Marcus Lafayette Byrn

The Mystery of Medicine Explained

A Family Physician and Household Companion

Marcus Lafayette Byrn

The Mystery of Medicine Explained
A Family Physician and Household Companion

ISBN/EAN: 9783337211059

Printed in Europe, USA, Canada, Australia, Japan

Cover: Foto ©berggeist007 / pixelio.de

More available books at **www.hansebooks.com**

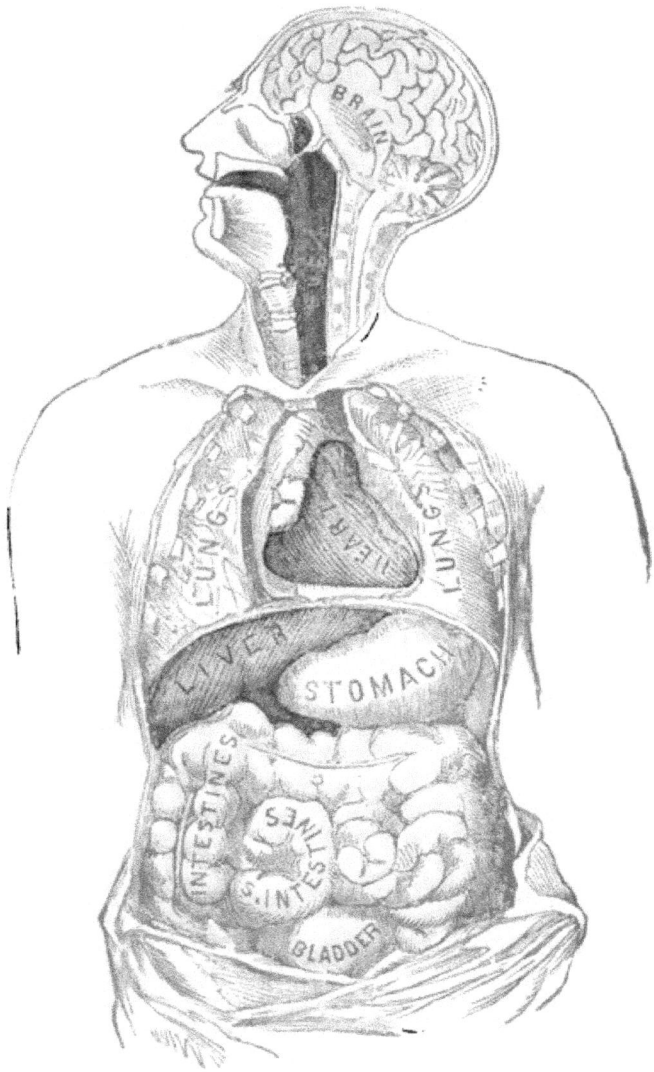

FIFTIETH EDITION—REVISED AND IMPROVED

THE

MYSTERY OF MEDICINE

EXPLAINED;

A

FAMILY PHYSICIAN,

AND

Household Companion;

PREPARED FOR THE USE OF FAMILIES,

Plantations, Ships, Travelers, &c.

By M. LAFAYETTE BYRN, M.D.,

GRADUATE OF "THE UNIVERSITY OF THE CITY OF NEW YORK," AUTHOR OF "POISONS
IN OUR FOOD," ETC. ETC.

NEW YORK:
PUBLISHED BY M. LAFAYETTE BYRN, M. D.
No. 49 NASSAU STREET.
1875.

PREFACE.

THIS book has been written for the "PEOPLE!"—the rich, the poor, the old, the young, male and female, the learned and the illiterate, those who are well and those who are sick; on land and on water, in the city and in the country, in the rural country-seat of the retired merchant or the log-cabin and camp-fire of the hardy pioneer or backwoodsman; for the clerk of sedentary habits, and for the farmer who toils in rain and in sunshine; for the young man far away from home, and for the mother who keeps watch over her loved ones through the long hours of dreary night in sickness;—in a word, for the million.

It is intended as a guide for preserving health and prolonging life, by giving that kind of information (couched in language free from medical technicalities), which has long been needed by the masses. It has been my aim not only to simplify the laws of health and physical education, but to give such plain descriptions of the various ailments which our bodies are subject to, that every one may know from the symptoms, each ailment or disease, and be enabled to give the best remedies, where a physician can not be had, or, in cases of emergency, to know what to do before the physician arrives, so as to alleviate suffering or be the means of saving life.

Also, to enable the reader to treat intelligently and successfully many of the "ills which flesh is heir to," which are liable to occur in a family at any time, without being under the necessity of sending for a physician.

All the works that I have previously met with, on this subject, have invariably *attempted too much*, by giving directions for treatment that the great mass of their readers would be unwilling to try. I have endeavored to give *simple, brief* and *safe* instructions, so no one need be afraid to follow them I have, of course, availed myself of all the sources of informa tion possible in the preparation of the work. The study of our own natures is, perhaps, the most elevating and ennobling subject which can engage the mind, and we ought to deem it as much a part of our sacred duty to promote physical health and happiness in this life, as to prepare our spirits for the fruition of a happy immortality.

Knowing the weight of responsibility resting upon me in issuing a volume like this, for the guidance, comfort, and benefit of my fellow-beings, I feel to implore the blessings of God, that my labors be not in vain.

M. LAFAYETTE BYRN, M. D

New York City, 1871.

THE FRAME OF THE HUMAN BODY

The Names by which the different Bones in the Human Body are Known.

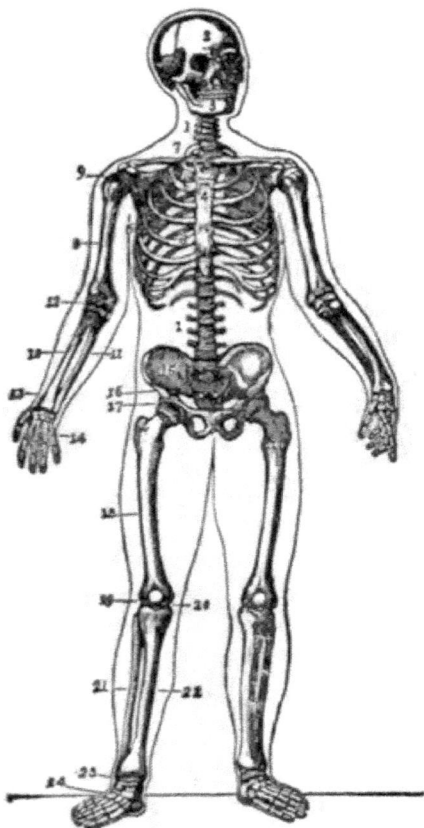

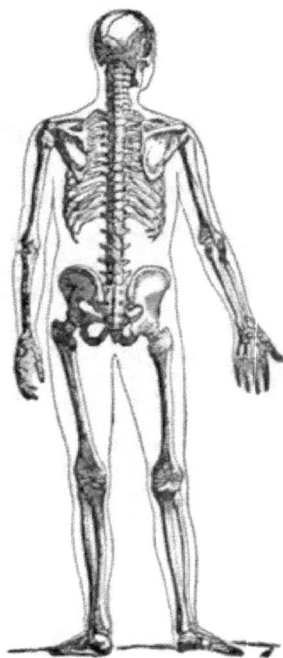

FRONT VIEW.   BACK VIEW.

1. The back-bone or spinal column.
2. The skull or head of the human body.
3. The under jaw.
4. The sternum or breast-bone.
5. The ribs.
6. The cartilages which connect the ribs with the breast-bone.
7. The clavicle or collar-bone.
8. The humerus, or bone of the arm.
9. The shoulder-joint.
10. The radius } two bones running parallel with
11. The ulna { each other from the elbow to the wrist.
12. The elbow-joint.

13. The wrist.
14. The bones of the hand.
15. The pelvis, or haunch-bone.
16. The sacrum, a wedge-shaped bone at the lower end of the back-bone.
17. The hip-joint.
18. The thigh-bone.
19. The patella or cap of the knee.
20. The knee-joint.
21. The fibula or lesser bone of the leg.
22. The tibia or shin-bone of the leg.
23. The ankle-joint.
24. The bones of the feet.

We present here two Views of the Human Frame; they will be found very useful to our readers as a matter of information and satisfaction, as they will be able to tell the exact location, formation and peculiarities of the various parts of the Human System; especially in cases of fractures, dislocations, wounds, etc., will it be found of value.

(*See Pages* 273 *to* 286.)

PLATE II.

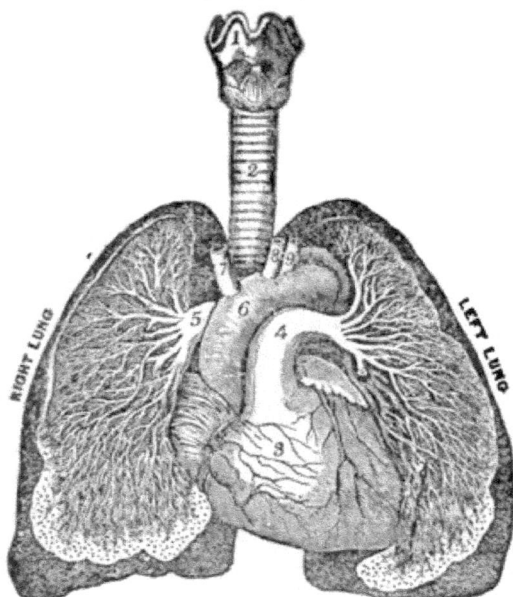

THE HEART AND LUNGS.

These are placed within the chest, at the upper end of the trunk.

By breathing we draw in *air*, which purifies the blood, and prepares it for nourishing and sustaining the body, and the refuse matter which it separates from the blood is carried out.

The air enters into the lungs where the oxygen contained therein comes in contact with the blood, impure from having circulated through the body. The oxygen of the air purifies this blood, which immediately after passes again through the body to return in like manner.

The blood is carried into the lungs by small blood vessels, or tubes. These small tubes are continually poring impure blood into the lungs from the heart, while others are carrying it back again into the heart after it is purified.

As the blood passes through the body, it gathers up the decayed and waste particles or atoms, and in this way it becomes impure. It then returns to the lungs to be purified, and thus made fit again to go out into the system, to perform its life-preserving work anew.

How necessary, then, that we should breathe pure **air**!

(*See Pages* 30, 125, 318.)

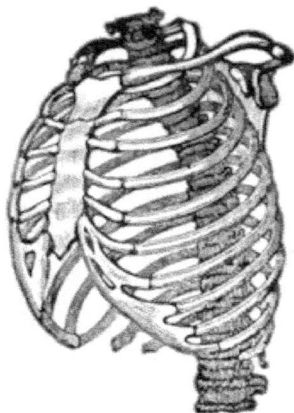

BACK-BONE, RIBS, AND COLLAR-BONES.

The Back-bone or Spinal Column, extends from the skull, behind, down the middle of the back, and is composed of twenty-four short, round, and perforated bones, called *vertebræ* by physicians. They are held together by the cartilage or gristle, which is between each bone, thus allowing the spinal column to be flexible.

There are twenty-four ribs, twelve on each side. They grow out of the spine, forming a hoop by meeting and being fastened to the breast-bone in front.

They are the framework of that part of the human trunk called the chest, in which are the heart and lungs; they serve as a protection to those vital organs. The collar-bones are two long slender bones passing over the ribs in front from the highest part of the shoulder-bones to the head of the breast-bone. The collar-bones prevent the arms from sliding too far forward.

(*See Pages* 276, 281, 286.)

PLATE IV.

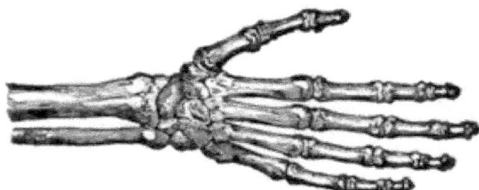

BONES OF THE HAND AND FINGERS.

The wrist has eight bones, all being wedge-shaped, and strongly united together by ligaments or gristle. In the hand and fingers there are nineteen bones, which are also strongly held together by powerful ligaments. It will be useful to examine this engraving carefully, so as to be well informed on the subject in cases of emergency, such as dislocations, sprains, etc.

(*See pages* 249, 250.)

BONES OF THE FEET.

There are twenty-six bones in the foot; they are of peculia. shape, and are all held together by very strong ligaments, forming the arch of the foot, called hollow of the foot on the under side, and the instep on the top.

(*See pages* 122, 283.)

13

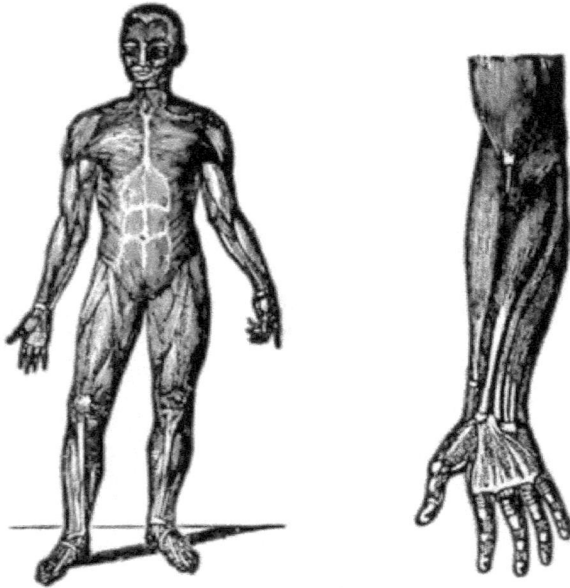

THE MUSCULAR SYSTEM.

There are about four hundred Muscles in the human body, each one having its peculiar office to perform. Muscles are the fleshy or lean parts of the body, formed of fibres or threads. They are fastened to the bones, and are the instruments by which we move the different parts of the body,—by them is every motion performed. They have the power of contracting or shortening, by which their two ends are drawn nearer to each other, constituting motion. This contracting power is subject to the WILL, and is exercised when we please, and yet we can not tell how. It is one of the great mysteries of our bodies, which we can not explain.

(*See pages* 286, 288.)

15

PLATE VL

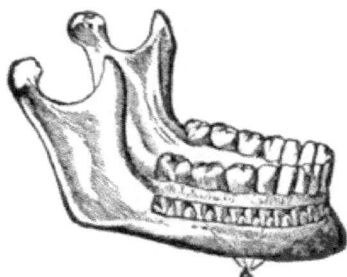

SIDE VIEW OF UNDER JAW-BONE AND TEETH.

Children have twenty teeth—ten in the upper, and ten in the lower jaw; when a child becomes six or seven years of age, the teeth loosen, and, if they do not drop out, they should be removed without delay, otherwise they will prevent the proper formation and regularity of the new permanent teeth, which are growing under them. Letter A, in the engraving, shows the position of the new teeth in the under jaw, pushing out the old ones.

(*See pages* 69, 241, 251.)

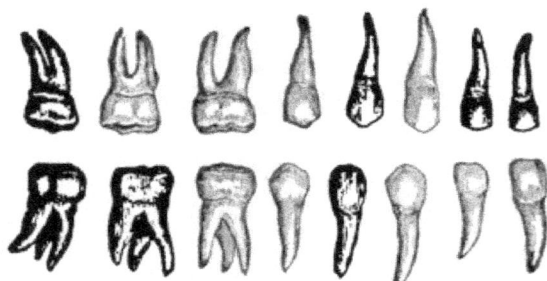

THE DIFFERENT KINDS OF TEETH.

The internal part, or the ivory of the teeth, is a more solid substance than bone, and this is covered with a smooth, white, and still harder substance, called *enamel*, which when once decayed or destroyed, is never again restored. This enamel gives the teeth strength, as well as hardness, for biting, chewing, and grinding the food; it also prevents injury from these operations, and from the action of acids on the bone of the teeth; and adds much to their beauty and durability.

Teething and the proper care of the teeth are matters of great importance. (*See pages* 57, 95.)

Fig. 1

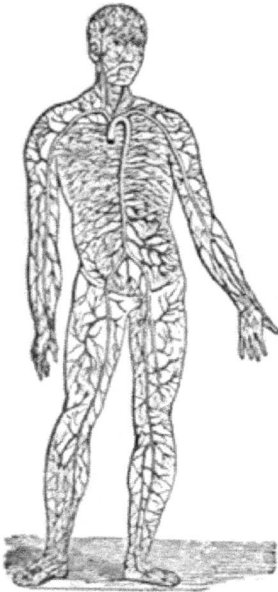

Fig. 2

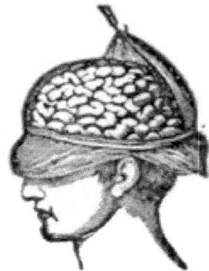

The "*Circulation of the Blood*" (Fig. 1) was discovered by Dr. Harvey. From the earliest of our being to life's latest hour, this life-giving process must go on. Beginning in the great centre of the system, the heart, the blood is propelled through the *arteries* to the remotest parts, giving nourishment and life to the system, and is then returned through the *veins* to the heart, and thence to the lungs, to be purified by the oxygen in the air.

The blood thus makes the rounds of the circulation once in about four minutes.

The "*Brain*" (Fig. 2) is the centre of the nervous system, and weighs from three to four pounds. Every nerve of the body is connected with the brain, and may be called the "Sentinels of the System," being the mediums of all our sensations, such as hunger, thirst, pain, etc.

(*See pages* 134, 194, 329, 332.)

Fig. 1.

Fig. 2.

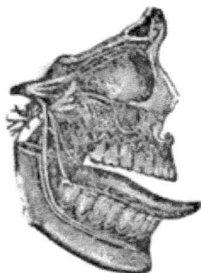
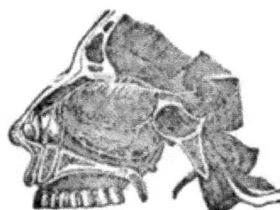

" THE TONGUE," and *Sense of Taste*, is represented in Fig. 1. The sense of taste lies chiefly on the upper surface of the tongue. Branches of the nerve of taste, are also spread over the palate or roof of the mouth, and on the inside of the cheeks and lips. The tongue, also, assists in various other little operations besides taste and talking ; by it we move the food in the mouth from side to side, or hold it in a proper position for chewing. In sickness the tongue becomes dry and hard, as there is no saliva or spittle secreted to moisten it, and at such times there is *very little taste*, as the tongue is covered with a fur, or coat. In such cases, the tongue should be frequently moistened. (*See pages* 45, 203, 208.)

" The Nose," or Sense of Smell, is seen in Fig. 2. The sense of smell is situated in the lining of the nostrils. The inner surface of the nostrils are lined with a thin mucous membrane, over which the branches of the nerve of smell are spread, and which are kept constantly moist in their natural state with a thin fluid called *mucus*, which keeps the branches of the nerve of smell in perfect order, and protects them from harm. It also catches the odorous or noxious particles which are drawn into the nostrils with the air, and are thereby prevented from being carried deeper into the lungs, where they would be injurious. By the act of sneezing these particles are ejected from the nostrils.

(See pages 97, 402, 121 *and* 231.)

PLATE IX.

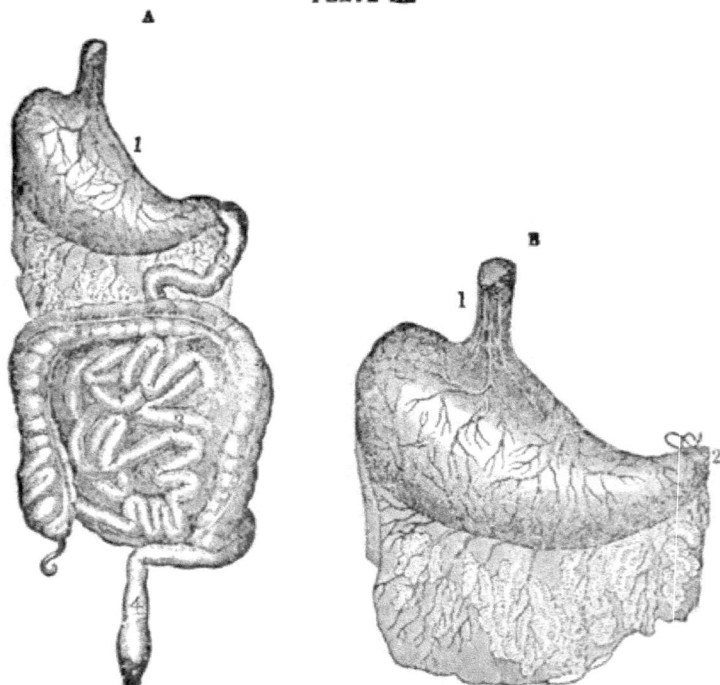

HUMAN STOMACH AND BOWELS.   THE HUMAN STOMACH.

We here give a fine representation of the Stomach and Bowels, united together, (A) in which "1" is the stomach, "2" is the upper portion of the small bowels, "3" the middle portion of the small bowels, around which, in the shape of an arch, is seen the large bowels; "4" is the rectum, or lower outlet of the bowels.

The food while in the stomach becomes *dissolved* or *digested*. It is then called *chyme;* this passes out of the stomach into what is termed the "*duodenum*" (or upper part of small bowels), where it is then separated into two classes, a milky fluid called *chyle*, or the part which enters into and forms the life of the blood, and the waste or useless part, which should be ejected from the bowels regularly each morning soon after rising. "THE STOMACH," detached from the bowels, is seen at "B," at figure "1" is a section of the Esophagus, or tube leading from the throat to the stomach; and figure "2" is the outlet into the upper portion of the bowels.

(See Pages 104, 158, 41, 52, 218, 320, 341, 64, 105, 157, 180, 349 352, 134, 340.)

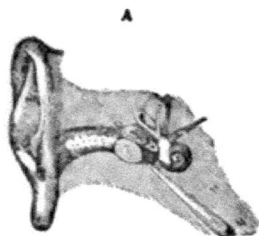

A   B

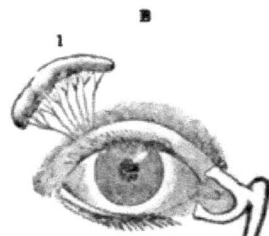

THE HUMAN EAR.   THE HUMAN EYE AND LACHRYMAL GLAND.

THE EAR (A), the organ of hearing, *external* and *internal*, is here beautifully delineated in the engraving. The very intricate structure of the parts concerned in the sense of hearing, with the Eustachian tube leading from the ear to the throat, are very nicely shown by the artist.

THE EYE is seen at B; the lacrymal gland which supplies the eyeball with moisture, and is the gland which secretes the tears, is seen at fig. 1. The lacrymal duct or canal, extending from the corner of the eye, into the nose (fig. 2, B), carries off any superfluity of moisture into the nose, which mostly becomes evaporated by breathing.

(See pages **103**, 309, 310, **116**, 295.*)*

THE
MYSTERY OF MEDICINE
EXPLAINED.

HOW TO PRESERVE HEALTH.

THE variety of temperaments or constitutions renders it possible for health to be very different in different persons, hence what would preserve the health of one would occasion disease in another. Persons of a *sanguine* temperament, whose vessels are full, and whose fibres are firm and active, easily excited to motion, and often to irregular actions, bear evacuations well, and have their health best promoted by abstinence and low living, by avoiding excess of every kind, and particularly guarding against cold after active bodily exertions. The *bilious* temperament, with a constitution more acutely sensible, always more irritable, requires the same precautions as the sanguine; but the evacuations best adapted, which are indeed almost indispensable to this kind of constitution, are the free and frequent use of the milder laxatives. To preserve the health of the *melancholic*, of those whose complexion is dark, and whose powers are torpid, whose mind is dull, but persevering, much exercise is required to assist digestion and to turn the circulating fluids to the skin. From the torpor of the bowels, to which persons of this temperament are peculiarly liable, they will demand the occasional use of purgatives of the aloetic kind; their occupations and amusements should be varied and interesting to the mind. The *phlegmatic* temperament is pale in complexion, languid in its exertions; the vessels, if full, are torpid, the constitution inactive; the mind not easily excited to exertion. The diet in this temperament, require to be nutritive and somewhat stimulating, though it ought not to go the length of what would be called high living.

The health of females has some peculiarities arising from the delicacy of their frame, the monthly discharge, the state of pregnancy, and of nursing. All these circumstances coustitute a condition very different from the robust and vigorous strength of man in the prime of life; yet equally perfect, relatively to the sex and the individual. The irritability of infants, and the nimble tricks of boyhood, are consistent with good health, though they would be unsuitable at a more advanced period of life.

Health varies in people of different occupations. The acuteness of the senses which is necessary in some employments, would be morbid in persons otherwise engaged. But some have various diseases or predispositions to disease, either derived from parents or acquired in the progress of life, which render health with them only a comparative term. The scrofulous can hardly be said in strictness ever to be in perfect health; but their disease may be dormant or undeveloped; and, in favorable circumstances, may permit the subject of it to enjoy an exemption from pain and inconvenience, to the end of a long life. Gouty and rheumatic patients may also enjoy good health during the intervals of their attacks.

AIR.

Atmospheric air, or that by which we are usually surrounded, is not a simple, but a compound body, consisting of at least four distinct substances, viz: *oxygen, azote, carbonic acid,* and *aqueous,* or *watery vapor.*

The two former substances, however, constitute almost the whole of the atmospheric air near the surface of the earth; the other two are variable in their proportions; the first exists only in minute quantities, which it is difficult to appreciate. Vital air, or oxygen, which constitutes about one-fourth of the atmosphere, is necessary to respiration and combustion, and an animal immersed in it will live much longer than in the same quantity of common air. The remaining three-fourths, called azote, or mephitic air, is totally incapable of supporting life or combustion for an instant.

The oxygen which is received into the lungs of animals from the atmosphere, communicates the red color to the blood, and is the principal agent which imparts heat and activity to the system. When animals die for want of oxygen in the air, their blood is always found black. Independently of its destruction by the respiration of men and other animals, there is a constant consumption of the oxygenous portion of atmospheric air, by the burning of combustible bodies; by the fermentation and putreaction of vegetable substances, etc.

A diminished proportion, therefore, of the oxygen of our atmosphere, and an increased amount of carbonic acid and other deleterious gases, is undoubtedly produced from the innumerable processes of combustion, putrefaction, and respiration of men and animals, particularly in populous cities, the atmosphere of which is almost constantly prejudicial to health.

In the open country there are few causes to contaminate the atmosphere, and the vegetable productions continually tend to make it more pure. The winds which agitate the atmosphere, and constantly occasion its change of place, waft the pure country air to the inhabitants of the cities, and dissipate that from which the oxygen has been in a great measure extracted. The air of any place where a numerous body of people is assembled together, especially if to the breath of the crowd there be added the vapor of a great number of candles, lamps, or gas-lights, is rendered extremely prejudicial, as these circumstances occasion a great consumption of oxygen.

The fact is well known, that when air has been long confined and stagnated in mines, wells, and cellars, it becomes so extremely poisonous as to prove immediately fatal to those who imprudently attempt to enter such places. No person should descend into a well or cellar, which has been long closed, without first letting down a lighted candle; if it burns clear there is no danger, but if it cease to burn, we may be sure that no one can enter without the utmost danger of immediate suffocation. It sometimes happens also, that when air is suffered to stagnate in rooms, hospitals, jails, ships, &c., it partakes of the same unwholesome or pernicious quality, and is a source of disease. It is obvious, therefore, that in all confined or crowded places, the correcting of vitiated air, by means of cleanliness and frequent ventilation, is of the highest importance to health, and the most effectual preservative from disease. No accumulation, therefore, of filth about our houses, clothes, or in the public streets, should on any pretence be suffered to continue, especially during the heat of summer.

It is a very injurious custom for a number of persons to occupy or sleep in a small apartment, and if it be very close, and a fire be kept in it, the danger is increased. The vapor of *charcoal*, when burnt in a close apartment, produces the most dangerous effects. Our houses, which are made close and almost air-tight, should be ventilated daily, by admitting a free circulation of air to pass through opposite windows·

and our beds ought to be frequently exposed to the influence
of the open air also.

Houses situated in low marshy situations, or near lakes or
ponds of stagnant water, are constantly exposed to the influence of damp and noxious exhalations.

Among the most powerful means furnished by nature for
correcting air which has become unfit for breathing, is the
growth and vegetation of plants. Animal bodies consume
oxygen, and give off carbonic acid; plants and vegetables
consume carbonic acid, and give off oxygen. The generality
of plants possess the property of correcting the most corrupt
air within a few hours, when they are exposed to the light of
the sun; during the night or in the shade, however, they
destroy the purity of the air, which renders it a dangerous
practice to allow plants to vegetate in apartments occupied
for sleeping.

MARSHES.—The neighborhood of marshes is peculiarly unwholesome, especially towards the decline of summer and
during autumn, and more particularly after sunset. The air
of marshy districts is loaded with an excess of dampness, and
with the various gases given out during the putrefaction of the
vegetable matters contained in the waters of the marsh. Persons exposed to this air are liable to various diseases, but especially ague, bilious fevers, diarrhœas, and dysenteries. They
who breathe it habitually exhibit a pallid countenance, a
bloated appearance of the abdomen and limbs, and are affected
with loss of appetite and indigestion. Health is best preserved in marshy districts by a regular and temperate life--
exercise in the open air during the middle of the day, and by
retiring as soon as the sun sets, within the house, and closing
all the doors and windows except enough for ventilation. The
sleeping apartment should be in the upper story, and rendered
perfectly dry by a fire, lit a few hours before going to bed,
and then extinguished. Exposure to the open air should, if
possible, not take place in the morning before the sun
has had time to dispel the fog, which, at its rising, covers the
surface of the marsh. Persons who are intemperate, or use
ardent spirits habitually, are those most liable to suffer from
the unwholesome air of marshes; such generally perish from
diseases of the liver and dropsy.

NIGHT AIR.—Many diseases are brought on by imprudent
exposure of the body to the night air; and this, at all seasons,
in every climate, and variety of temperature. The causes of
this bad property of the night air, it is not difficult to assign
The heat is almost universally several degrees lower than it

the daytime; the air deposits dew and other moisture; the pores of the skin are open, from the exercise and fatigues of the day; the evening feverishness leaves the body in some degree debilitated and susceptible of external impressions; and from all these concurrent causes are produced the various effects of cold acting as a check to perspiration; such as catarrhs, sore throats, coughs, consumptions, rheumatisms, asthmas, fevers, and dysenteries. In warm climates, the night air and dews, with their tainted impregnations, act with much malignancy. In civilized life, and in crowded towns, how many fall victims to their own imprudence, in exposing themselves to the cold, the damp, and the frostiness of the night air! Issuing from warm apartments with blazing fires, or from crowded churches, theatres. or ball-rooms, with exhausted strength, profuse perspiration, thin dresses, and much of the person uncovered, how many are attacked with a benumbing cold and universal shivering, which prove the forerunners of dangerous inflammations of the brain, of the lungs, or of the bowels, which either cut them off in a few days, or lay the foundation of consumption or other lingering illness. Never stand to talk, even for *a moment*, in the open air, after coming out of a heated or crowded room, or after active exercise. Such being the dangers of exposure to the night air, it ought to be inculcated on all, both young and old, to guard against them, by avoiding all rash and hasty changes of place and temperature, by hardening the frame by due exercise and walking in the open air in the daytime; and on occasions where the night air must be braved, taking care to be sufficiently clothed; and to avoid drawing in the cold air too strong or hastily with the mouth open. Always breathe through the *nostrils* for a short time after going out of a warm room into the cold air, keeping the *mouth shut.*

Sea Air.—The air upon the sea and in its neighborhood is generally distinguished by its greater coldness, purity, and sharpness; and is therefore in many cases directed to patients whose complaints do not affect their respiration, and who have vigor of constitution enough to derive benefit from the stimulus which such air occasions. A residence by the sea-side is beneficial to persons of a scrofulous habit and debilitated constitution, provided they take care not to expose themselves to cold and damp; and in the fine season, when there is no reason against it, they ought to bathe. In complaints of the chest, the use of sea-bathing, and a residence near the sea, are more questionable; and by such an inland rural situation, in a mild equable climate, is to be preferred. A sea voyage has long been famous for its good effects at the commencement of consumptive complaints; and these good effects may be ascribed partly to the good air at sea, partly to the affection of the stomach and skin induced by sea-sickness, and to the excitement of the mind, caused by change of scene and occupations.

VENTILATION.—The air, as we have already remarked, cannot become stagnant or unchanged for even a short period without its becoming unfit for breathing, and destructive to the health of those who breathe it. The streets of a city should, therefore, be so laid out as to insure a constant and free circulation of air; hence the unwholesomeness of a residence in narrow alleys, courts and passages. Not less important is the continued renewal of the air of our apartments—the ventilation of which, however, should be so conducted as to prevent a current of air from blowing directly upon the persons within them. Our bed-chambers, in particular, should be freely ventilated during the day; and even at night, when the windows are closed, the chimney should be left open, or, if the room is small, and the weather sultry, a door, opening into another room, or a window partly open, or the sash pulled down to admit fresh air. No consideration of economy should prevent the most constant attention being paid to proper ventilation, so essential is the latter to health and comfort.

CELLARS.—It is important that cellars should be perfectly dry, kept strictly clean and freely ventilated. The damp and foul air so frequently generated in cellars, where dryness, cleanliness, and ventilation are not properly attended to, is often the cause of disease, not only in the persons who inhabit the house to which the cellar is attached, but in others residing in the immediate neighborhood. No house can be considered a healthy residence, in the cellar of which water is allowed to stagnate: this may easily be obviated, in most situations, by a sink dug to gravel. The air of cellars can be preserved sufficiently dry and wholesome by free ventilation, the removal of all filth and corruptible materials, and frequently white-washing the walls. Cellars, especially when entirely under ground, are improper places of residence; appropriating them as places of residence for the poor, or as workshops, should be prohibited by law.

CLIMATE.—Climate is considered by physicians, not with reference to geographical situation, but to the state of regions as to the warmth and steadiness of their temperature, or the dryness or moisture of their atmosphere. The interior of continents and islands is generally mountainous, and, in conse quence, cold. From the bracing qualities of the prevailing winds, the inhabitants are robust, and disposed to inflammatory diseases: invalids, or persons coming from warm climates, should, therefore, prepare themselves gradually for mountainous regions, by not coming abruptly into those colder parts.

CLOTHING.

CLOTHING possesses no warmth in itself, but merely prevents the heat of the body from being carried off by the air, and other surrounding bodies, faster than it can be supplied by the blood. The essential requisites for clothing are, that it be soft and pliable, so as not to obstruct the free and easy motion of the joints, or occasion inconvenience by its weight or tightness; that it be adequate to protect the body from the external influence of the atmosphere, and preserve it in that degree of temperature which is most agreeable, as well as best adapted to the exercise of its different healthy functions and motions; and that it does not produce any detrimental effects, occasion any unnecessary degree of perspiration, or absorb the vapors of the atmosphere. Clothes of a light color, have the least attraction for heat; those of a black, the greatest; the first mentioned are, therefore, most proper in hot, the last in cold weather.

But besides these general properties of commodious and comfortable clothing, it should be suited in quantity and material to the climate, the season of the year, the period of life, the constitution, and the habits and mode of living. Thus, a person who is engaged in a sedentary employment, will always require warmer clothing than one who is actively engaged in manual, or other labor demanding considerable muscular exertion; and the latter will always require an addition of clothing, the moment he has ceased from his active labors, to what is proper whilst engaged in them. Neither do young persons, or those in the prime of life, and in robust health, require clothing in the day, or covering in the night, of so warm a nature as persons advanced in years; because the performance of their functions is more equal and vigorous, and of course, the generation of heat in the body is quicker, and of greater extent, than is the case in old age.

One of the safest rules in the regulation of dress, is to adjust it to the vicissitudes or fluctuations of the season; and this rule should be carefully attended to by the invalid, the delicate, the infirm, and the old. The winter clothing should not be left off too early in the spring, nor the summer clothing worn too late in the autumn. Neither should this rule be disregarded by the young, and those in the enjoyment of perfect health The grand rule is, so to regulate the clothing, that, when ex posed to the external air, the difference of temperature experienced, shall not be such as to produce any unpleasant impression, whatever may be the inclemency of the weather, when we go abroad. Thinner clothing are necessary within doors than without, and a greater warmth of clothing after night, and during cold, damp weather, than during the day, and when the air is perfectly dry.

Persons of delicate and irritable constitutions, whose powers of life are feeble, and whose circulation is languid and irregu lar, are very apt to suffer severely by a very slight diminution of the temperature of their skin. This is also the case with invalids. All such persons, therefore, ought rather to exceed, than be deficient in the quantity and warmth of their clothing.

But while clothing should not be too light, or too small in amount, neither should it be too heavy, or too much in quan tity. The effects are equally mischievous. By over-clothing, too much perspiration is drawn out of the body, by which the frame is greatly weakened, and coldness and numbness of the extremities are occasioned.

Tight clothes are invariably detrimental to the health, com fort, and symmetry of the body. By the pressure they make upon the muscles, and the impediment they offer to their free exercise, they produce in them an emaciation and debility which prevent them from supporting properly the natural and graceful position of the body, or effecting its active movements with sufficient vigor. They prevent also, the free circulation of the blood, and cause it to accumulate in the veins of the head, lungs, or abdomen. When the pressure of the clothes, or any part of them, is around the neck, it is apt to produce head ache, discoloration of the face, giddiness, and apoplexy, or other diseases of the brain; when upon the chest and waist, it prevents the full development of the lungs, impedes breathing, and interferes with the proper action of the heart, in conse quence of which, the health of the whole system suffers; when around the abdomen, the stomach, liver, and bowels are affect ed, and indigestion is produced, or the nutrition of the whole body is rendered imperfect. The clothes, therefore, should be perfectly loose, leaving to every part the fullest liberty, and to all their natural and unconstrained motions. Avoid muffling up the neck, head, ears, &c., when in good health. This is all important at every period of life, but particularly so, during infancy and childhood.

The skin of a child, from the neck downwards, ought to be

kept warm by proper clothing; bare legs and chests, thin and insufficient clothing cause croup, inflammation of the lungs, &c. Keep flannel or woolen next their skin in cold weather.

FLANNELS OR WOOLENS worn next the skin, in addition to the ordinary clothing, are of very great service in preserving the health. They produce a moderate warmth of the surface, promote perspiration, readily absorb the perspired fluids, and easily part with them again by evaporation, on account of the porous nature of their texture. Woolens should be worn *at all seasons* by the *aged*, and all subject to diseases of the chest or bowels, and by invalids.

Flannel is also well adapted for infants and young children, especially in autumn, winter, and spring. Older children do not require it, excepting during the cold weather, and all persons under forty, in good health, should reserve it as a resource for their declining years, during which it becomes every year more and more useful and necessary. Flannel that has been worn during the day, ought not to be habitually worn at night, but exchanged for a woolen night shirt. Always wear the same kind of materials next the skin at night as you do in day time. Many persons who wear woolen under-garments during the day time, sleep in a cotton night shirt, and thereby greatly impair their health. Flannels need to be washed often.

Such persons as find flannel too irritable to their skin, may obviate this, by having it lined with thin muslin. We especially recommend the use of flannel drawers to females.

COTTON.—Cotton, as an article of clothing, especially when worn in contact with the skin, is far better adapted for general use than linen, but is inferior to woolens. In warm weather, and in hot climates, it is the most comfortable article for an inner dress. It is cooler than linen, inasmuch as it conducts more slowly the excess of external heat to our bodies, and when a sudden reduction of atmospherical temperature occurs, on the other hand, it abstracts more slowly the heat from the body, and thus preserves the surface of a more steady and uniform temperature.

LINEN.—Whatever may be said in favor of the comforts of linen, and the greater ease with which it is kept clean, it is by no means a substance well adapted for the dress worn next to the skin, at any season of the year, nor by any class of persons.

HEAD DRESS.—Whatever covering is worn upon the head should be light, sufficiently large, and adapted in its form to the shape of the head. Too heavy or warm a covering, or one which compresses unduly the head, is productive of pain and inconvenience. In summer, the color of the hat or bonnet should be white, or at least some shade approaching to white, in consequence of the tendency of all dark colors to absorb and transmit the rays of heat. The brim of the hat should also be sufficiently broad to protect the face and eyes from the sun Although the nature of a head dress may appear to be a sub ject of very little importance in regard to health or comfort, yet every one has perhaps experienced more or less of the pain and inconvenience, occasioned by wearing a new hat, too small in the crown, and unfitted in shape to the head, and the almost immediate relief which results from exchanging it for one of more ample dimensions.

CRAVAT.—It is important that of whatever it is composed, it be very light and loosely applied. When the neck is kept too warmly covered, it is very injurious; the throwing off of the cravat for a few moments, or exchanging it for one of lighter materials, will often give rise to a violent inflammation of the throat. Cravats and neck ties must not be worn too tightly; this is often the cause of a "horrible headache."

CORSETS AND TIGHT-LACING.—Of all the whims of fashion, no one is more absurd, or more mischievous in its effects, than that which condemns the female, under the pretence of improving the grace and beauty of her shape, to the torture of a tightly laced corset. Equally detrimental to comfort and to health, this portion of female attire cannot be too severely censured. It is productive of not the least advantage, real or imaginary, to compensate for the injury it produces, nor to excuse the folly of females in persisting in its use. The immediate effect of tight-lacing is, by compressing firmly the chest, to prevent its free expansion in the act of breathing; a less amount of air is taken into the lungs, and as a consequence, the blood is less perfectly changed. The impediment to breathing is increased when the corset extends so low as to compress the abdomen; by the bowels being then forced upwards against the diaphragm, (or partition between the chest and abdomen,) the latter is prevented from descending, and the dimensions of the chest are thus contracted from below. A sense of oppression and weight is always experienced about the breast when the corset is drawn very tight around the body; the breathing is short, quick, and panting; and not only is the blood prevented, in a great measure, from undergoing that change in the lungs by which it is adapted for the healthy nourishment of the various organs, but the actions of the heart are also impeded; violent palpitation is not unfrequently produced, accompanied with a sense of giddiness and

Whalebone, Corsets, and Tight Lacing.

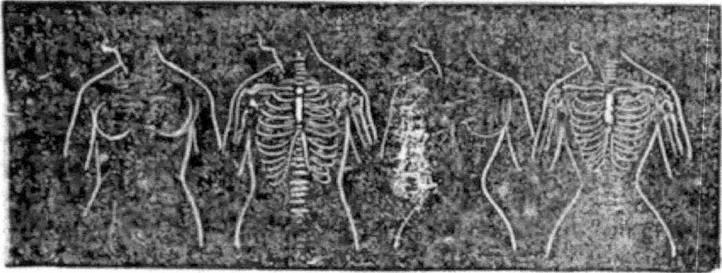

NATURAL FORM AND THE RIBS.   FASHIONABLE FORM AND THE RIBS.

A dress, tight over the chest, not only binds the ribs together, and thus prevents the free play of the lungs, but it crowds all the vital organs upon each other, so as to derange their proper action, and obstruct the circulation in all parts of the system. Thousands die annually, the victims of consumption, produced by tight lacing.

Such casing and confining are deadly foes to health and life, as everything is which prevents perfect freedom of action to the vital organs. The ribs, perhaps, may be more easily changed than any of the other bones of the body. Their very structure is such that the constant pressure or tightness of clothing, day after day, needs to be but slight to bend the ribs downward or inward.

During childhood, the bones are soft and pliable, and readily accommodate themselves to any position which is habitual. Tight-fitting dresses on a young lady from the age of fourteen to twenty, are the only appliances needed to make her sadly deformed in chest for the remainder of her life, which cannot be of long duration in consequence.

It is well known that a loose and easy dress contributes much to give the sex the fine proportions of body that are observable in the Grecian statues, and which serve as models to modern artists, Nature being too much disfigured among us to afford such models now. The Greek women were ignorant of the use of whalebone stays, by which our women distort their shape, instead of displaying it. This practice is carried to so great an excess, that it must in time degenerate the species, if not abandoned. It is only a *habit*, a *fashion*, the females have "got into." Now who will confer a blessing on the human race, b introducing the fashion of "common sense," instead of tight-lacing? **Try it** ye millions, **try it**

occasional fainting. When the corset is worn constantly from early youth, the growth of the ribs is prevented, and the whole capacity of the chest is permanently contracted; and hence spitting of blood, difficulty of breathing, or even more dangerous and fatal diseases of the lungs and heart are induced. Consumption is a very common complaint, the production or aggravation of which may be traced to tight lacing. But it is not merely to the chest that the injurious effects of the corset are confined; it likewise compresses the whole of the upper portion of the abdomen (or bowels,) and by the yielding nature of this portion of the body, the pressure upon the organs within is even more considerable than that experienced by the heart and lungs. The liver, the stomach, and the bowels in particular, experience this pressure to a very great extent; in consequence, the free and healthy secretions of the liver are prevented from taking place, the stomach and the bowels can no longer perform their functions with proper vigor and regularity; the digestion of the food is impeded, and the bowels become costive and distended with wind. In this manner, in connection with the injury inflicted upon the lungs, the vigor of the whole system becomes prostrated, the skin assumes a sallow hue, the countenance a haggard and wrinkled appearance, and all the functions of life are performed imperfectly. It is a fact, that nothing is better adapted to produce the premature decay of beauty, and the early appearance of old age, than tight-lacing.

There are two other effects produced by this article of dress, which would be sufficient of themselves to induce every prudent and sensible female to abandon it. The first is the great injury inflicted upon *the breasts*, by which their proper *development is prevented*, and the nipple is almost entirely obliterated, so that, when called upon to fulfil the sacred office of nurse towards her offspring, the mother finds, to her sorrow, that, from her folly, she has totally incapacitated herself from performing its duties, or experiencing its pleasures. The second effect is that produced by the pressure of the corset upon the pelvis (hips) and the womb, more especially when worn in early youth, or during the first stages of pregnancy. From this cause barrenness, miscarriages, or a stunted and deformed offspring may result, or the pains, the difficulties, and the dangers of child-birth, may be increased to a frightful degree. Let no American woman talk about the Chinese women compressing their feet to prevent them from growing, so long as she continues the life-destroying custom of tight-lacing.

Garters.—Tight garters are injurious by impeding the cir

culation of the blood in the leg. Swelling and numbness of the leg, and permanent enlargement of the veins of that limb, are consequences of wearing tight garters.

EXERCISE.

By this means disease may often be prevented, and not unfrequently cured, even when it has taken a very strong hold upon the constitution. Generally speaking, a slothful and sedentary life is the source of all those diseases which are termed slow or chronic, the number of which is in our day very considerable.

The exercise which is necessary to the maintenance of the health, vigor, and the perfect and full development of the human frame, is such as will bring into action every limb and muscle ; this is termed *active* exercise, and is produced by the exertions of the body in walking, running, and various species of labor.

The chief kinds of *passive* exercise are, riding, swinging, and rowing. To derive all the advantage resulting from exercise, it must be regular. Little benefit need be expected, when, to occasional exercise of the muscles, a long period of inaction succeeds. Exercise, to be beneficial, must also be in the open air, and should never be carried to the length of inducing *undue fatigue*. The other general rules in regard to exercise, may be laid down as follows :—

The effect of exercise should be as general as possible, and not confined to any particular limb or part of the body, as walking, running, riding on horseback, etc.

Little benefit is to be expected from exercise, unless it be performed in a pure air.

The higher and drier the situation, and the more varied the air in which exercise is performed, the more beneficial will be its effects.

On commencing any exercise, we should always begin with the more *gentle*, and then proceed to the more laborious : and as sudden transitions are always wrong, the same rule should be followed when exercise is given up.

A good appetite after exercise, is a proof that it has not been carried to any improper excess.

It is a good rule, frequently to vary the exercise.

Lord Bacon correctly observes, it is requisite to long life, that the body should never abide long in one posture, but every half hour at least, should change it, except during sleep.

Muscular motion is most agreeable and healthful, when the

stomach is neither too empty, nor too much distended. Active exercise is improper, therefore, immediately after a meal, or after long fasting.

Nothing can be more injudicious than to sit down to a substantial dinner or supper, immediately after a fatiguing walk, ride, or other violent exertion. Every man should rest for some time after exercise, before he sits down to eat.

In taking exercise, the dress should be free and easy, particularly about the neck and joints.

In violent exercises, a flannel waistcoat ought to be worn next the skin, to obviate the possibility of injury from a sudden chilling of the surface of the body.

It will always be found very refreshing, after fatiguing exercise, to wash the feet in tepid water, before going to bed.

Serious thinking, when we are walking or taking other exercise, soon fatigues us; but if we give ourselves up to amusing thoughts, or the conversation of agreeable and intelligent friends, the good effects resulting from exercise are increased.

It is very desirable to have a certain object or spot by which the exertion is to be bounded. Exercise undertaken merely as a task, or without being connected with some purpose by which the mind is agreeably occupied and excited, is seldom productive of much advantage.

WALKING.—There is no exercise so natural to us, or in every respect so conducive to health, as walking. It is the most perfect kind of exercise in which the human body can be employed; for by it every limb is put in motion, and the circulation of the blood is effectually carried on, throughout the minutest veins and arteries of the system.

Walking is of two kinds, either on plain ground or where there are ascents. The latter is in every respect greatly preferable. Walking against a high wind is very severe exercise, and not to be recommended.

As, from various circumstances, persons residing in large towns, and engaged in sedentary occupations, cannot take all that exercise abroad, which is necessary for their health, they ought, at least as much as possible, to accustom themselves to walk about, even in their own houses, instead of sitting constantly at a desk or table.

The following rules are recommended to the attention of those who make use of this excellent species of exercise:

The most proper walk, for health, is in a pure and dry air, and in rather an elevated situation, avoiding marshy and damp places.

In the summer season, the walk should be taken morning and evening, but by no means during the middle of the day, unless the person be guarded from the oppressive heat of the sun. In winter, the best period of the day for walking is usually after breakfast, or from ten to one o'clock.

It is advisable, occasionally, to change the direction of the walk.

We ought to accustom ourselves to a very steady and regular, but not to a very quick pace; in setting out, it should be rather slower than what we afterwards indulge in.

An agreeable companion during a walk contributes much to serenity of mind; but unless the manner of walking of both is similar, as well as the taste and character congenial, it is better to walk alone.

To read during a walk is highly detrimental to the eyes, and destroys almost all the good effects that can be derived from the exercise.

SWIMMING.—For the young, the robust, and healthy, swimming is an excellent recreation. It combines all the advantages to be derived from bathing, with active exercise of nearly every part of the body. Swimming, however, as well from the powerful and constant exertion it demands, and the coldness of the water in which the body is immersed, is improper for the debilitated, or those exhausted at the time from fatigue.

RIDING.—Next to walking, riding on horseback is the most salutary and useful species of exercise, especially for invalids.

Persons laboring under ill health, whether occasioned by too long continued sedentary habits, or from defective digestion, as well as those predisposed to consumption, will experience from the exercise of riding the most decided advantage.

In riding to preserve health, eight or ten miles a day are sufficient to answer all the purposes; but, in riding to restore health, these little excursions will avail nothing. To attain the latter object, the mind, as well as the body, must be roused from its languor. Upon this account, long journeys are recommended to such people, in order, by the variety or novelty of the scenes through which the invalid passes, to awaken or divert the mind. Many have, by these means been surprised into health.

ROWING.—Rowing a boat, to those who are not daily accustomed to the task, may be ranked among the most active species of exercise. To the robust and those in perfect health this exercise, when not carried to the extent of producing very considerable fatigue, is one admirably calculated to impart strength to the arms, and breadth and development to the chest. When, however, it is too frequently repeated, to the neglect of other species of exercise, it is very apt to produce a partial and ungraceful expansion of the frame.

FRICTION.—Friction of the skin, in conjunction with regular bathing, forms a very important means of preserving and improving the health of the body. It removes thoroughly from the surface every species of impurity which may accidentally adhere to it, and promotes the freedom of the blood's circulation in the minute vessels of the skin. It promotes the growth and development of the muscles—invigorates the digestive organs, and imparts a comfortable glow and an increased energy to the whole system, by which it is rendered less liable, during cold and changeable weather, to become affected with disease.

Though useful to all, frictions are peculiarly adapted to increase the health and vigor of persons of debilitated habits, who lead a sedentary life, are subject to dyspepsia, gout, and rheumatism, or who are particularly liable to be affected by cold by slight variations of the weather. Their whole bodies, more particularly their limbs and the front part of the body, should be rubbed for a few minutes, morning and evening, with a flesh-brush or coarse towel, until the surface begins to grow red, and assumes an agreeable glow. In many cases, sponging the body with cool or tepid water, will be found to increase the good effects to be derived from the practice. It is preferable to the *too-often repeated* and *much abused plunging* into a bath-tub, which is practised to a great extent at present, in cities especially. Frictions are highly useful in the case of delicate females; and in children they promote their growth and activity, and prevent many of the diseases to which they are liable.

The best time for using friction is in the morning, when the stomach is not distended with food. They who are subject to wakefulness and disturbed sleep will find, in addition to a properly regulated diet and active exercise in the open air, that sponging the body with tepid water, followed by brisk frictions of the surface, will more effectually induce quiet repose than any other means.

APPETITE.

THERE are three kinds of appetite: 1st. The *natural* or *healthy* appetite, which is stimulated and satisfied with the most simple food; 2d. The *artificial* appetite, or that excited by condiments, liquors, pickles, high-seasoned dishes, variety of food, wine, &c., and which remains only so long as the operation of these stimulants continues: 3d. The appetite *of habit*, or that by which persons enjoying no inconsiderable health, accustom themselves to take food at stated hours, but frequently without relishing it. The *true* and *healthy* appetite alone can ascertain the quantity of food proper for the individual. If we were seldom to trespass the due limits of temperance, our natural appetite would be able accurately to determine how much food we may consume with satisfaction and benefit, but the age we live in is one of intemperance in eating as well as everything else. If, after a meal, we feel ourselves refreshed, and as cheerful as before it, or more so, we may be assured that we have taken no more than a proper quantity of food; for, if the right measure be exceeded, torpor, heaviness, and relaxation, are the necessary consequences; our digestion will be impaired, and a variety of complaints gradually induced.

HUNGER.—As a general rule, the sensation of hunger should as seldom as possible be allowed to occur; for, although the old proverb, "hunger is the best sauce for our food," is true, if the term hunger be used merely to signify *keenness of appetite;* yet, the moment it becomes a painful sensation, the stomach and other organs suffer, and the energies of the system are, to a certain extent, prostrated.

ABSTINENCE.—Abstinence from food, for a limited period, is often, during health, of very great importance; it is one of the most powerful means of obviating the effects of any accidental excess, of warding off an impending attack of disease, and of removing those disorders of the stomach, incident upon the introduction into it of food of an improper kind. Occasional abstinence from food, by omitting a meal or two, or substituting for an animal diet, a bowl of gruel, or a slice of bread and tea, restores the force of the digestive organs, by diminishing their action, and giving them rest, and time to resume their healthful energies; while, at the same time, when the system is rapidly verging into disease, or the vessels are overloaded with blood, it removes from the first a stimulus which might increase its deviation from health, and upon the second, it acts as an evacuant, by allowing the secretions time to remove from them their excessive amount of fluids. The studious, as well as they who lead sedentary lives, are especially benefited by occasional abstinence; as such persons, from the want of sufficient exercise, are generally the severest sufferers from diseases of repletion, and from a disordered state of the digestive organs.

Food.—Excess of food, even of the lightest and most whole some kind, interrupts digestion, oppresses and irritates the stomach, produces a feverish heat of the surface, loads the vessels with an excess of blood, and when sufficient exercise is not taken, renders the body unwieldy, by the accumulation of fat beneath the skin, and around the internal organs. Partaking of a *great variety* of food at one meal, is injurious; it causes more to be eaten than is proper, and impedes the digestive powers of the stomach. With respect to the solid or fluid nature of our food, a certain degree of solidity assists its digestibility; soups, jellies, gravies, and the like, are more readily digested, when bread or other solid substance is added to them, than when they are eaten alone. A sufficient *bulk* of food in the stomach, to give it a gentle stimulus and distension, is absolutely necessary for healthy digestion: it is on this account, that all condensed articles which contain much nutriment in a very small space, are unwholesome. In regard to the concentration of food, very erroneous and injurious opinions generally prevail. It is supposed, by most persons, that by extracting the nutritious principle or principles of any given article of food, they are able, with greater certainty and effect, to nourish the body of the sick and delicate; thus, we continually hear of strong beef-tea, pure arrow-root jelly, and the like, being prepared with great care for such persons. But many of our readers will be much surprised to hear, that dogs and other animals, fed on the strongest beef-tea, or pure jelly alone, rapidly emaciate, and die within a short period, and that precisely the same consequences would ensue, were the strongest man confined to the same food. A certain *bulk,* therefore, of food taken into the stomach, is essential to nutrition; and all attempts to combine *too much nutriment in too small a mass,* materially impairs the wholesomeness of our food.

Vegetable Food.—Although vegetable food requires a longer time to digest in the stomach than animal, and notwithstanding the latter presents a larger amount of nutritive matter in a smaller bulk than the former; yet the human system can derive from vegetable food as great a quantity of suitable nourishment as from animal, while the former produces much less excitement and heat, and is far less liable to produce over-fullness of the blood-vessels, or to predispose the organs to disease. As a general rule, it will be found that they who make use of a diet, consisting *chiefly* of vegetable substances, properly cooked, such as rice, oatmeal, potatoes &c., have a manifest advantage in looks, strength and spirits, over those who partake largely of animal food; they are remarkable for the firm, healthy plumpness of their muscles, and the transparency of their skins. The diet of children, and young persons generally, should consist almost exclusively of vegetables and milk. In summer, and in warm climates, a greater proportion of vegetable food is required than in winter and in cold climates. They who, with a sufficiency of daily exercise in the open air, to preserve the activity of the digestive organs, nevertheless spend ordinarily a life of ease and comparative inaction, will find their health and comfort better promoted by a diet principally vegetable, than by animal food. Towards the decline of life, though, the amount of animal food should be gradually increased.

ANIMAL FOOD.—Man is destined to live upon both animal and vegetable food, and a proper combination of both constitutes the aliment which, generally speaking, is best adapted to his taste, and the one by which the health and vigor of his system is under most circumstances best sustained.

The nourishment communicated by both animal and vegetable food is much the same; but the animal product is the most easily separated by the digestive organs, and is afforded in the greatest amount. The blood of the individual who partakes largely of animal food, is richer than the blood of those fed principally upon a vegetable aliment. The first gives, likewise, a greater tendency to inflammatory affections than the latter. For those who are accustomed to active and laborious employments, a greater amount of animal food will be proper than for the sedentary and inactive. Infants require less animal food than children, children than adults, and women than men. In summer, the quantity of animal food should always be diminished, whatever may be the habits or occupations of the individual. In winter, and in the more northern climates, a more permanent and stimulating nourishment is required than under opposite circumstances: this is best afforded by animal food. The different *kinds* of animal food differ in the degree of nourishment they afford, as well as in the ease with which they are digested. Thus, the flesh of *full-grown* animals is much more digestible and nutritious than that of their young; and as it respects the larger animals, this rule is without exception. Beef and mutton, for example, are more easily digested, and more wholesome than veal and lamb. The *sex* of animals, too, influences the nature of the food; the flesh of the female being more delicate than that of the male. The mode of killing, too, gives a tenderness to the flesh. Hunted animals are more tender than those that are killed on the spot. The flesh of animals which are allowed to range freely in the open air, is more wholesome and nutritious than of such as are stall-fed. In general, the flesh which is dark colored, and which contains a large proportion of fibrin, is more digestible and nutritious than the white flesh of animals. The *black meat* of fowls, so called, is more nourishing than the white meat, (the breast, &c.) Thus, the flesh of domestic fowls is not so readily dissolved in the stomach as that of the different kinds of game. By cooking, animal food is changed in its texture, being generally rendered softer, and easier of digestion; but by improper modes of cooking, a reverse effect is produced, the food being rendered indigestible, unnutritious and unwholesome.

VARIETIES OF ANIMAL AND VEGETABLE FOOD.—*Gelatine*, or animal jelly, is highly nutritious; but in its separate or concentrated state, it is difficult of digestion; hence, the impropriety of the dyspeptic, and persons of weak stomachs generally, being fed upon strong soups, calves' feet jelly, and similar articles of food.

CALVES' FEET JELLY.—Plain calves' feet jelly, or that which is sweetened, is grateful to the palate, very nutritious, and not very difficult of digestion; it is sometimes a useful article of diet for convalescents; it may be taken cold, or dissolved in warm water, according to circumstances. It should, however, only be given occasionally, or in moderation; for jelly, like all other concentrated food, is not so readily converted into blood, as many other articles which contain a less amount of nutriment. The addition of wine and spices to the jelly, renders it an improper article of diet under most circumstances.

ALBUMEN.—The purest example of albumen is that presented by the white of the egg; it nevertheless enters largely into the composition of many of the animal fluids and solids. As an article of food, it is at once readily converted into blood, it being taken up by the absorbent vessels, without its being required to undergo digestion, while at the same time it is highly nutritious. The injurious effects resulting from the eating of hard boiled eggs, are occasioned in a great measure by the effects of the heat upon the oily matter of the yolk.

MILK.—This is one of the most valuable presents which a bountiful providence has bestowed upon man. In many instances, either alone, or in combination with vegetables, it has formed the sole sustenance of life—maintaining fully the health and robustness of the system, without any of the disadvantages which result from an excess of animal food on the one hand, or the diminished strength and vigor which have been supposed to be the effect of a purely vegetable diet, on the other.

Incalculable would be the benefits which would result to the working and laboring classes of our country, were they to substitute this wholesome and nourishing food in their families, for the expensive and unnutritious slops, which, under the name of tea or coffee, constitute the chief of their morning and evening meals; or, to substitute a tumbler of milk for the pernicious dram of ardent spirits, beer, porter, or ale.

For children, milk with bread, or a simple preparation of milk with rice, or with eggs and sugar, is perhaps the best and most wholesome food that can be devised : it should, at least, form the principal part of their nourishment for the first few years of their life.

Milk, to be perfectly wholesome, should be drawn from *sound, young* animals, supplied with a sufficiency of their natural food, and allowed free exercise in the open air. The best mode of using it, is, undoubtedly, in its *raw* state, and when it has stood about two hours after being drawn.

Eggs.—Eggs contain a great deal of nourishment in a small bulk ; and when perfectly fresh, and *soft* boiled, they constitute a species of food of very easy digestion. When hard boiled, and especially when fried, they are indigestible and stimulating.

Cheese.—All kinds of cheese are of difficult digestion ; and can with safety be made use of, only in very small quantities, as a condiment along with other food. The idea entertained by many, that a portion of old cheese taken with the desert aids digestion, is perfectly absurd. When cheese has advanced very near to a state of putrefaction, though eaten by certain epicures, it is at once disgusting to the senses, and injurious to the stomach. Certain changes which cheese occasionally undergoes, impart to it poisonous properties. Roasted or cooked cheese, is very indigestible, and liable to occasion painful sensations in the stomach, headache, sour belchings, feverish heat of the skin, and disturbed sleep. When eaten, cheese should always be combined with a large portion of bread.

Butter.—Butter is used as a sauce to many articles of food, and is frequently added to flour to be baked into cakes and pastry, and it is in both these forms injurious, for though it does not produce effects that are immediately apparent, it lays the foundation of stomach complaints of the greatest obstinacy. Its use in this form is also very apt to give rise to diseases of the *skin*, very difficult to cure. Persons laboring under stomach complaints should not use much butter in any form. It is also very unwholesome when heated. It is a bad part of the management of children, to pamper their palates by frequently indulging them with butter; as it is apt to give rise to a gross and unhealthy habit of body, characterized by the frequent appearance of boils and other sores, discharges from behind the ears, &c., or eruptions on the head, and other parts of the skin.

Fat affords a rich nutriment, requiring, however, strong powers of digestion, and adapted only to the healthy and laborious; it is more wholesome, however, when eaten with a proper quantity of lean, or with a considerable addition of farinaceous aliment in the form of potatoes, bread, rice, &c., &c.

Beef.—Beef affords a strong, easily digested, and wholesome nourishment; it should be tender, fat, and well mixed, (lean and fat) and taken from a bullock of middle age.

Of its different parts, the *fat* is less easily digested than the lean; the *tongue* and also the *tripe*, being of a more dense texture than the other parts, are more indigestible. The best mode of preparing beef, is by roasting, or boiling. Beef-steaks appear to be the form, however, in which its nutritious qualities are best retained.

The excessive body of fat which is accumulated upon what is called *prize beef*, adds nothing to its goodness, but on the contrary, renders it less wholesome and nutritious.

Mutton.—Mutton is a highly nutritious and wholesome meat. It appears to be the most digestible of all animal food. The flesh of the male animal, however, has in general so strong and disagreeable a taste, and is, besides, so exceedingly coarse, and difficult of digestion, that it is only adapted to persons of strong digestive powers. *Ewe-mutton*, if it is more than between three and four years old, is likewise tough and coarse. *Wether-mutton*, or the flesh of the castrated animal, is most esteemed, and is by far the sweetest and most digestible.

Lamb being less heating, and less dense than mutton, is better suited to persons convalescent from acute diseases; but by the majority of patients laboring under indigestion, or any other severe affection of the stomach, it is not found so digestible or proper a diet as wether-mutton. It is, however, to persons in health, a light and wholesome food, especially when the lamb is not killed too young. A lamb that has been allowed to suck five or six months, is fatter and more muscular, and in every respect better, than one which has been killed when two months old, and before it has had time to attain its proper consistency.

Venison.—The flesh of the deer is reckoned a great delica cy ; it is nutritious, savory, and easy of digestion.

Veal.—The flesh of the calf, like that of all young animals, abounds in gelatinous matter ; it is far less easy of digestion than the flesh of the ox, or beef. For persons in health, the most proper mode of cooking veal is by roasting or baking.

Veal Broth produces a laxative effect upon the bowels, and is a very suitable food for persons troubled with costiveness.

Pork.—Good *pork* is unquestionably a very savory food, and affords strong nourishment, well suited, as an occasional diet, to persons who lead an active or laborious life ; but it is not easily digested, nor can it be considered so wholesome as beef or mutton. The too frequent and long continued use of this meat favors obesity or fatness, and is apt to disorder the stomach and bowels, and occasions pimples or boils upon the skin.

Bacon.—It is a strong, very indigestible, and stimulating food, adapted only to persons of robust frame, and accustomed to laborious occupations. The best mode of cooking bacon is by boiling it with vegetables. When fried with eggs, it is decidedly unwholesome.

Ham.—When properly cured, and when *boiled*, ham is a very palatable and wholesome food. It is, however, stimulating and difficult of digestion, and only suited to such persons as are in full health and exercise much in the open air. Fried ham is still more indigestible than that which is boiled.

Sausages.—In whatever form they are eaten, sausages are an indigestible and unwholesome food, fitted only for the stomach of the most robust. When sausages have been long kept, particularly in a damp place, they are apt to undergo certain changes, in consequence of which they become poisonous.

Poultry.—Poultry, in the common acceptation of the term, includes all the domesticated birds used as food, as the common fowl, turkey, duck, and goose. In point of digestibility they rank nearly in the order we have enumerated them.

Chicken Soup.—Chicken soup, when properly prepared, is a light food, adapted to many invalids and to persons convalescent from fevers. For their use it should be prepared from the fleshy or lean parts of the chicken, well boiled in water with a little salt, the scum and fat being taken off as it rises.

Fish.—Fish are less nutritious than the flesh of warm-blooded animals, while to most stomachs they are more difficult of digestion. When used habitually, they are apt to induce dis eases of the skin and disorders of the bowels. The fat of fish is still more indigestible than that of other animals, and readily turns rancid on the stomach. When not in season, all kinds of fish everywhere, are very indigestible and unwholesome. The best mode of cooking fish is by boiling; stewed or fried fish are very indigestible. Salted and dried fish are a still more unwholesome food than such as are eaten fresh. Butter and the acid fruits form improper sauces for fish, causing it almost always to oppress and irritate the stomach : nor should *fish and milk ever be taken at the same meal;* this combination has frequently occasioned severe bowel complaints.

Salt-water fish are the best. Those fish which have scales are, in general, the most easily digested, and the best.

SALTED MEAT.—Salted meat is more difficult of digestion than that which is eaten fresh ; it is also less nutritious, both from the pickle in which it is immersed washing out, as it were, a considerable amount of its nutritive parts, and from the chemical change which it always undergoes to a greater or less extent. When used as food, salted meat should always be well boiled, and eaten with a large quantity of vegetables.

CRABS AND LOBSTERS, in whatever manner cooked, are indigestible and decidedly unwholesome.

THE MUSSEL, a shell-fish often used as food, is highly indigestible and unwholesome.

OYSTERS, when taken raw or after being slightly cooked by roasting, are a light, nutritious, and easily digested food. The hard white part, or eye, should always be rejected. When *thoroughly* cooked, particularly when stewed or fried, oysters constitute, on the other hand, one of the most indigestible and pernicious articles of food in ordinary use. When out of season, oysters are always unwholesome. The juice of the oyster, thickened with grated biscuit and warmed, is sometimes an excellent diet for persons laboring under great delicacy of stomach.

SOUPS.—For the laboring classes generally, there is scarcely a more wholesome and economical article of diet than soup. We allude now to the ordinary domestic soups, prepared from beef, mutton, or veal, with the addition of various vegetables. In the preparation of soup, the meat and vegetables should be well boiled, and whatever seasoning is added to increase the flavor, care should be taken that it be not thereby rendered too stimulating. The combinations of flour and butter, which are sometimes met with in soups, under the denomination of dumplings, are highly indigestible and improper. Soup should always be eaten with plenty of bread; this gives it that degree of consistency which, in all our food, appears to cause it to be the most readily acted upon by the stomach.

Many suppose that soups generally are calculated only for those whose powers of digestion are weak; but this is a mistake, the reverse being generally the case. When the digestive powers are weak or deranged, it will almost always be found that solid food agrees the best, particularly solid animal food; this the stomach seems to digest with ease and in a very short time; whereas, liquid food is apt, in such cases, unduly to distend the stomach and to require a greater strength of digestive power.

RICE.—When mixed with other food, it furnishes a wholesome article of diet. Rice is supposed to be in some degree astringent; and in looseness of the bowels, the water in which it has been boiled forms an excellent drink. By its mild mucilaginous properties, it aids greatly also in allaying irritation in all diseases of the bowels.

OATS.—The meal obtained by grinding the grain of oats affords a wholesome and nutritious food, used boiled with water, in the form of gruel, or made into thin cakes.

GRUEL.—By gruel is generally understood oat-meal or Indian meal boiled in water. Thin plain oat-meal gruel, or a gruel made in the same way from Indian meal, is a useful diet for convalescents from fevers, and for those who have committed an excess in eating.

RYE affords a meal, the food prepared from which, though less nutritious than wheat, is nevertheless wholesome and sufficiently nourishing. Rye bread is more difficult, however, of digestion, and is apt to turn sour in the stomach and to irritate the bowels.

BARLEY.—Barley forms an excellent article of nourishment when boiled in water, or made into cakes. Barley bread is not, however, a very pleasant nor wholesome food.

BARLEY-WATER.—The water in which barley is well boiled, forms one of our best drinks, in various fevers and other diseases.

MAIZE, OR INDIAN CORN.—The meal made by grinding Indian corn, prepared in various ways, but especially when made into mush, or with the addition of wheat flour baked into bread, furnishes a most wholesome, nourishing, and palatable food, and one well adapted for the support of the active and laborious generally. Indian bread, properly prepared, were it not from habit and fashion, would recommend itself to every palate by its agreeable flavor, and the beauty of its appearance; it is far preferable to the ordinary bread made from wheat alone. To make this bread, a mush should be made of the Indian meal in the usual way; into this, when cold, with the addition of a very small quantity of warm water, and a little salt and yeast, is to be kneaded a sufficiency of wheat flour to make it into a paste; when sufficiently raised, it is to be again kneaded, and baked in the same manner as bread.

BREAD.—New bread is particularly unwholesome and indigestible. The only apparent exception is in the case of new rolls, which healthy stomachs manage to digest pretty well, provided they be well baked, and the crust bears a considerable proportion to the whole.

Bread slightly *toasted*, but not burned, is a wholesome diet, especially for persons upon whose stomachs most articles of vegetable food, including bread in its ordinary state, are apt to turn sour. In eating toast, the butter should not be spread upon it until it is cold.

PANADO.—The crumb of wheaten bread softened with boiling water. It forms an excellent diet for children; for those affected with fevers, and for women during the first days after delivery.

GINGER-BREAD.—When well baked, and eaten in moderation, it affords, under many circumstances, a useful stimulus to the stomach. It is an excellent article for individuals going to sea; it being frequently, in cases of sea-sickness, retained on the stomach, when every other article is immediately rejected. Travelers, also, on setting out early in the morning, will find, that eating a small portion of it, will afford a grateful stimulus to the stomach, when they have been obliged to commence their journey without breakfasting.

PASTRY, or dough mixed with butter, is used in a great variety of forms, and though grateful to the taste, is highly indigestible, and injurious to health. At dinner, in the shape of pies and tarts, pastry is thrown into the already loaded stomach, and the over-taxed powers of that organ are unable to digest what is difficult to manage when they are the most vigorous. To children, pastry is peculiarly unsuitable; they who use it much, are subject to runnings from the ears, disorders of the bowels, eruptions on the skin, and inflammatory complaints of various kinds.

PUDDINGS, when composed of flour, or crumbs of bread, combined with suet and dried fruit, are extremely indigestible, and constitute one of the most unwholesome dishes served at meals. Puddings and dumplings made of batter, baked or boiled, are also indigestible, and unwholesome. Bread and milk pudding, as well as rice pudding, is readily digested, and may be eaten in moderation, without injury.

Sago, boiled with water, or milk, furnishes an agreeable and nourishing jelly; it is easy of digestion, and excites the system but little; and is an excellent article of diet for convalescents and for children.

Potatoes constitute an article of diet, which, whether we have reference to the nourishment it affords, the agreeableness of its flavor, its wholesome qualities, and the extent to which it is consumed, is certainly of the greatest importance to man. Potatoes are the lightest and most nutritious of those vegetables which are served at table in their natural state; and, next to bread, the very best accompaniment to every kind of animal food. The dry, mealy kinds are the best, and should always be preferred to those which are hard and waxy. The best manner of cooking the potato, is by boiling in two waters, or by roasting. Finely mashed, or fried potatoes, are indigestible and oppressive to the stomach.

Cabbage affords but little nutriment, is very flatulent, and where the stomach is delicate or irritable, it is very apt to produce uneasy sensations, cholic, &c. Boiling in two waters deprives it, in a great degree, of that unpleasant taste and smell, which are so disagreeable to many palates.

Sourcrout forms an excellent and wholesome vegetable food for the crews of ships destined for long voyages; and for all persons so situated as to be deprived of a sufficient supply of fresh vegetables. In regard to its effects upon individuals, whose powers of digestion are impaired, the same remarks will apply as to cabbage in its recent state.

COOKERY.

When meat is boiled too long or too fast, if it contains much albumen, as in beef, we shall obtain a hard and indigestible mass, like an over-boiled egg; or in young meats, such as veal, where there is more gelatine, the result will be a gelatinous substance, not easily digestible. Young and viscid food, therefore, as veal, chickens, &c., are more wholesome when roasted than when boiled, and are easier digested. Boiling is very properly applied to vegetables; as it renders them more soluble in the stomach, and deprives them of a quantity of air and other particles which are pernicious to weak stomachs. The *quality* of the water used in boiling requires some atten tion; mutton boiled in hard water is more tender and juicy than when soft water is used, while hard water renders vegetables harder and less digestible.

ROASTING.—By this process, the fibre of meats is made crisp the fat melted, and the water evaporated. When underdone roasted meat may be more nourishing; but, from the closeness of its texture, it will not be so easily digested. Meat loses more by roasting than by boiling; by boiling, mutton loses one-fifth, and beef one-fourth; but by roasting, they lose one-third of their weight.

FRYING is, perhaps, the most objectionable of all the operations of cookery. The heat is applied through the medium of boiling oil or fat, which is rendered scorched, and therefore extremely liable to disagree with the stomach.

BROILING.——By this operation, the sudden browning or hardening of the surface prevents the evaporation of the juices of the meat, and imparts a peculiar tenderness to it. But the over-excited health brought on by eating meats thus cooked, is peculiarly liable to become changed into disease from very slight causes.

BAKING.—Baked meats are not so easily digested, on account of the greater retention of their oils. Such dishes, accordingly, require the stimulus of various seasonings to increase the digestive powers of the stomach. As there is often much pastry, made with butter, used to confine the juices of the meats baked, such accompaniments render meat pies of all kinds of food the most difficult of digestion.

STEWING has a similar effect to boiling in depriving the meat of much of its nourishing juices; but as the fluid in which the meat is stewed is made use of as food in connection with the latter, little nourishment is absolutely lost by this mode of cooking. Stewed meat is less easily digested than that which is boiled; it is also more stimulating. Simple stewing is a mode of cookery well adapted for the food eaten by those of robust frames and laborious habits.

MEALS.—Regularity in the number of meals, and the periods at which they are taken, is of the first importance; on it much of the equable and pleasant enjoyment of health depends. In general, three meals, in the course of the day, seem the most desirable, and the best adapted to the wants and constitution of the human frame; while, at the same time, this number is best suited to the powers of the digestive organs.

The practice which leaves the great bulk of the day without a meal, and then crowds two or three together, is manifestly bad, as it produces in the body a state of exhaustion and fatigue, which strongly tends to enfeeble the powers of digestion To confirm and preserve health, whatever may be the number of meals taken, they should be eaten at regular times and stated periods. Six o'clock dinners, are only another way of destroying health, and shortening life. The extremes of too long fasting, and too frequent repletion, should be carefully avoided ; for the langor of exhaustion and the fever of repletion, are equally injurious to the healthy state of the stomach : its muscular fibres are debilitated by excess ; while a collapsed state of the organ occasions its loss of tone and energy, and superinduces constitutional weakness. And it should be remembered, that one meal should be duly digested before the introduction of another into the stomach.

BREAKFAST.—During sleep, the whole of the food taken the previous day has probably been digested ; but, in general, it is proper to interpose some time between rising and taking breakfast, and take some light exercise.

DINNER.—The period for dining appears to be well chosen for the active classes of society more especially. Dinner should always consist of one kind of *meat*, plainly cooked. Variety of food, like too much seasoning, keeps up the appetite after the wants of the system are satisfied ; the stomach is oppressed by too great a quantity of food, and digestion is impeded even to a greater extent than were the same amount to be eaten of a single dish. Let it be recollected, also, that dishes compounded of a number of ingredients, the natural qualities of which are completely disguised, by the refinements of cookery, are altogether unwholesome : many of them are little better than poisons. It is all-important that sufficient *time* should be allowed for this meal, in order that the food may be properly *chewed*, without which its digestion will be greatly retarded. If the food be sufficiently plain and juicy, thirst will seldom be experienced ; but when a desire to drink is experienced, a moderate draught of water will be proper. But no other liquor should be taken—water is the only natural diluent of our food, every other liquor impedes its digestion. The custom in use among some people of taking drams or bitters before meals, for the purpose of whetting the appetite, is highly pernicious, and has quite a contrary tendency to that designed, as it relaxes the stomach, and consequently enfeebles t for the operations it has to perform. Nor is the fashion of taking wine, or brandy and water, during dinner, less reprehensible. The use of bottled cider, porter, or beverage, during this meal, is also injurious, as it unnecessarily distends the stomach, and thus prevents its muscular contractions, at the

rery time when it is necessary they should be brought into
action, and preserved in their full vigor. To say the least of
all these vulgar errors in diet, they check the process of digestion, and paralyze the powers of the stomach.

SUPPER.—As the powers of the body, and digestion among
the rest, are diminished in their activity during sleep, it is an
unsafe measure to load the stomach at bed-time with a quantity or various kinds of food. Do not eat meat for supper.
Under no circumstance should food of any kind be taken for
two or three hours before retiring to rest.

DRINKS.—In warm weather a much greater quantity of drink
is demanded, than when the atmosphere is temperate or cold.
This arises from the stimulating effects of heat upon the system; but chiefly by the waste of the fluid portion of the blood,
occasioned by the increased perspiration. For the same rea
sons, active exercise or labor augments the thirst. Salted,
high-seasoned, and all stimulating food increase the demand
for drink, by stimulating the lining membrane of the mouth,
throat, and digestive organs, and increasing the viscidity and
exciting properties of the blood. Nature calls for water to
take out the salt or other stimulating substances contained in
the food, so that digestion may be more easily accomplished.

Persons in good health, generally, take a great portion of
their drinks, especially at dinner, of the temperature of the
atmosphere; but in weaker stomachs, the drinks may be required to be a little warmed, though it is seldom safe to take
them habitually *very hot;* and far less is it proper to chill the
energies of the stomach, by cold or *iced drinks.* The quantity
of drink taken, is also of much consequence to good digestion;
a large volume of fluid will prevent the food from being properly acted upon by the stomach; and if there be too little,
the mass will be dry and hard. Different kinds of food require
different quantities of liquid: animal food requires more than
vegetable; roasted, more than boiled; and baked meat, more
still than roasted. To drink much *before* a meal, is unwise;
but to drink more or less, during a meal, according to the nature of the food, assists digestion.

TOAST-WATER is perfectly wholesome, and agrees frequently
with persons whose stomachs do not relish pure water. It has
a slightly nutritive quality, and may be allowed in all the feverish and other cases, where cooling drinks are proper.

WHEY affords a bland, easily assimilated nourishment
increasing the secretions, and tending to produce a beneficial
change in the fluids of the body It contains a considerable

amount of sugar, which renders it sufficiently nutritious. As a drink, whey is well adapted to allay thirst in hot weather.

Buttermilk contains but little nutritious matter; but, in warm weather, it forms an excellent cooling drink, and, with bread, may constitute a considerable part of the diet of children.

TEA.—The properties of tea seem to be those of an astringent and narcotic; but like some other narcotics, in small quantity, its first effect is that of a very gentle stimulant, and certain kinds of it, when taken pretty strong, and near the usual time of going to rest, have the effect of keeping off sleep; but when weak, and taken moderately, and tempered with cream and sugar, it acts merely as a grateful diluent, and produces a slight exhilaration.

Tea and coffee, when used in moderation, are beneficial, by preventing the waste of the tissues of the body. Of course, where they disagree with the system, as we observe in some temperaments, they must be discontinued.

The green and high-flavored teas are those which are the least wholesome. Tea should not be taken soon after dinner.

The following rules, respecting the use of tea, will be found useful:—Carefully avoid the high-priced and high flavored teas, more especially if green. Take with it, at all times, a good proportion of milk, and some sugar, as correctives to any possible noxious qualities present. Make the infusion properly, with water, soft, and otherwise of a good quality, and in a boiling state. Take less tea in the morning than in the evening.

COFFEE.—The infusion of coffee acts as a stimulant upon the stomach, the heart and the nervous system, increasing the circulation of the blood, augmenting the heat of the skin, and exhilarating the mind; these, its immediate effects, are followed, however, by an equal degree of depression in the functions of those several organs: the excitement and subsequent degression being in proportion always to the strength of the infusion, and the quantity drank. Coffee bears a strong analogy, in its effects upon the system, to wine, ardent spirits, and opium; from the latter, its effects, however, are very different in degree. Coffee, therefore, when drank very strong, or indulged in to excess, is unquestionably injurious; it seldom fails to disorder the stomach, impair its digestive powers, and in delicate habits it often occasions watchfulness, palpitation of the heart, headache, and many of those complaints, vaguely denominated nervous. To the dyspeptic and sedentary especially it forms a very improper article of diet. When taken weak and with plenty of cream or milk and sugar. It may however be indulged in sparingly, by persons in health, and who lead active lives, without much inconvenience. Coffee should never be taken late in the evening, in consequence of its tendency to prevent sleep.

CLEANLINESS.

Personal cleanliness consists in the careful removal of every impurity from the surface of the body, whether generated by itself and attached to the clothing in immediate contact with it, or contracted from the air and other surfaces with which the body is accidentally or necessarily surrounded. Allowing impurities to accumulate upon the surface not only gives rise to a disgusting effluvia, by which the air the individual breathes is contaminated, but occasions various eruptive diseases of the skin, and the internal organs sympathize with it. Whatever suspends or impedes the functions of the surface, whether cold or dirt, a derangement to a greater or less extent of the internal organs invariably results. By the action of the numerous blood-vessels of the skin, there is removed from out the system, in the form of an insensible perspiration, an immense amount of matter, the retention of which would be productive of uncomfortable feelings or disease.

The means of preserving the purity of the skin is frequent ablutions with water, with the addition occasionally of soap and friction. The ablution of the body should be frequent and general, and not confined simply to the face, neck and hands. Washing the entire surface in water of a proper temperature, and at short intervals, would greatly increase the strength, health and pleasurable feelings of all, whatever may be their sex, age, or condition in life. Frequent change of clothing is equally necessary. Domestic cleanliness is of scarcely less importance to the preservation of health than that of the person and clothing. From every apartment of our dwellings, as well as from the yards, cellars, vaults, and enclosures attached to them, during every season of the year, every thing which by its decomposition is liable to contaminate the air, be immediately removed.

We should prevent all accumulations of filth and stagnant water in the streets, courts, and alleys of towns, drain marshes, fill deep ditches and ponds with stones, etc., and frequently, in warm weather more especially, dissolve one pound of chloride of lime, or copperas, in a pail of hot water, and throw same down the sinks, privies, &c.—

Never leave slops, or any thing offensive, in sleeping apart
ments, but let them be removed at once, more especially in
cases of sickness. Many families have been prostrated with
various forms of fevers, dysentery, or other diseases, by the
slops from the kitchen being thrown into the yard, and there
decomposing, and generating poisonous gases.

Unless their debility be very great, and unless it be
productive of much pain and suffering to move them, the bed
and body linen of the sick should be kept very clean, and frequently changed; their apartment should be cleaned and wel'
aired, and all offensive discharges should be very carefully and
speedily removed.

Soap.—In addition to the perspiration which is thrown out
by the skin, a portion of which always remains upon the surface, the latter is constantly lubricated by an oily fluid. This
oily exudation greases the linen when it is worn for too long a
time—catches the dust floating in the air, and causes it to adhere to the skin, and likewise retains in contact with our
bodies, a portion of the dead matter, which it is the office of
the skin to discharge from the system. The removal of this
deposit, which is constantly accumulating, is absolutely necessary, as well for personal comfort as for the preservation of
health. It cannot be effectually removed without the occasional use of soap, with which it combines without difficulty.
Washing all over, with soap and water, occasionally, is very
necessary.

Mechanics, and they who, from any cause, are peculiarly
liable to have deposited upon their skin, dust, dirt. or any
foreign matters, should wash with soap and water often, and
also rub afterwards well with a rough dry towel, as well for
the preservation of the skin as of their health generally.

If you can only get a bowl of cold water, some good soap
and a rough towel, you can have all the advantages to be
derived from the most fashionable bath-tub—only being careful to avoid getting cold, by thorough rubbing of the surface
with a dry towel, after the washing all over in soap and water.

The ordinary brown and yellow kinds of soap are altogether
unfitted for cleansing the skin, as they irritate it, and when
frequently used, most generally cause it to become rough,
chapped, or covered with painful and unsightly pimples. Most
if not all, of the colored and variegated soaps, prepared express-
ly for the toilet, are equally objectionable. Pure *white soap*
ought, therefore, to be invariably used in ablutions of the face
and hands, or of the surface generally.

A Curved, Round-Shouldered, and Erect Spine.

How very distressing, and yet how common it is to see curved or deformed spines. The habits of children, especially of girls, if not corrected in time, create a fearful frequency of this spinal defect. Nature has given to all, both male and female, a sufficiency of bone and muscle to sustain them in the most graceful and healthy position, and when these are correctly and faithfully used, and their strength developed, they fulfil their intended purposes, and keep the form erect. Look at the following illustrations of improper positions of the body, and you will be able to tell the cause of these deformed spines.

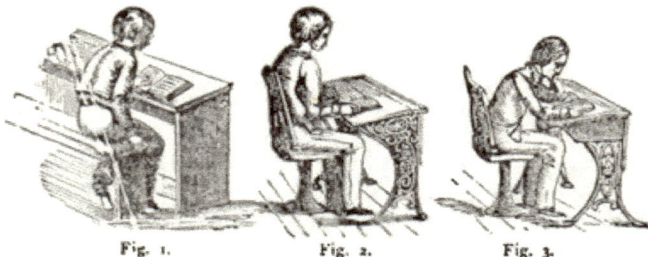

Fig. 1.   Fig. 2.   Fig. 3.

Old Style and New Style of Desks and Seats for Schools.

Too many schools are furnished with seats of the same uniform height. If they are high enough for the larger scholars, they are too high for the smaller children. (See fig. 1.) In sitting, a child should find a support for the back, and rest for the entire thigh-bones and feet, otherwise the bones of these, being soft and growing, are liable to become distorted, or out of shape. Fig. 2 represents a proper position, and fig. 3 an improper position, for sitting.

Proper and Improper Positions to lie in Bed.

Curvatures of the spine may be caused by too many pillows upon which the head rests while in bed, as represented above. Young persons should lie as nearly level as possible, with the head but slightly raised if at all. As they advance in life, a more elevated position of the head may be desirable. Most people lie upon the right side; some lie upon the back, but this latter position is not favorable to those who are liable to nightmare. A frequent change of position is very desirable. If you awake during the night, change your position.

SLEEP.

It is highly important that every body should understand that sound, refreshing sleep is of the utmost consequence to the health of the body, and the vigor of the mind.

Among the marks and symptoms of long life, that of being naturally a regular and sound sleeper, is considered to be one of the surest indications. Great watchfulness, by accelerating the consumption of the fluids and solids, abridges life, and a proper quantity of repose must tend to its prolongation.

QUANTITY OF SLEEP.—What number of hours are necessary to be passed in sleep, is a question that has occasioned much discussion. The opinion generally entertained by the ablest physicians, is, that from seven to eight hours, in the four-and-twenty, constitute, generally speaking, the proper time, and that this period should scarcely ever be exceeded by adults, in the enjoyment of health, though the delicate require more than the vigorous, women more than men, and very young children more than either; but it is worthy of particular remark, that the sick and weakly seldom require more than eight hours, or at the most, nine hours, and will rarely, if ever, fail to be injured by a longer indulgence.

PROPER TIME FOR SLEEP.—Nature certainly intended exercise for the day, and rest for the night. Working at night and sleeping in the day time will, sooner or later, destroy the best constitution. Another point to be considered is, that by the custom of sitting up late at night, the eyes suffer severely, day-light being much more favorable to those delicate organs, than any artificial light whatsoever.

The plan of going to bed early, and rising betimes, has been called the golden rule for the preservation of health and the attainment of long life, and it is a maxim sanctioned by various proverbial expressions.

Indulging in sleep during the day-time, and more especially after dinner, is always productive of more or less injury to health, while it is never found to produce even that temporary feeling of refreshment which results from the same amount of repose taken at night.

BEST MEANS OF PROMOTING SLEEP.—The principal circumstances to be attended to, in order to procure refreshing sleep, are, the nature and quantity of our food and exercise; the size and ventilation of the bed-chamber; the quality of the bed and of its coverings; and the state of the mind.

It is certain that a full stomach almost invariably occasions restless nights, and it is, therefore, an important rule to make a very light supper.

With some persons, the most effectual methods of procuring sleep will fail, unless exercise be resorted to in the open air. Pure air has of itself an exhilarating and soothing effect on the mind, conducive to sound repose. It is an excellent plan when the exercise of the day has been limited, to walk up and down a large room or passage for half an hour, or more, before going to bed, and the use of the dumb-bells for a part of the time will augment its good effects.

If, notwithstanding an adherence to the preceding rules, sleep is still found to be unsound and unrefreshing, a brisk use of the flesh-brush, before going to bed, or rising from the bed, and freely ventilating it, will often produce a very favorable change.

Another excellent practice, in case you have gone to bed, and cannot sleep, is to rise, shake the bed well, draw the upper clothes down to the feet, and walk about the room, warmly clad, till both you and the bed are aired. Opiates and sleeping draughts should never be resorted to, to procure rest—once resorted to, their habitual use will become necessary, as sleep will not occur without their aid; while by their prejudicial influence upon the stomach and other organs, their employment will never fail, gradually, to undermine the health of the system.

The following miscellaneous rules respecting sleep deserve to be recorded in this place: Many real or imaginary invalids lie long in bed in the morning, to make up for a deficiency of sleep in the night time; but this ought not to be permitted, for the body must necessarily be enervated by long continuance in a hot and foul air. By rising early, and going to bed in due time, their sleep will become sound and refreshing, which otherwise they cannot expect to be the case. It is an indispensable rule, that fat people should avoid soft beds, and should sleep little and rise early, this being the only chance they have of keeping their bulk within due bounds. Such persons as are subject to cold feet, ought to have their legs better covered than the rest of the body, when they are in bed. We should never suffer ourselves to doze, or fall asleep, before we go to bed. Reading in bed at night is a most pernicious custom; it strains the eyes, prevents sleep, and injures the health. Remember, sleep is sound, sweet, and refreshing, according as the mind is free from uneasiness, and the digestive organs are easy, quiet and clear.

Beds.—The use of feather beds is very common in this country, especially in the rural districts, yet there can be no doubt that they are highly injurious to health. To the invalid, and to young persons who are disposed to distortion of the spine and shoulder, they are particularly hurtful. Such as consider them a necessary luxury in the winter, should invariably exchange them for a mattress in the spring and summer. The injury resulting from feather beds is occasioned, principally, by their accumulating too much heat about the body, and in this manner causing a profuse and debilitating perspiration, and predisposing the system to the influence of slight changes of temperature. By yielding unequally to the pressure of the body, the latter is thrown into a distorted position, which being resumed regularly almost every night, is liable to cause in the young and weakly a permanent deformity. Corn husk or shuck mattresses are superior to every other kind of bed, and it is highly desirable they should be generally adopted. By those whose means will not permit the purchase of hair mattressess, those of moss or straw will be found an excellent substitute. Feather beds are more injurious to the health of children, than even of adults, and especially if they are weakly.

In very cold climates feather beds are often necessary, and the aged may often require them, in order to preserve or increase their heat, which is sometimes inconsiderable, and if lessened would prevent their sleeping.

Young people and invalids, in particular, ought to avoid many, and heavy, bed-clothes. The use of curtains to the bed should be avoided: they are injurious, by preventing the proper circulation of the air breathed by those who occupy the bed, and by accumulating dust, cause it to be inhaled into and irritate the lungs.

Beds and bed clothes are apt to become damp for want of proper airing when not constantly used. Colds, rheumatisms, and even more fatal complaints may be caused by occupying a damp bed. Beds, instead of being made up soon after the persons rise from them, should be turned down, or their coverings thrown separately over the backs of chairs, and thus exposed to the fresh air from the open windows during the day.

BED-ROOMS.—A bed-room ought not to be situated on the ground floor: an elevated apartment is particularly recommended. It should be airy, large, and lofty. The more airy a bed-room is, the better; and it will be still better if it be also exposed to the influence of the sun. A bed-room ought to be well ventilated in the day time, as it is principally occupied in the night, when all the doors and windows are shut. The windows should be kept open as much as the season will admit of, during the day.

Keeping open the windows of bed-rooms during the night ought never, however, to be attempted, but with the greatest caution, except a small space for ventilation, by lowering the top sash, when practicable, or raising the lower one slightly.

Do not sleep in a very warm room.

Unless there is an apprehension of damp, a bed-room should rarely have a fire in it. They who live in hot countries ought to be very particular regarding the place they sleep in. The apartment should be roomy, dark, shaded from the rays of the sun and moon; temperate as to heat and cold, and rather inclined to coolness than heat; while a free admission of air is allowed during the day time, the windows should be carefully closed as soon as the night sets in.

DREAMING.—As a general rule, dreaming may be prevented by whatever causes perfect and uninterrupted sleep; such as sufficient exercise during the day, temperance in eating and drinking, a cheerful and contented mind, and the avoidance of late or heavy supper, or of strong tea or coffee during the evening. Many of the sudden deaths which take place during the night, in persons apparently in the full enjoyment of health, are to be attributed to night-mare.

The *night-mare* is a certain uneasy feeling during sleep, as of great anxiety and difficulty of breathing, and of strong but ineffectual efforts to shake off some incumbent pressure, or to relieve one's-self from great inconvenience. It commonly arises from an imperfect and unhealthy digestion, from heavy suppers, and from a constrained uneasy posture of the body. Such persons as are subject to the night-mare should take no food whatever in the evening, should keep the bowels open, and should sleep upon a mattress with the head and shoulders raised.

THE PASSIONS.

JOY.—Instances are not wanting, in which this passion when unexpectedly excited and violent, has produced disease, or even immediate death; but when moderate, and existing only in the form of cheerfulness, it has a beneficial effect in preserving health, as well as in the cure of disease.

HOPE.—Of all the passions, *hope* is the mildest; and, though it operates without any visible commotion of the mind or of the body, it has a most powerful influence on the health of the one, and the serenity of the other: it contributes, indeed, so much to the welfare of both, that if it were extinguished, we could neither enjoy any pleasure in this life, nor any prospect of happiness in the life to come; but by the beneficent will of Providence, it is the last of the passions that forsakes us.

Love is one of the strongest and most absorbing passions with which the mind is affected, and has at its commencement when happy, and properly guided by reason, a favorable influence on all the functions of the body; but being often in its progress attended with other passions, such as fear and *jealousy*, it is liable to become the source of infinite disquietude. No passion undermines the constitution so insidiously, as violent and unreasonable or misplaced love. While the whole soul is occupied with the thoughts of a pleasing attachment both the mind and the body become languid from the continuance of vehement desire; and should there arise any prospect, real or imaginary, of being frustrated in its gratification, the person is agitated with all the horrors and pernicious effects of *despair.* Love, when violent and unsuccessful, frequently produces a wasting of the body, terminating sooner or later in death.

Fear —When intense or habitually indulged in, it destroys the energies of both mind and body, retards the motion of the blood, obstructs digestion, and prevents the proper nutrition of the body. Violent terror has been known, in an instant, to turn the hair perfectly white, and in other instances, to produce loss of mind, or even instantaneous death. By weakening the energies of the system, this passion disposes greatly to disease during the prevalence of epidemics.

Grief.—There is no passion more injurious to health than *grief* when it sinks deep into the mind. By enfeebling the whole nervous system, it depresses the motion of the heart, and retards the circulation of the blood; it disorders the stomach and bowels, and ultimately every other organ of the body, producing indigestion, consumption, and other chronic diseases. Grief long continued, often gives a shock to the constitution that nothing can retrieve. Grief, like fear, predisposes to an attack of epidemical diseases.

Anger is a passion suddenly excited, and which often no less suddenly subsides. The nerves are unduly excited; the pulsation of the heart and arteries, and with them the motion of the blood, are sometimes so much increased, as to occasion the *bursting of some of the minute vessels of the brain or lungs.* The stomach, liver and bowels, are often violently affected by intense anger—digestion is always disordered, a violent colic is sometimes produced, and very often all the symptoms of jaundice. Thus it is often the immediate agent in the production of fevers, inflammations, spitting of blood, apoplexy, and other acute disorders. An essential means for their subjection, is a regular, active mode of life, a mild and moderate diet, and the abandonment of all intense excitements and stimulating drinks.

Anxiety of Mind, when constantly indulged in, destroys the digestive powers of the stomach, impairs the functions of the lungs, disturbs the regular circulation of the blood, and impedes the nutrition of the system. It is a fruitful source of chronic affections of the stomach, liver, heart, lungs and brain. Even the anxiety induced, in a sensitive mind, by the ill-humor, caprice and unkind treatment of others, is deeply felt, and proves highly injurious to health.

CARE OF THE HAIR.

Whatever has a tendency to impede the passage of the fluids by which the hair is nourished, from the root along the cavity which exists in the centre of each hair, must necessarily prevent its proper growth—render it thin, and deprive it of its soft and glossy appearance. There can be little doubt that this is the effect. to a certain extent, of the practice of twisting the hair from its natural position, and of plaiting or firmly braiding it, pursued in obedience to the dictates of fashion, by most females.

Whenever the hair becomes thin and irregular, or its beauty is otherwise impaired, nothing is better calculated to restore its proper growth than cutting it short. Frequently cutting the hair also prevents it from splitting at the ends.

In children, keeping the hair short is a circumstance of no little importance. Nothing is more common than to see a luxuriant head of hair accompanied in children by paleness of complexion, weak eyes, and frequent complaints of headache.

The hair of children should be cut short until they are eight or nine years old. There is good reason for believing, that children who have a great quantity of hair, are those most liable to eruptions, as scald head, &c.: it is at least certain, that in them eruptions are very difficult to remove.

Mothers, whose vanity may be alarmed lest repeated cutting the hair for so many years should make it coarse, may be assured they have no cause for this apprehension.

When there is any tendency to sores or eruptions on the head of children, fine combs are very apt to promote them. The seldomer a fine comb is applied to the head of an infant the better. However, the head should be well washed with a good lather of soap and cold water (once a week), afterward washing the soap out with clean water, rubbing with a dry towel, and then giving the hair a good brushing. Washing the hair in cold water every day, is a great advantage to it

PRESERVATION OF THE SIGHT.

The following are the general rules for preserving the sight unimpaired for the longest possible period.

All sudden changes from darkness to light, and the contrary, hould be avoided as much as possible.

Avoid looking attentively at minute objects, either at dawn or twilight, and in dark places.

Avoid sitting near a dazzling or intense light, as of a lamp or candle, and facing a hot fire.

Avoid reading or sewing much by an imperfect light, as well as by artificial lights of any kind.

Avoid all dazzling and glaring sunshine, especially when it is reflected from snow, white sand, or other light colored bodies.

Avoid rubbing or fretting the eyes in any manner, and wiping them with cotton handkerchiefs.

Avoid all spirituous and heating liquors, rich and highly seasoned food, and every species of intemperance, all of which invariably injure the eyes and impair their sight.

CARE OF THE BOWELS.

An evacuation once in the twenty-four hours is the best standard of frequency—this, in general, takes place whenever the digestive organs are in a state of health.

Confinement to a diet composed chiefly of dry animal food, or of food highly seasoned—the use of fresh bread, and of warm rolls and cakes, very generally induce a costive state of the bowels. Costiveness is very common also in persons who use little exercise, or who pass the greater part of the day within doors in occupations of a sedentary character—females are much more subject to it than males. Lying in bed to a late hour in the morning is unfavorable to a regular condition of the bowels.

Early risers, who pass several hours of the morning walking abroad in the open air, seldom complain of any want of regularity in their stools.

The daily use of wine, especially the red or astringent varieties, retards very materially the natural discharges from the bowels. The same effect takes place in persons who past the greater part of their time in company, and who from a false delicacy resist the calls of nature.

In costiveness, besides early rising, daily exercise of the body in the open air, and abstinence from wine and ardent spirits, the diet should be composed principally of vegetable food. Plain soups, especially of veal and mutton, with the addition

of the ordinary vegetables, well boiled and not too highly seasoned, will be found a very excellent diet for those inclined to
costiveness. Fresh fruits, perfectly ripe, or fruit cooked, with
or without the addition of sugar or molasses, are gently laxative. Spinach, when in season, and properly boiled, is also a
very pleasant and wholesome vegetable for persons of costive
habits. Bran bread, or wheaten bread with an admixture of
rye or Indian meal, is better suited to the habitually costive
than bread composed entirely of fine wheat flour. For drink,
those troubled with costiveness should make use of water,
either alone, or with the addition of a small quantity of sugar
or molasses. A very pleasant drink is made by dissolving
currant jelly in water, or by pouring boiling water upon sliced
apples or peaches, and allowing it to stand until cold. This
acts gently upon the bowels. Buttermilk, or sweet whey, may
likewise be occasionally drank with advantage by those whose
discharges are defective : all ardent spirits and wines, especially those of an astringent nature, should be carefully avoided.
The method recommended by the celebrated Locke, for procuring a regular discharge from the bowels, is founded on
correct principles, and should not be neglected ; it is, " to
solicit nature, by going regularly to stool every morning,
whether one has a call or not."

    To remove costiveness, individuals should be extremely
cautious in resorting to purgatives, or those medicines, under
whatever name they may be sold, which have the effect of
inducing evacuations from the bowels. The frequent use of
these articles, however mild their operation may appear to be,
tends to disturb the stomach and bowels ; and consequently to
vitiate or retard digestion.

THE FEET.

    The feet are extremely subject to the impression of cold, and
when chilled, in consequence of the close sympathy between
them-and other parts of the body, disease is apt to be occasioned in some one of the internal organs. They should be
protected always from cold and damp, but when accidentally
wet, the shoes and stockings should be immediately changed,
and the feet bathed in cold water and rubbed perfectly dry
with a coarse cloth. Tight and misshapen shoes are injurious,
by preventing the circulation of the blood, causing coldness of
the feet, and producing corns. The feet are generally much
neglected. They should be washed at least *once every day,*
and rubbed well afterwards with a rough towel, or flesh brush.
Thin shoes are killing thousands of females every year—away

with them at once. Females should wear thick, warm shoes, at all times during cold or damp weather.

PHYSICAL EDUCATION AND DISEASE. OF CHILDREN.

GENERAL SYMPTOMS OF DISEASE.—It is often difficult in very young children to determine the disease with which the little sufferer is afflicted,—for as we can obtain no information from the patient relative to his feelings or the history of his malady, we are obliged to form an opinion from external symptoms alone. The symptoms of disease are nearly the same in childhood as in maturer age. The sleep, motions, breathing, evacuations, pulse, appetite and mental manifestations, all afford important appearances. The *countenance* is in most cases an important index of disease; when the expression is calm or lighted up by a smile, it indicates a state of ease and a regular performance of all the functions: frowning or contraction of the features, pale, red, or blue tint of the face, rolling up of the eyes, swelling of the upper lip, twitching of the muscles, dilatation or spreading of the nostrils, are all evidence of approaching or confirmed disease. The *sleep*, in a healthy infant is quiet and profound, and indicates a state of ease and comfort; but if there is sleeplessness, sudden starting during sleep, slight spasms, screaming, fright, or deep and laborious sleep, there is some disturbance of the brain or stomach and bowels.

Crying is the natural language of infancy, and the only means by which very young infants express their wants and sufferings. *Healthy infants* cry but seldom, and then only to express some slight uneasiness or vexation: but violent paroxysms of crying, (unless from anger,) plaintive moaning, short and suppressed, a hoarse or shrill cry, indicate disease. The *breathing* in health is full, easy and regular,—but it sometimes becomes slow, irregular, difficult, short, laboring, rattling, shrill "crowing" intermittent, or attended with cough, sneezing or hiccup in sickness.

The *tongue* and *mouth* sometimes show important characteristics of disease: paleness or redness, brown or white fur, dryness, swelling, trembling, ulcers or cracks in the tongue are signs of disease. Unusual redness of the gums and throat, increased or diminished secretion of saliva, (spittle,) bitter taste or brown scum on the teeth, also point to diseased action in some part of the system. The *skin*, in health is soft, slightly moist and warm, and has a slight crimson tint of the blood but when it becomes harsh, dry, hot, shriveled, pale, cold, clammy, blue, red, or yellow, some disease exists. A cold profuse perspiration shows debility or other disease: many of the eruptions of the skin indicate disease of the digestive organs; itching and tingling of the skin are signs of irritation from worms or other causes. The *breath*, if rancid, sour or unpleasant, is the result of fever, indigestion or ulceration. The *evacuations* from the stomach, bowels and bladder, are in most cases peculiarly characteristic, and are important symptoms: frequent vomiting of curdlike, green, bilious, sour, bloody or black matter, is evidence of disease of the brain, stomach, liver, kidneys or bowels. When the evacuations from the bowels are frequent, slimy, bloody, green, black, gray, or profuse and destitute of odor, or mixed with flakes of mucus or pieces of undigested food, or when great costiveness is present, there is disorder of the digestive organs. If the *urine* is white, red, of unpleasant odor, or deposits a brown or gray sediment, (or settlings,) is scanty, profuse, or passed with difficulty, there is morbid action of the kidneys or some other part. The development of the *bones* is often indicative of some disease: narrowness of the chest and very prominent breast bone, great length of body and limbs, large joints, curvature of the bones of the legs, brittleness of the bones, large head, weak joints, open seams, (sutures,) in the skull, and crooked back, all indicate a rickety, scrofulous or debilitated state of the system.

The symptoms above enumerated are sufficient to enable the common observer to detect the existence of disease.

MANAGEMENT OF CHILDREN.

AIR.—A constant supply of pure air is indispensable to the health of every human being, from the first moment of existence to the end of life. This is even more necessary for infants than for adults, on account of the rapidity of the circulation and breathing, and the weak and irritable state of the nervous system. Children confined in badly aired rooms become pale, feeble, irritable and finally consumptive. The air in their sleeping rooms should not be too warm, as this causes oppressive breathing, too great perspiration, feverishness and oppression of the head. Neither should it be *too cold*, for this checks the insensible perspiration which is constantly going on during health. Very cold air also closes the pores on the mucus membrane of the nose, throat, windpipe and lungs, and inflammation and fever ensue. Beware how you attempt to "harden" your children by putting them to

Illustrations of Healthy and Unhealthy Positions of the Body.

CORRECT POSITION.   INCORRECT POSITION.

See how that round-shouldered youth is sitting with his shoulders against the back of the chair, and the lower portion of his spine several inches from it, giving his body the shape of a half-hoop. Parents should regard such a position in their children with apprehension as to the result, and should rectify it at once. The other young gent. has learned a thing or two worth knowing, as you will see by his manly, dignified position in his chair. He may be President yet— who knows?

HEALTHY POSITION.   UNHEALTHY POSITION.

An upright position, in either sitting or walking, favors a healthy action of all the various organs of the system, and besides it gives a graceful and dignified appearance to the human form. Children and adults are more or less inclined to lean forward with their heads upon their elbows, even when their seats are provided with backs ; such a position oft-repeated must in time result unfavorably. There is a very "*don't care*" kind of look about one of these young ladies; if she lives long enough, she *will care.*

IMPROPER POSITION.   PROPER POSITION.

Learn to sit up, young man, and to imitate your opposite neighbor ; for the unnatural position which you have assumed will never make you a good writer A bold upright position, with the pen held loosely between the fingers, and determined purpose to imitate some definite copy as nearly as possible, is the only true road to success in the art of writing or good penmanship. This young-ster on the left looks as if he was "bowing his back for a heavy burthen "—and so he is ; if he don't quit it, he will have burthen enough.

sleep in a very cold room; croup and inflammation of the lungs are often thus produced.

Children should be carried or permitted to go into the open air *often*, and always with their faces bare unless the weather be inclement. The face should never be muffled or covered during sleep. The temperature of the nursery and sleeping room should never exceed 65° or 70°; older persons should not judge of this by their own feelings, as a degree of cold which to them would be comfortable, might be injurious or dangerous to an infant. Keep a thermometer, and go by that.

BATHING.—The infant may be bathed every morning during the first two or three months, in tepid water, after which the bath may be nearly or quite cold. The idea that the cold bath is best for *all* children, is erroneous,—it is equally wrong to suppose none but the *warm* bath is safe and beneficial. Some infants have not sufficient vitality and strength to bring about reaction so as to recover from the shock and regain their natural warmth, but become weak, chilly and sick, by the use of the cold bath. Those of a robust and *full habit of body* on the contrary, by the use of the warm bath are affected by congestions of the stomach, lungs and brain, and suffer from indigestion, headache and convulsions. For pale, feeble and irritable children, the *tepid bath* is usually the best adapted: the temperature of this bath is from 80° to 90°. For those of active circulation, good digestive powers and robust health, the temperature may be 60°, — constituting the *cold bath*. Immediately after bathing, the surface of the body and limbs should be well dried and rubbed with a linen napkin, and the child dressed and allowed to exercise. The cold bath should not be used during profuse perspiration, during a chill, nor sooner than *three hours after eating*. If there is roughness or smarting of the skin after bathing, it may be rubbed with a very little sweet oil, or a little starch powder. Children are easily bathed by a cloth, or soft sponge and basin of water.

In bathing infants, the head should always be *wet*, but not *washed with soap, as this fades the hair*. *Too frequent* use of the cold bath produces eruptions on the skin, debility and disease of the heart.

CLOTHING.—Every part of the person should be covered except the head, face and hands. Bare arms and low-neck dresses may be very pretty to look at, but it is a dangerous practice; many times little children are thus sacrificed to the whims of folly and fashion.

The habit of putting caps on infants is useless and injurious;

the habit of allowing them to go barefoot is vulgar and cruel; the practice of swathing or bandaging the bodies of children, is totally useless, and very pernicious to their health and comfort. Soft, white woolen flannel should be worn next to the skin during winter and the colder part of spring and autumn. When the flannel is laid off, cotton may be substituted by degrees: this may be done also, at any time when woolen ppears to be too irritating, as sometimes happens. The night clothes should be light and perfectly loose: the whole dress should be fastened by means of buttons, hooks and eyes and tapes, to avoid *wounds* from *pins*.

SLEEP.—The sleep should be voluntary, and not forced or induced by medicines or rocking; neither should it be disturbed for the purpose of washing, nursing or dressing. Young children require a soft, warm couch in winter, but during warm weather they should lie upon beds filled with straw, cotton, curled hair, moss or corn-husks. The pillow should be of the *same material* as the bed, to prevent the head becoming too much heated, and avoid taking cold, earache, catarrh and snuffles. During the first three or four months it is better to allow the child to sleep with the mother,—after which it may sleep alone, (if the weather is warm,) in a cradle, cot or couch. The habit of *rocking infants*, if frequent or long continued, is injurious,—but if gentle and only occasionally during waking hours, it is both harmless and pleasing to the child. The bed or cradle should be high and without curtains: that old fashioned contrivance called a "trundle bed," is a vile relic of barbarism, and deserves to be totally banished from civilized society. The *position* should be occasionally changed during sleep: this prevents too much pressure on any part, accumulation of heat, deformity of the head, and fatigue. The room in which the child sleeps should be partially darkened : during waking, it requires the stimulus of a mellow light, this conduces both to its health and cheerfulness. The eyes, however, should not be long exposed to the intense glaring light of the sun, fire or lamp.

All perfumery, flowers, medicines or food, or anything exhaling a strong odor, should be excluded from sleeping apartments. Children ought never to sleep with *old* or *sick* persons : neither should they be *fondled or kissed by old, diseased* or strange persons, for fear of incurring some disease Young children should be protected against *loud noises strong odors*, and *sudden frights*. Children of more advanced age should retire early, and rise early in the morning.

EXERCISE.—During the first few months of infancy, **but**

little exercise is required,—nor does the organization admit
of more than a small amount of passive and gentle movement.
The infant may be carried about in the arms within doors.—
or when the weather is pleasant it may be carried in the
arms or drawn in a wagon, laying on a pillow, in the open
air, a few minutes at a time, several times daily. All rough
tossing, jolting and *dandling* are injurious. For the first two
months, the infant cannot be placed in the *erect posture*
without the risk of deformity.

When the infant indicates some desire to sit alone and
move about, he may be allowed to sit, lie or roll about on the
floor with the utmost freedom. When riding, the position
should be *often changed*, and nursing should be done equally
upon the *right* and *left* side, to avoid deformity of the head
or spine.

It is better that the child should not attempt to stand or
walk before the ninth or tenth month: it should rather be
encouraged to creep until it acquires sufficient strength and
firmness to walk voluntarily. Children should not be confined to *little chairs*, "baby jumpers," or any apparatus for
restraint; although it may relieve the mother or gratify the
laziness of the nurse,—it is still unnatural and injurious to
the delicate growing system of the child. After the age of
two years, girls as well as boys should be permitted to roam
free and unconfined over the wide field of nature, and inhale
the "pure breath of heaven." The first six or eight years of
childhood should be passed in various kinds of exercise and
amusements. Confinement in school rooms or shops, or at
desks, or to any laborious occupation previous to that age, is
always injurious and unkind. Nothing is *gained* and much
is lost by sending a child to school too young. No child
under six years should be sent to school, and even then only a
few hours at a time.

Children should be indulged in pursuing little mechanical
operations, and in learning to build and construct whatever
their tastes may incline them to: they should be taught to
admire the beauties of nature, rather than be supplied with
the various little toys and expensive contrivances of art.
They should be permitted to take short rambles for the purpose of collecting flowers, insects, minerals or fruits, or to
observe the habits of animals and birds, and enjoy the prospect of natural scenery.

DRINKS.—Infants feel the sensation of *thirst*, as early as
that of hunger, and are highly gratified and benefitted by a
small quantity of cold water several times daily.

Restlessness and crying are often caused by thirst, but mis
taken for colic or hunger, and the infant is dosed with cordials
or opiates, or forced to take the breast, which only increases
its distress, perhaps surfeits the stomach and causes nausea
and vomiting. This leads to the apprehension that he is sick
and dose succeeds dose, till disease is often produced,—when
a spoonfull of cold water would have removed all unpleasant
sensations. Many people think it is a "healthy sign" for a
child to throw up its milk, when the whole trouble is that
the child has *too much milk* and too *little water !* (*Very cold,*
as well as very *hot* drinks should be avoided.)

Diet.—As a general rule the mother's milk should be the
only food for the first nine or ten months ; no solid food
should be allowed until a sufficient number of teeth are
developed, to enable the child to chew.

When it becomes necessary to increase the amount of food
derived from the breast,—or to "bring up the child by hand,"
cows' milk is the best substitute : it should be mixed with an
equal quantity of warm water and a little loaf sugar added.
All preparations of gruel, panada, broth, soup or solid food,
are unfit for the diet of an infant. Animal food is too strong
and stimulating for infants before the first set of teeth are
complete. After teething is completed, a more solid and
nutritious diet may be allowed.

Food.—Care is requisite that too much of one kind of food
be not taken,—but rather a due proportion of several kinds :
the food should not be taken too fast or swallowed without
being well chewed. Sugar and other sweet substances are
not injurious when taken in *due quantity* and with *other articles of diet ;* all sweet and ripe fruits, when freed from the
seeds, stones and skins, are wholesome and nutritious in small
quantities. Sour or green fruits are decidedly pernicious,—
cherries of most kinds, and also some kinds of grapes, are
peculiarly unwholesome : ripe fruits, either dried or recent,
when cooked or preserved and made palatable with sugar, are
not objectionable for children.

NOURISHMENT OF INFANTS.

The foundation of incurable chronic diseases, and of constitutional debility in after-life, is often laid within the first
month, or even first few days, after birth, by improper management ; and a great amount of the suffering and mortality
which occurs during infancy must be ascribed to the same
prevailing source. The custom of feeding children with inap-
propriate articles of food, very soon after birth, is wrong. Nc

sooner is the infant washed and dressed, than the nurse is ready with her spoon and cup of gruel, pulverized crackers dissolved in water, or some such preparation, to fill its stomach to the utmost of its capacity ; and this process of stuffing is continued with a ruinous degree of diligence and perseverance. The digestive organs of the new-born babe are thus often seriously injured during the first twenty-four hours. Nature herself seems to point out the impropriety of this practice. She withholds the nourishment which she provides (the milk) until many hours after birth. I do not mean to inculcate that nourishment is to be entirely withheld from the infant until the milk is secreted *under all circumstances;* but I am persuaded that, with healthy infants, several hours, at least, should be suffered to pass immediately after birth before any food is introduced into its stomach, and not, under any circumstances, give food to the child until it has *first been put to the mother's breast* several times, to see if the milk will not be excited to flow. In nine cases out of ten, perhaps, the griping, flatulency, diarrhœa, and colic, which so frequently harass infants, during the first half year after birth, are the results of indigestion, brought on by errors in diet. To relieve the colic, griping, flatulency, diarrhœa, &c., which ensue, recourse is had to cat-mint tea, aniseed tea, Godfrey's cordial, paregoric, or some other palliative or nostrum, and thus an additional cause of indigestion is brought into operation. The screams and restlessness of the infant occasioned by the griping and colic, are frequently regarded as manifestations of *hunger.* To appease this supposed craving, the stomach is almost constantly kept in a state of distension with food ; and thus the helpless babe has no chance of escaping from the torments and ruinous consequences of its unfortunate situation. Great distress and suffering are sometimes witnessed during the early period of infancy from indigestion, and consequent irritation of the stomach and bowels, even where the child is wholly nourished by the *breast.* For when, during the time which intervenes between the secretion of milk and the birth of the child, crude articles of nourishment are introduced into the infant's stomach, the digestive powers are often at once so deranged and impaired, that even the wholesome and congenial fluid furnished by the mother will not be easily digested and acidity, flatulency, and colic, will continue to harass the child until the digestive powers gradually acquire a greater degree of vigor.

Let the child's stomach be once or twice filled during the *first twenty-four hours with gruel,* or any of the ordinary preparations employed by nurses for this purpose, and the chances will probably be as ten to one that sourness of the stomach, vomiting, colic, griping, &c., will supervene. There is no period throughout the whole course of life in which the observance of caution, in relation to the food, is of greater moment than in the comparatively short interval which passes between the birth of the infant and the secretion of its natural aliment (milk). Do not be in too great haste to give purgative medicine to a child, soon after it is born, to open its bowels. Active purgatives are sometimes given for this purpose ; and there is much reason for believing that the infant's digestive organs are often injured in this manner.

The very best thing to stimulate a secretion of milk is *applying the child to the breast*, for a few minutes at a time, (commencing as soon as the mother is comfortably settled in bed,) every half an hour or so. If the child is hungry, his efforts will be the more vigorous to procure nourishment. A mixture of two parts of fresh cow's milk and one part of warm water, approaches nearer to the nature of human milk than anything else that can be conveniently procured. Of this a few teaspoonfuls may be given in cases where the secretion has been long delayed, carefully avoiding overcharging the stomach until the mother's breasts are ready to yield their more congenial nutriment. In order to excite the early secretion of milk, it will be proper to let the child draw the breasts, for a few minutes, soon after the mother is comfortably fixed in bed, provided her health and strength will admit of it. After the secretion of milk is once fully established, and furnished in sufficient quantity, the infant should be nourished *exclusively by the breast.* Not even the mild and simple fluid just mentioned should be allowed, unless some special reason exist for the use of additional nourishment. It seldom occurs in healthy mothers, that the quantity of milk supplied by the breast is not sufficient to afford adequate nourishment to the child for the first two or three months, and, in general, much longer, without the necessity of any additional artificial food.

The infant should be nourished exclusively by the breast *until the first teeth make their appearance.* No other kind of nourishment whatever should be allowed before this period, unless from deficiency of milk or some other cause, the use of additional nourishment becomes necessary. After the first teeth have come out, small portions of barley water, thinly prepared arrow-root, or a mixture of equal parts of cow's milk and water, may be given two or three times daily, in addition to the nourishment drawn from the breasts. I do not mean to say that, when the child arrives at this stage, it becomes *necessary*, or even proper, as a *general rule*, to exhibit any additional articles of food. In general, however, the simple and mild liquids just mentioned may be given at this period with very little risk of unpleasant consequences, for the digestive organs have by this time acquired a degree of power and activity sufficient to obviate the painful and disturbing effects which would arise from the use of such food during the first four or five weeks after birth. It is also of much consequence that the food should be introduced into the stomach as *gradually* as practicable. This can be most conveniently done by causing the infant to suck the fluid from a bottle, furnished with the usual tube, the mouth-piece of which is pierced with a *small* opening. By this contrivance, the child will receive its food in the same *gradual manner* as when nourished at the *breast*, and it will rarely take more than its appetite calls for, an error which is frequently committed when *fed with a spoon* After the *seventh* month, small portions of the preparations of food just mentioned (such as corn starch, pulverized crackers with milk and water to soften them, gruel, made of oatmeal or wheat flour, &c.,) should be given at *regular periods*, three or four times daily. This will prepare the infant for the sudden change which it has to undergo, in the character of its food, when it is weaned, and thereby tend to lessen the liability to unpleasant consequences from the change. Infants who have been *moderately* fed with suitable articles of food some time previous to weaning, almost always accommodate themselves much more readily, and with much less uneasiness to the change, than such as have seldom or never received any other nourishment than that which they draw from the mother's or nurse's breasts.

Mothers ought never to delegate the suckling of their infants to others. This sacred office should rest with the mother alone. The mother who submits the suckling of her infant to another, while her own breasts are ready to furnish an ample supply of milk, can scarcely possess an amiable and moral heart.

It can scarcely be doubted that the *mother's* milk is, in general, better adapted to the constitutional temperament of her offspring than that furnished by others. Besides, when the suckling of the infant is submitted to a nurse, it is liable to various sources of injury and disorder, which are, in a great degree, if not entirely, obviated when this important duty is performed by the mother. No *hired nurse* can be depended on to give the tender care, and have the affectionate regard, for a child that its own mother has.

Unfortunately, however, mothers are not always in a condition that enables them to suckle their own infants, and the employment of a wet-nurse, or recourse to artificial nursing, is unavoidable. The causes which may prevent a mother from nursing her child are : A decided deficiency or total failure in the secretion of milk, in consequence of disease or torpor of the breasts ; a bad state of milk, rendering it decidedly prejudicial to the health of the child ; the presence of a morbid *taint*, or some *communicable chronic diseas* in the mother's *system ;* when suckling gives rise to painful or dangerous affections in the mother, as colic, cough, distressing nervous affections, great weakness, epilepsy, &c.

When causes of this kind render it improper or impracticable for the mother to nurse her child, it then becomes a question whether a wet-nurse should be employed, or artificial nursing resorted to. It would, in general, be much better to nurse the child artificially, *under the eye of its mother*, than to place it entirely at the mercy of the wet-nurse. Nurses, doubtless, are sometimes found to whom a child may be safely intrusted ; but experience has but too often shown that the reverse is the case.

Attention must also be paid to the previous and present health of *the nurse.* No woman who has led a debauched course of life, even though reformed, can be regarded as a perfectly safe nurse, however careful and attentive she might otherwise be. Females of this description are apt to have their systems contaminated with some morbid taint, which may give an unwholesome quality to the milk, and injure the child's constitution. The existence of scabby or scaly eruptions on the skin, unless they are of transient character, and of chronic ulcers, particularly on the legs, should be regarded as sufficient objections to a nurse. A manifest scrofulous habit, also, is decidedly objectionable. The *age of the milk* is another point of considerable importance. Milk that is six or seven months old seldom agrees well with infants during the first two or three months after birth. In general, the milk becomes much more rich and nutritious after the fourth month than it is previous to this period ; and milk of this kind, from its requiring stronger digestive powers than younger milk, often gives rise to much disturbance of the stomach and bowels in new-born infants. As a general rule, therefore, the age of the milk should not vary much from that of the child, up to about the fourth month. After this period, such a relation between the ages of the milk and child is not of much importance—for a child five or six months old and upwards, may be nourished by a fresh breast with entire safety.

The occurrence of the menstrual evacuation during nursing, is almost invariably attended with diminution and deterioration of the milk, and constitutes a well-grounded objection to a wet-nurse. This is more especially the case during the first three or four months of infancy. When a child at this early period is put to the breast of a nurse who menstruates, it rarely fails to experience derangements of the stomach and bowels. After the seventh or eighth month of age, there is much less inconvenience and disorder to be apprehended from this source; but, even at this advanced period of infancy, the milk of a nurse, thus circumstanced, may give rise to disturbances in the digestive organs, and should, if possible, be avoided. Nature here, as elsewhere, is a safe guide. We perceive that menstruation is almost universally suspended during the period of suckling; and we may presume that this arrangement of nature is designed for some useful purpose— for the well-being, doubtless, of the infant. Nature, therefore, as well as experience, indicates the propriety of withholding the breast from the child when, from constitutional peculiarity, or some accidental influence, the monthly turns make their appearance in the nurse or the mother.

A nurse who has but one good breast should never be selected. A child suckled by one breast only, is apt to contract the habit of *squinting*, from having its eyes constantly directed to one side. Some attention should also be paid to the nurse's nipples. If they are very small, the child will be apt to fatigue itself in sucking, without being able fully to satisfy its wants. This defect can seldom be properly remedied. The practice of drawing out the nipples by suction, with a pipe or bottle, will be of advantage; but when the nipples are very small, and deeply imbedded in the breast, it can scarcely remedy the evil. In some instances the nipples yield the milk so freely, that the child is continually harassed by a sense of strangulation, while suckling, from inability to swallow as rapidly as the milk issues into its mouth. This may, in general, be remedied by passing a piece of fine tape or elastic pretty firmly round the base of the nipple; or the nurse may compress the nipple moderately between the first and second fingers, while the child is suckling.

Finally, particular regard should be had to the temper and moral habits of the nurse. An irritable, passionate, and sour tempered female, is but illy suited for this important duty Not only is the child liable to be maltreated by a nurse of this character, during the fits of ill-nature and passion ; but the most serious and alarming effects may be produced on its tender organization, by the milk of such a nurse. It is well known that violent anger, and habitual sourness of temper are peculiarly apt to give a pernicious quality to the milk. Children have been thrown into convulsions, by suckling soon after the nurse has been agitated by violent anger, rage or fright ; and alarming vomiting and purging is particularly apt to occur from this cause. Indeed, every kind of inordinate excitement, or depression of the mind is unfavorable to the secretion of healthy milk. Protracted grief, sorrow, or mental distress and anxiety in the nurse, seldom fail to exert a prejudicial influence on the health of the nursling. This circumstance ought not to be overlooked, in choosing a wet-nurse.

ARTIFICIAL NURSING.—Under *judicious management*, infants will, in general, experience no particular inconvenience from a course of artificial nursing ; and, as a general rule, this mode of nourishing children, when properly conducted, is upon the whole preferable to the employment of a wet-nurse, whose competency and fitness is doubtful. This preference, however, is founded rather on the greater risk which the child incurs of being maltreated and neglected, when submitted to the exclusive care of a wet-nurse, than when nursed artificially, under the immediate superintendence of a parent.

There are circumstances, in relation to the condition of the child, which render the employment of a wet-nurse, notwithstanding all the risks that have been mentioned, preferable to artificial nursing. *Very young*, and peculiarly *delicate* and feeble infant children do well when raised by the hand. Fresh and wholesome milk from the breasts of the mother, or a healthy nurse, is almost indispensable to the well-being of an infant thus circumstanced. The same observations apply to infants, whose stomachs and bowels are peculiarly weak and irritable. Finally, if upon trial, the slightest and most appropriate kinds of artificial food are found to disorder the stomach and bowels, the life of the infant will very probably depend on its being nursed by a fresh and wholesome breast.

Sometimes the mother, though incapable of supplying a sufficient *quantity* of nourishment by the breast, is still able to furnish small portions of *wholesome milk*, and when this is the case she ought, by all means, to continue suckling the child, in conjunction with the use of artificial nourishment. Particular care should be taken to keep the bottle perfectly clean and sweet. It should be well washed, both inside and outside, with hot water, every morning and evening. The same food should not be suffered to remain in the bottle more than *three hours.* After the child has satisfied its appetite, no *new supply* of food should be added to what may have *been left.* The quantity of nourishment put into the bottle, should not be much greater than what may be deemed fully sufficient for *one nursing.* By these precautions the food will always be sweet, and free from offensive or irritating qualities. Nursing bottles are now easily obtained in almost every part of the country. When the child uses the bottle, it should be taken up and supported in an easy position, on the lap or arms of the nurse, imitating the position of the mother's breast. The child should be kept quiet for at least thirty or forty minutes after having received its nourishment. Rest is particularly favorable to digestion.

Children, who are entirely nursed by artificial diet, should be restricted to the use of the milk-and-water mixture mentioned above, until several teeth have made their appearance. They will, in general, enjoy more perfect health and thrive better, when nourished exclusively with this simple food, than under the use of any other nourishment that can be made. After the third month, however, the proportion of milk should be somewhat increased : namely, three parts of milk to one part of water. After the first teeth are protruded, the food may be a little more varied and substantial. Grated crackers dissoled in warm water; oat-meal gruel; liquid preparations of arrow-root, tapioca, or cago; milk thickened with rice flour, and thin pap, may be allowed in moderate quantities along with the ordinary milk-and-water mixture. When these preparations do not agree with the child's stomach, they should be used with an equal portion of weak mutton, chicken, or beef broth, clear and well freed from fat. A mixture of this kind is, in general, easily digested, and rarely causes any unpleasant effects, when used *after the first teeth have made their appearance.*

After the first grinding teeth are protruded, weak broths, slightly thickened with oat-meal, rice flour, arrow-root, or grated crackers, mixed with milk, constitute, in general, the most appropriate articles of nourishment. A small portion of stale bread may also be allowed, two or three times daily, at this stage of infancy. The animal food given to young children should be plainly *roasted or boiled.* Fried and broiled meats, and all food *heated a second time,* by hashing or mincing, being less digestible, should be avoided. Many people, from a mistaken expectation of strengthening weakly children, give them more animal food, and sometimes twice or thrice a day : but it will be found much more frequently to add to debility than to the increase of strength. Those children, on the whole, who eat the *least animal food*, are the most healthy. Nothing is more absurd than the notion that, in early life, children require a variety of food.

The peculiarly excitable state of the system during teething, and the consequent tendency to feverish irritation, render the *free* use of animal food decidedly objectionable during this stage of childhood. Small portions of the more digestible meats may be allowed to healthy children, once daily, with little or no risk of injury ; but they should never be permitted to form the principal part of the food. The lean parts of mutton, lamb, tender beef, game, and fowl, should be selected. Veal, pork, pig, goose, duck, and all kinds of salted meat, being of much more difficult digestion, can seldom be used without impeding digestion, and finally injuring the tone of the stomach. Veal is decidedly the most objectionable of all the meats in common use for children. Fresh fish, boiled, and taken in moderate portions, seldom disagrees with the stomachs of children, and may be used, occasionally, with perfect propriety. Soft boiled eggs, too, form an appropriate article of nourishment for children after the first teeth have come out. When fried, or boiled hard, they are altogether unsuitable Strongly seasoned meats, compound dishes, hashes, meat pies, and pastry, are to be *wholly rejected*.

The introduction of fresh food into the stomach before that which was previously taken is *entirely digested*, seldom fails to operate injuriously. As a general rule, from three to four hours may be regarded as a suitable interval between the meals. If the child requires nourishment between the regular meals, small portions of liquid food, such as milk, &c., should be used. When solid animal food forms a part of the diet of children, it should be taken at noon, or in the forenoon.

Pure water, with or without small portions of milk, constitutes the best drink for children.

CANDIES AND SWEETMEATS.—Indulgence in the use of sweetmeats is a copious source of disease and mortality during childhood. Fruits preserved with their skins, as *raisins*, are particularly pernicious. The skin of all fruits is of difficult digestion.

The conduct of parents, in relation to this subject, is often extremely irrational and pernicious in its consequences. They would not themselves venture on the frequent and free use of confectioneries of this kind ; and yet will indulge their chil dren without scarcely any restraint, in the use of these pernicious luxuries. The *sicklier* and *weaker* the child is, the more apt, in general, is it to be allowed these destructive gratifications. The pale, feeble, and sickly child, whose stomach is hardly able to digest the most simple and appropriate food, is sought to be appeased and delighted by the luscious and scarcely digestible articles of the confectioner. Indigestion, bowel irritation, terminating often in ulceration and incurable diarrhœa, are the frequent consequences of such conduct; and at best, such indulgence must prolong the feeble and sickly condition of the child, and not unfrequently eventuate in permanent debility.

With regard to the use of fresh fruits, writers, on this subject, have expressed different opinions. Apples, peaches, and apricots, (freed from the skin,) when *perfectly ripe* and mellow, may be occasionally allowed to children, in moderate portions, with entire safety, unless the stomach and bowels be very weak and irritable. In children of a costive habit, the temperate use of these fruits may even have a beneficial effect, by their tendency to excite the action of the bowels. Pears, even of the tenderest kinds, appear to be much more indigestible than ripe apples or peaches. Stewed or roasted fruits, particularly the two latter kinds, are, in general, well adapted to the digestive powers of young children, and may be allowed, occasionally, with perfect propriety, provided they are not very sour. When the *acid* or *sourness* prevails to such a degree as to require the addition of sugar to render them sufficiently palatable, stewed or roasted fruits of this kind rarely agree well with weak and delicate stomachs, and cannot be allowed to young children without considerable risk of injury.

In general, all fruits having a firm cuticle or skin, such as grapes, whortleberries, &c., are improper articles of food for children. The *pulp* of grapes, *freed from the seeds*, rarely causes disorder in the bowels when taken in moderation..— Fruit that contains small, hard and insoluble seeds—such as strawberries, blackberries, currants, &c., are particularly apt, when taken *freely*, to disorder the stomach and bowels. The seeds, resisting the digestive powers, irritate the mucous membrane of the bowels; and when, from previous causes, this membrane has become enfeebled and irritable, they may readily excite dangerous irritation. Small insoluble bodies of this kind, frequently remain lodged in the *folds of the bowels* for many days and even weeks, and give rise to severe and unmanageable disorders of the bowels.

Cherries are among the *most pernicious fruits* in common

use, and ought to be wholly excluded from the list of articles
with which children may be occasionally indulged. Even
when eat without the stones, they are peculiarly apt to derange
the bowels; and when swallowed with the stones, which, with
children, is not unfrequently the case, they are capable of producing violent and even fatal diseases. Most alarming and
fatal consequences have resulted from the irritation of cherry
stones *lodged in the bowels.* Convulsions, inflammation, and
harassing diarrhœa are among the affections which are apt to
arise from this cause. All fresh fruits have a tendency to excite, more or less strongly, the action of the bowels. As a
general rule, therefore, every kind of fresh fruit is improper
for children whose digestive organs are weak and irritable, or
who are habitually liable to disorder of the bowels.

Exercise.--Uncertain and awkward motions of the arms—
stamping with the legs, and drawing them up. are the first
feeble attempts which the infant makes in the use of his muscles. But even these muscular exertions appear to be *indispensable* to the preservation of its health and the proper
development of its powers; and it should be an especial object
of care to allow entire freedom of motion, *several hours daily*,
by avoiding all modes of dress and position tending to restrain
the free use of the extremities. With this view, the infant
should be taken from its bed two or three times a day, and
laid on his back upon a soft mattress. or any other level and
slightly resisting surface, and divested of every thing calculated to restrain the motion of its limbs and body. Confining
an infant's feet in long under-clothes is decidedly objectionable, after they get to be a few weeks old. Children who are
frequently permitted to exercise their muscles in this way, will
learn to use their limbs and walk earlier than those who are
seldom allowed this freedom of voluntary action.

Carrying.—This should be commenced as early as two
weeks after birth, provided the infant be not unusually feeble;
and it should be daily attended to, as one of the regular and
indispensable duties of nursing. The manner, however, in
which very young children are usually carried or exercised, is
extremely reprehensible, as it is calculated to give rise to very
unfortunate consequences in relation to the health and regular
conformation of the child's body. We allude, particularly, to
the common practice of carrying infants with their bodies in
an *erect position*, before the backbone and muscles have acquired a sufficient degree of firmness and activity to support
the body and head in this posture. The child is usually carried by the nurse pressing its thighs and hips, with the left

forearm, against her body, whilst its body is balanced in an upright posture, by resting lightly against her bosom. Thus the whole weight of the infant's body rests upon the feeble and yielding backbone, while the unsupported head is, in general, suffered to lean constantly to one side, or *to roll about in every direction.* This mode of carrying infants must interfere, very materially, with the regular and symmetrical development of the body. The feeble backbone, yielding to the weight of the head and body, is always curved outwards while the infant is held or carried in the erect position ; and, when this is daily repeated for several hours, as is frequently the case, the back is liable to become *permanently bent* or distorted. A habit, too, of leaning the head to one side is sometimes contracted by the child ; and, from the violent manner in which the head is liable to fall from side to side, serious and even fatal injury may be inflicted on the spinal marrow of the neck. But even after the spine and its muscles have acquired a sufficient degree of firmness, to enable the child to support its head and body in an erect position, without difficulty, it incurs considerable risk of injury from the usual practice of carrying it almost exclusively *on one arm.* When the child is carried almost wholly on one arm, it is apt to acquire the habit of *leaning to one side,* which it is always very difficult to correct. The child, also, when carried in this manner, usually throws one of its arms around the neck of the nurse, in order to support itself more steadily in the erect position ; and of course always with the same arm, when the side on which it is carried is not changed by the nurse. In consequence of this position, the shoulder-blade and side of the chest are liable to be *forced upwards and outwards,* which may result in permanent distortion.

The backbone and its muscles seldom acquire sufficient strength and firmness before the end of the third month, to enable the child to support its body in an upright position, without inconvenience or risk. Until this power is acquired, the infant should not be carried, or suffered to sit, with its body erect, without supporting it in such a manner as to lighten the pressure made on the spine, and aid it in maintaining the upright posture of its head and body. But even when thus supported by the nurse, it should not be kept in an erect position more than one or two minutes *at a time,* until it is two months old. At first (a few days after birth) the infant should be taken from its cradle or bed, two or three times daily, and laid on its back, upon a pillow, and carried gently about the chamber. The best mode of carrying very young infants is to

lay them into a small, oblong basket. By this contrivance a
gentle and agreeable swinging or undulating motion will be
communicated to them ; and the sides of the basket being three
or four inches higher than the child's body, a cover or netting
may be thrown over it, without restraining the free motion of its
limbs. After the third or fourth week, the child may be carried
lying in the arms of a careful nurse, in such a way as to afford
entire support to the body and head. It is painful to see the
violent and generally abortive efforts which the infant makes to
steady its little head, when raised into a sitting posture. It will
sometimes succeed in balancing its head for a moment, to the
great delight of the fond mother ; but the effort is almost invariably speedily followed by a sudden and often violent
rolling of the head from side to side, which cannot but be injurious.

All rapid, *whirling* and *jerking* or *jolting* motions are calculated to injure the health of infants. Running or jumping
with an infant in the arms, descending rapidly a flight of stairs,
whirling round, etc., ought to be rigidly forbidden. The practice of supporting very young infants in a sitting posture on
the knee and jolting them violently cannot be to severely censured. It is not uncommon to see mothers and nurses jolt infants in this manner, with a violence that threatens dislocation. *Tossing* them rudely on the arms, is equally reprehensible. These *violent* agitations " powerfully affect the delicate
organization of infants, and may be productive of spasms, epilepsy, and apoplectic fits." *Gentle* and cautious tossing on
the arms affords an agreeable exercise of the body, and may
be beneficial by the moderate agitation which it causes in the
internal organs.

With infants predisposed to diseases of the head, strong
rocking should be particularly avoided.

RIDING IN A CARRIAGE.—This is an excellent mode of affording suitable exercise to infants, and may, with great propriety,
be employed as an occasional substitute for carrying in the
arms.

The body of the carriage should be long enough to permit
the infant, when quite young, to lie down at *full length ,* and
the sides ought to be sufficiently high to prevent its falling or
rolling out. Like carrying in the arms, this mode of exercising infants is liable to be conducted very improperly. This
duty is usually entrusted to children or young girls, who being
generally more disposed to consult their own sportive inclinations than the comfort and safety of their charge, are apt to
draw the carriage along with great rapidity, paying little or

no attention to the roughness or unevenness of the ground over which they pass. After the child has acquired some degree of strength, it should be placed in a half sitting posture, with its head and back well supported by pillows, etc.

WALKING.—After the infant has acquired sufficient strength to support itself in the sitting posture, it should be placed on a soft carpet several times daily, and surrounded with its toys. When thus left to the free use of its limbs, it will soon learn to crawl. The common practice of teaching children to walk by supporting them prematurely on their legs, and leading them forward w'thout allowing them the advantage of having their muscles *previously strengthened*, and in some degree brought under the commands of the will, by *crawling*, is objectionable on various accounts. It seldom fails to produce more or less unnatural curvature of the legs ; and in infants of a scrofulous or ricketty habit, it may readily give rise to distortion of the spine and round shoulder. Children who are permitted to exercise their muscles by *crawling*, generally acquire a much firmer step, and enjoy more robust health than " those who have been taught to walk before the crawling exercise."

If we are earnestly desirous of training up our children in such a manner that they may acquire a firm step and well-formed limbs, we shall gain our purpose much more certainly and safely by pursuing this *gradual* and *cautious* mode of teaching them *the use of their legs*, than by the more common practice of placing them prematurely on their feet, without permitting them first to learn to crawl.

Leading-strings and go-carts, formerly so much in use, are now, very properly, almost universally abandoned. The very common practice of teaching infants to walk by holding them by *one of their hands*, is very wrong. When led in this way, the child's arm is continually, and often forcibly, extended upwards : if it happen to lose its balance, or trip, or if its legs are yet too feeble to support itself long in the erect posture, the whole weight of its body is often suspended by one arm. Frequently, too, it is entirely *raised from the ground by one arm*, in order to help it over some obstacle, or to hasten its progress over a rough and difficult piece of ground. It is easy to perceive that this practice must necessarily, and in no inconsiderable degree, tend to draw the shoulder and side of the chest out of their natural position ; and when frequently repeated, to give permanent *deformity* to these parts.

Nursery-maids seldom exercise sufficient care in this respect. Too indolent to carry the infant in their arms, as they are directed and *supposed to do*, they are apt, as soon as they are no longer observed, to place the child on the ground, and to hurry or rather *drag it along*, in the most careless and unfeeling manner.   Of a similar, but still more reprehensible character, is the practice of raising infants from the ground by both arms and swinging them about in the air.

After children have acquired the entire use of their legs *walking* is decidedly the best exercise they can take.   Parents ought not to intimidate their children by inspiring them with a constant dread of falling or hurting themselves.   The custom of exaggerating the dangers incident to their usual sports—and of plying them continually with admonitory injunctions against accidents when they are engaged in their amusements, is calculated to favor the occurrence of the very accidents which they are meant to obviate, by the timidity which these perpetual lessons of caution and fear almost inevitably inspire.   When the ground is soft, it is much better to let the child take the chance of two or three falls, and give it full scope for the exercise of its limbs, by running about until it is satisfied.   When children fall or hurt themselves, they should not be soothed by expressions of extreme pity and sorrow ; for plaintive words and expressions of great sorrow tend very effectually to render them effeminate and timid.   Children who are thus accustomed to excessive commisseration, seldom fail to acknowledge this tender sympathy, by straining their little lungs to the utmost by crying on every slight injury they receive.

After children have passed through the period of teething, they should be encouraged in the pursuit of active amusement out of doors, as an essential and regular part of physical discipline.   The practice of obliging children to remain within doors, and to con over their lessons between or after school hours, is a barbarous " march of civilization."   These intervals should be devoted to innocent amusement and bodily exercise.

Exposure.—Infants ought to be early accustomed to the fresh and open air.   The practice of confining them, during the first five or six weeks, to close and heated rooms, has a direct tendency to impair the energies of the system, and to impede its healthful development.   Pure air is most grateful to the feelings of children.   After having been carried out, but a few times, they evince, even at a very early age, a strong desire to return to the open air.   While yet on the arms of the nurse, they anxiously point to the door, and make efforts to approach and open it.   When they can scarcely crawl, they instinctively advance towards that part of the room from which they have a prospect of escaping.

When the weather is clear and of a mild temperature, infants should be carried into the open air once or twice daily, as soon as they are three or four weeks old. During *cold* and *damp* weather, they should be occasionally conveyed into an adjoining well-aired room: avoiding, however, strong currents of air, or sitting with them near an open window. Important as the enjoyment of fresh air is to the health and comfort of infants, care should be taken to accustom them *gradually* to the impressions of the external air—more especially when the atmosphere is cold and damp. The practice of exposing children, soon after birth, at once to the open and cold air, with the view of "hardening them," as it is called, is attended with considerable risk of injury, and should not be permitted by parents, except when the weather is clear and very mild. Even in summer, the infant should not, as a general rule, be carried at once into the external air, without having been previously accustomed to the air of a well-ventilated chamber. After the child is three or four days old, it ought to be conveyed, several times daily, into an adjoining room having, at first, only the windows open, and in four or five days afterwards, the doors also, so as to admit a free circulation of the air through every part of the room. This having been practised for ten or twelve days, the child may then be carried out of doors, and permitted to enjoy the pure and open air. At first, it should not be allowed to remain out of doors more than five or ten minutes at a time, but gradually extended. *Hanging up the linen* of children, or *drying their diapers* in the place where they sleep, is very improper.

WEANING.—The only thing that is usually regarded by mothers, in fixing on the time for weaning, is the *age* of the infant. The child is suckled until it attains a certain age, without any regard to the *development of its digestive powers*, or the state of its health and constitutional vigor. By this course, children may be kept at the breast, long after the vigor of the digestive functions, and the demands of the system require a more substantial and nutritive diet; and on the other hand, they may be separated from the breast before the stomach has acquired sufficient energy to digest with due facility a stronger and less congenial food. The progressive development of the digestive powers, and the demands of the organization in relation to nourishment, are very various among different infants. It is particularly important that the condition of infants, with regard to these circumstances, should be consulted in regulating the period of nursing. The obvious correspondence which exists between the successive appearance of the teeth, and the development of the digestive powers, afford us a safe guide in relation to this sub ject.

The progress of teething is, doubtless, our safest guide in re gulating the nourishment of infants, and in deciding on the period at which they may with propriety be put on the ex clusive use of artificial food.   Not unfrequently, however, circumstances of an irregular or morbid character render it expedient, or even indispensable, to wean the child, before it has attained the age and development which, under ordinary circumstances, would be deemed requisite to justify its final separation from the breast.

The mother may be affected with some constitutional disease, which may so contaminate her milk, as to render it highly injurious to the child's health, if she continues to nourish it at the breast.   Mothers, affected with *scrofula*, or ulcerated *can cer*, should, on no account, suckle their infants.

The mother may also be so exhausted and debilitated by an attack of some acute disease, and the measures requisite to subdue it, that she cannot continue to suckle her infant, with out increasing her prostration and superinducing a train of alarming and highly distressing affections.   The same difficulty is apt to occur in mothers of a feeble, delicate and nervous habit of body, particularly when the digestive powers are weak, or so disordered that nourishing and substantial aliment cannot be taken.   Under these circumstances, suckling can seldom be continued without producing the worst effects.

Many young ladies, on becoming mothers, are incapable of supporting the constant drain to which the wants of their in fants subject them.   They lose their good looks, become gradually weaker and paler, and, as their strength declines, they become more and more afflicted with a variety of harass ing nervous affections.   Medicinal means are of no permanent advantage.   They may procure more or less temporary miti gation of the symptoms, but they are wholly inadequate to the removal of the malady.   *Nothing but weaning will suffice* —and the entire separation of the child from the breast is generally soon followed by a progressive subsidence of the sufferings of the patient.

On the part of the mother, the effects of unduly protracted nursing are sometimes extremely pernicious.   We not unfre quently see women pale, debilitated, and constantly tormented with dyspeptic and nervous affections, suckling their infants for eighteen or twenty months, and occasionally much longer without suspecting that their sufferings and ill-health are the result of exhaustion from the constant drain of nursing.

Many mothers are able to suckle their children until they arrive at the *proper period of weaning* without the least inconvenience, who, nevertheless, will suffer very serious derangements of health when the nursing is extended considerably *beyond the time* which nature points out as the proper period for terminating it.

On the part of the infant, also, suckling, when continued much beyond the proper period, is apt to exert a highly injurious influence. It is well known that after the eleventh or twelfth month the milk almost invariably becomes *diminished in quantity*, as well as more or less deteriorated in *quality;* and, in proportion as the nursing is protracted, so will it lose more and more its nutritious and wholesome character. In many instances, indeed, the milk begins to deteriorate as early as the ninth or tenth month, corresponding in this respect with the proper period of weaning as it is usually indicated by the progress of teething. Children who are suckled an undue length of time generally gradually lose their fresh and healthy appearance. The countenance becomes very pale, and acquires a languid, fretful, and sickly expression.

In some instances the milk loses its wholesome properties at an early period, without any very serious or obvious derangement of health in the mother's system. When this occurs, the infant often throws up the milk, soon after nursing, and becomes harassed with colic, griping, acidity, and diarrhœa, attended with paleness, debility, emaciation, and frequently with scabby eruptions about the face and head. If the child becomes affected in this manner, when nourished *exclusively at the breast*, we may presume that the milk has become depraved and injurious to its digestive organs. If any doubt exist as to the agency of the milk in the production of the disorder, the breast should be withheld from the child as long as can be done without any particular inconvenience to the mother, and artificial nourishment, or the milk of a nurse, substituted. If the mother's milk has been the cause of the child's illness, an obvious abatement of the symptoms will soon take place ; and should this occur, the child ought to be gradually entirely separated from the mother.

The recurrence of the menses, during nursing, exerts, in many cases, a decidedly prejudicial influence on the properties of the milk, and often renders weaning necessary before the usual period of separating the child from the breast. When the mother finds the child becoming sickly, feeble, and annoyed with disorder of the stomach and bowels, after *her monthly sickness has returned*, or after she finds herself in a

etate of *pregnancy*, and relief is not obtained, in due time,
from the use of appropriate remedial means, the child ought
to be gradually weaned. Should a woman with an infant at
her breast, again become pregnant, one of two things will
usually take place : either she will miscarry, or her milk will
become impoverished in quality and diminished in quantity.
It was not intended by nature that the processes of pregnancy
and nursing should go on simultaneously; but, on the contrary, that the one should commence when the other had terminated ; and experience sufficiently proves, that they will
not proceed well together.

Attention should also be paid to the *season* of the year, in
fixing on the period of weaning. In general, weaning may be
accomplished with less inconvenience and risk of unpleasant
consequences to the child, during the mild months of April,
May, September, October, and the early part of November,
than whilst the weather is inclement. Exercise in the open air
is always highly beneficial to the child at the time of weaning.
It tends to fortify the system of the child, and to enable its digestive organs to bear, without inconvenience, the change of nourishment. In consequence of the peculiar tendency of warm
weather to excite summer complaint, particularly in cities or
large towns, it is in general inexpedient to separate children
from the breast during the months of June, July and August ;
for the transition from the mother's milk to an exclusive artificial nourishment during this season, has a decided tendency
to favor the occurrence of this dangerous disease. Nevertheless, should the child be suffering from a deteriorated state of
the milk, it ought to be separated from the breast without any
regard to season : for a bad condition of the milk would doubtless be more injurious in this respect than a suitable artificial
nourishment. The child should, at the same time, have the
proper bathing, out-door exercise, etc.

DISEASES OF CHILDREN.

DIFFICULT TEETHING.—Teething is not *usually* attended with
much suffering or danger ; yet when there is much predisposition to disease during this process, any exciting cause may
produce violent and dangerous symptoms. The first teeth
usually begin to penetrate the gums about the seventh month
of infancy—they sometimes, however, appear as early as the
third or fourth, and in some cases as late as the twelfth or fif-
teenth. In difficult teething there is redness and tenderness of
the gums, increased flow of saliva or spittle, thirst, looseness of
the bowels, slight fever, restlessness and sometimes eruptions

on the skin.   In the more severe cases there are often ulcers of
the gums, diarrhea or dysentery, inflammation of the brain or
bowels, spasm of the windpipe, convulsions and death.   These
cases require perfect cleanliness, quiet, **pure air,** vegetable diet,
cool drinks, mild purgatives, and lancing of the gums.   Be
careful not to give anything to check the bowels *suddenly*, in
cases of looseness during teething, as the head is apt to become
the seat of **very** serious disease in such cases, producing spasms,
and sometimes inflammation of **the brain.**   Giving the child,
three or four times a day, a spoonful or two of blackberry root
tea (cold), is about as good, in cases of diarrhea during teething, as anything else. Flannel should be worn next the skin, and
let the child live on the mildest food, or the breast-milk, according to the age.

TOOTHACHE.—This may occur from decay of the tooth and exposure of the nerve, from inflammation of the nerve, gums or
membrane lining the socket, or from ulceration at the root of
the tooth.   If the tooth is much decayed, dark colored, or
ulcerated, it should be extracted : if the pain is caused by inflamed gums or socket, the gums should be freely lanced, warm
fomentations, such as hops and vinegar, with hot water, renewed every half hour, applied to the face, and a gentle purge
administered—castor oil, one teaspoonful, syrup of rhubarb,
one to two teaspoonfuls, or the same quantity of Rochelle salts
in a gill of cold water.

When the tooth has a cavity in it so as to expose the nerve
and cause pain, the application of a piece of cotton wet in some
stimulating medicine, such as oil of cloves, or cinnamon, or
paregoric and camphor, or a mixture of fine salt and alum, put
into the tooth on a piece of wet cotton, and renewed every
half hour, will usually give relief.

INFLAMED GUMS.—During the first teething the gums are
very liable to become inflamed ; in some cases it is slight, and
in others severe, and productive of serious consequences.   The
gums first become red, or dark-colored, swelled and painful,
child languid, feverish, thirsty, tongue furred, appetite impaired, and sleep disturbed.   When the inflammation occurs
before the double teeth appear, it often destroys the new teeth :
and when the inflammation proceeds to ulceration—if this is
not speedily checked, the other teeth become black, loose, and
decayed.   There is a flow of spittle, sometimes mixed with
blood, the breath is unpleasant, countenance pale, and sometimes severe attacks of diarrhea.   This condition is caused by
too much, or improper food, filthiness of the teeth, neglect to
lance the gums in difficult teething, biting hard substances, and

disorder of the stomach. The gums should be freely scarrified, (lanced) the bowels regulated, and some astringent medicine, such as strong green or black tea (cold), tea made of white oak bark, etc., applied frequently to the gums, the teeth cleaned and all decayed ones extracted : the diet should be very light, and the general health improved by gentle tonics ; a tea made from a mixture of equal parts of bruised gentian root, wild cherry bark, and orange-peel or sassafras bark—say one table-spoonful of the mixture in a pint of boiling water, cover up and let stand for one hour and a half ; of this one or two table-spoonfuls may be taken before each meal (cold.)

THRUSH.—Four or five varieties of inflamed mouth are de-scribed by authors ; but the most common of these are simple inflammation and thrush. The symptoms of the first are, red-ness and dryness of the mouth, the infant manifesting pain when attempting to nurse — caused by teething, bad diet, sharp acrid substances, cold, or over exertion of the muscles of the tongue and mouth in attempting to nurse from a badly-formed nipple. By removing the cause, and the use of simple washes and mild purges, a cure is soon effected. Thrush is confined in its attacks mostly to nursing infants. At the be-ginning of an attack the child is restless, the mouth red, dry, and hot, digestion is disturbed, and there is difficulty in nurs-ing : after one or two days, small white spots appear on the tongue and mouth, and sometimes spread over the entire sur-face. In the course of the disease, patches of curdlike matter fall off, and the spots are again covered as before ; it sometimes extends backwards into the throat, or ulcerates and becomes both tedious and troublesome—in some cases it proves fatal.

It is caused by improper diet, filthiness, impure air, disorder of the stomach and bowels, sudden stopping of diarrhea, and nursing from a sore nipple or a diseased nurse. In mild cases, pure air, proper-diet, cleanliness, mild purgatives—same as those recommended in treating toothache, and soothing washes (hop tea and sage tea mixed, is very good,) for the mouth, will remove the complaint.

The *first thing* to be done when an infant is affected with thrush, is to correct the acid state of the bowels by a few grains of calcined magnesia—or if the bowels be relaxed, by chalk, following the magnesia by a half teaspoonful of castor oil. This may be repeated every second day. The quality of the milk, and the state of the nipple of the mother are to be exam-ined. Milk and water—two parts of the former to one of the latter—in which a little isinglass should be dissolved if there is diarrhea, is to be the sole addition to the mother's milk ; all sugar is to be avoided.   If the state of the bowels be corrected, the thrush will generally get well, but it is expedient to assist the cure by the use of a solution of borax in water—one teaspoonful to half a pint—used to wash the mouth.   When the case is mild, the curd-like patches will separate in seven or eight days, leaving a healing surface below, and the mouth soon gets well, if it be not injudiciously scrubbed ("cleaned") daily by the urse.

BLEEDING FROM THE NOSE.—This is sometimes a frequent and troublesome disease with children, caused by injuries of the nose, fullness of the blood vessels of the head, &c.   In robust persons troubled with dizziness and headache, it is often beneficial, and, unless excessive, need not be restrained ; but in those of a pale and weak habit, it may, if long continued, produce debility and dropsy.   It may usually be restrained by the application of cold water to the head and neck, snuffing cold water, or alum water, up the nose, or stopping the nostrils with lint or cotton.   When these means fail, more efficient ones must be employed.

Gargling a strong tea, made of white oak bark, when cold, in the throat, then suddenly closing the mouth, and *stooping forward, to make the liquid come out of the nostrils*, repeated every few minutes, if necessary, will usually stop it.   Raising both hands above the head, while the nose is kept closed by an assistant, is a good and simple remedy.

To prevent a return, bathe the head in cold water, night and morning, live principally on a *vegetable diet*, keep the bowels regular, and avoid exposing the head to the heat of the sun. Using a rough towel or a flesh brush, night and morning, to rub the surface of the body and lower extremities, is advisable.

CANKER OF THE MOUTH.—This occurs in children of weak, scrofulous constitution, who are ill-fed and exposed to the influences of unhealthy habitations; and most generally immediately after acute disease, particularly measles.   The first symptom of the disease is a red, hard, angry-looking spot on the cheek, which quickly opens into a gangrenous, (mortified) ulcer inside the mouth, the gums become affected, the teeth drop out, the breath is very unpleasant, and the extending ulceration goes on destroying the cheek and contiguous parts, till it is either stopped or death ensues.

As the first cause of this fearful affection is traceable to poverty of constitution, the first remedial measure is to nourish.   The strongest meat-soup—beef-tea is the best—must be given in small quantities, frequently repeated ; milk and eggs,

,f the litt.e patient wil! take them. Wine may be allowed if he debility is extreme, but scarcely, if at all, should fever run nigh, and there is much heat of skin. A drachm of chlorate of potash is to be dissolved in six ounces of water, and to this added twenty drops of muriatic acid. A tablespoonful of the mixture to be given to a child of six years of age every four hours; it may be slightly sweetened. Half-grain doses of quinine, or an ounce of infusion (or tea) of Peruvian bark, may be given twice or three times in the twenty-four hours. A wash made of one teaspoonful of salt, dissolved in half a pint of water, should be frequently applied. The case ought to be seen by a medical man as soon as possible.

CROUP.

This is recognized as one of the most dangerous diseases of childhood. Its progress is rapid, and its treatment, to be successful, admits of no delay. Fortunately, if taken in time, it is greatly under the control of well-directed treatment. Its dangerous nature must ever make proper medical advice a necessity, but the importance of early active remedial measures renders it, at the same time, highly desirable that treatment should be resorted to without the slightest delay. The great danger in croup arises not only from the possibility of the narrow chink in the larynx, or upper part of the windpipe, through which the air passes, becoming closed by swelling, but also from the remarkable product of a peculiar inflammation which is formed upon, or thrown out by, the lining membrane of the parts. This formation, " false membrane" as it is named, resembles thin leather of an ash color. It takes the form of the tube which it lines, and, indeed, is sometimes coughed up in perfect tubular portions. However, when this false membrane forms, death is the result usually.

Croup may begin very suddenly. A child goes to bed, to all appearance perfectly well, and in the course of two or three hours comes a cough, which strikes even the most unobservant as peculiar, which falling upon the ear of the anxious parent, who has ever heard it before, tells at once of danger. The child seems as if it coughed through a *brazen* tube. Perhaps at first the little invalid is not awakened, and if now visited is found flushed and fevered, moaning slightly, perhaps, and restless, the breathing slightly quickened; the cough comes again, the child awakes, or is awakened; if it speaks, the voice is hoarse; if it cries, hoarser still. Should the disease be neglected at this time, or go on uncontrolled, the cough, still retaining its peculiar character, becomes more trequent; the breathing, quickened, is also accompanied by the characteristic dry wheezing occasioned by narrowing of the passage through which the air is drawn; the head is thrown back in the efforts to breathe, respiration is insufficiently performed, and the blood being insufficiently changed begins to evince its deteriorated character in the blue color of the lips, the dusky coldness of the skin, and the affection of the brain which gives rise to partial insensibility or delirium. The pulse, previously quick, becomes still quicker, but at the same time feebler, and at last the child dies in a state of almost unconscious suffocation. There may, however, in the progress of the disease, be intervals of comparative ease, alternating with paroxysms of spasmodic obstruction to the breathing, threatening, and sometimes causing, immediate suffocation. The average duration of a fatal attack of croup is from three to four days, but it may, and does, terminate much more speedily. When under proper treatment the disease is checked, the first best sign is the cough beginning to "loosen," the breathing at the same time becoming tranquil, and the skin moist; the pulse changes from its *hard quick beat* to one of a *softer* and a *slower* character. Croup does not, however, invariably begin *suddenly*—frequently the child has been suffering, apparently, from common cold in the head, and the attack of croup seems to be a consequence of the inflammatory affection of the membrane of the nose and throat extending into the windpipe, and taking on the peculiar character of the more fatal disease. At other times there has been slight drowsiness for some days previously, but not sufficiently well marked to attract attention, although at the same time, from hoarseness not being common among children, its occurrence should always rouse suspicion, especially if the child itself, or any of the family, have suffered from croup. Sometimes a child will have a croupy cough for some nights in succession before the attack of the real formed disease; and parents are apt to be lulled into security by the fact, that in children susceptible of croup any cough partakes more or less of the shrill croupy sound. Another, and highly dangerous, form of croup is that in which the inflammation commences in the throat, the tonsils, and soft palate, which quickly become covered with an ash-colored membrane. At first the child is supposed to be merely suffering from sore throat, for there may be little or no cough, or embarrassment of breathing, but the inflammation extends downward into the air passages, and the croupy symptoms become developed; by the time this stage is reached the case is all but hopeless. Fortunately this dreaded disease, the most distressing, perhaps, by which a parent can lose a child, is, in every form but the last, amenable to proper remedies, if adopted at once. So strik- ingly, indeed, is this the case, that it is very common to find parents taking the matter in their own hands after they have seen a child treated for the disease once or twice, keeping, *as they ought to do*, a supply of the proper medicines constantly at hand, and by their prompt application nipping the first at- tack in the bud ; the medical attendant is either not sent for, or, if he is, it is only to find that the proper treatment has been followed and the disease checked.

The great remedy in croup is *emetics*, or medicines which sicken the little sufferer at the stomach, and cause vomiting. A teaspoonful of the syrup of ipecac, or four grains of the powder, given at the very beginning of the disease, will usually produce vomiting in a short time ; if not, let the dose be repeated until that effect *is produced*. If ipecac is not to be had, as is often the case, at that hour of the night, one tea- spoonful of powdered alum, mixed with a tablespoonful or two of sweetened water, given, will usually produce vomiting ; or, if nothing better can be had, from ten to thirty drops of antimonial wine, (according to the age of the child,) given every fifteen minutes, until it has the effect of vomiting.

If the child is not better within an hour after the first vomiting, the emetic should be repeated. Besides this, apply some stimulating liniment to the throat, and around the upper part of the chest, front and back, and also keep a piece of flannel bound around the throat. A mixture of turpentine, sweet oil, spirits camphor, and whisky, (or spirits of any kind,) in equal parts, will make a good liniment, applying it with a woollen cloth or the hand, and afterwards putting on flannel next the skin. If all these ingredients are not to be had, use such as you have, even one of them alone. Let the child drink freely of toast water or thin gruel.

Besides this treatment, keep the child carefully wrapped up, to prevent checking the perspiration ; and also produce a *moisture in the room* as soon as possible, by means of hot water poured over mullein leaves, hops, sage, and horehound, all mixed together, or either one, if all cannot be had, using a large pan, pail, or dish, so that the steam arising therefrom will impart its moisture readily to the air of the room. And if the child be large enough, let it inhale frequently from an old tea-pot the vapor of hot water and mullein leaves, or some of the other ingredients mentioned above.

Also, give a good dose of casto＝ oil or other suitable medi

ne, to purge the bowels. When the child is getting better, be careful of a *relapse.* Do not suffer it to be exposed to the cold air out of doors till entirely well.

If a case of incipient croup be thus treated, it will, in all probability, and may be, subdued without medical assistance, though it is certainly safer to have it; but if the fever is extremely high, and if the breathing has any approach to a crowing sound, medical attendance *must* be procured if possible, and with the shortest possible delay. Always bear in mind that this disease must be *promptly treated.*

The causes of croup are almost invariably connected with cold and moisture, and particularly during east winds; but it may also be occasioned by the removal of wrappings from the throat, and exposure to a cool air when a child is heated. Children liable to croup are still more so after attacks of acute or debilitating disease.

The prevention of croup is, of course, of the highest importance, and, therefore, the causes of it must be avoided in every way. Slight colds should never be neglected in children or families thus predisposed, but should be treated by confinement to the house, or to bed if requisite, by milk diet, diluent drinks, and by the tolu ($\frac{1}{2}$ oz.) and mucilage (2 oz.) cough mixture, with the addition of wine of ipecac ($\frac{1}{2}$ oz.), one teaspoonful of the mixture every four hours; paregoric should also be given to allay troublesome cough, and, in fact, those measures recommended in *cold* carried out. The susceptibility may also be lessened by not clothing the throat too warmly, and by the regular practice of bathing the throat and chest well with cold water every morning, rubbing afterward with a rough towel, till thorough reaction ensues. This practice is, of course, better commenced in warm weather, and not too soon after an attack of the disease. Flannel should always be worn next the skin, and care taken particularly that bedchambers and rooms children habitually live in are not too warm, and never occupied while the floors are wet after washing. A residence a distance from water is to be preferred.

SPASMODIC OR CROWING CROUP.—This disease differs very much from the membranous croup. It is a species of convulsive or spasmodic affection of the muscles of the larynx (upper part of windpipe), which, by narrowing and closing the chink in that organ, through which the air passes, occasions the sound of the breathing to resemble that of the true inflammatory disease. This spurious croup is often an alarming, and sometimes a fatal disease; it generally occurs before the end of the third year of life, and in consequence of irritations acting more or less at a distance from the affected parts which receives the impressions through its nerves. Enlargement of the glands of the neck, affections such as eruptions of the scalp, the irritation of teething, or the presence of irritating matter in the bowels, may any of them give rise to this affection. It comes on suddenly; the child is seized in a moment with "catching at the breath," struggles, the face changes color, and the veins are full. If the spasm be not relaxed after a few ineffectual efforts at breathing, the child must die; but if the spasm gives way, the air is drawn into the chest with a crowing, croupy sound. It is of much importance that this spasmodic disease should be distinguished from real *inflammatory croup,* on account of the very different treatment required; it may be known by the *absence of fever,* the stopping of the breath being much *more instantaneous* than that which occurs in the real disease. In an affection presenting symptoms so sudden and so alarming, *immediate* remedies must be used; a little cold water should be dashed on the face at once, and, as recommended by Dr. Watson, a sponge dipped in hot water applied to the fore part of the throat, and after removing it apply a flannel bandage saturated with some stimulating liniment, medical assistance being of course procured if possible.

In the meantime, set the child in an upright position, with the head leaning forward, and exposed to the fresh air for a few moments, the body being at the same time well wrapped up. If not relieved, rub the spine (back bone) thoroughly with the open hand, moistened with the liniment previously mentioned, or any other which may be on hand.

When the spasmodic fit is over, examine the gums, and if red and inflamed let them be lanced. Also give a mild purgative every day or two until well. A teaspoonful of syrup of rhubarb, or castor oil, will answer.

Colds and Snuffles.—During the first month, most children are affected with colds, commonly in the nose, called snuffles. Warming the feet at the fire, will often be sufficient to cure them. But when the disease is attended with fever, it is best to administer three grains of ipecac, mixed in four table spoonfuls of warm water, and one table spoonful to be given every twenty minutes, until vomiting is produced. The bowels should be kept open with magnesia, rhubarb, manna, or castor oil, in small doses. Repeat the ipecac next day if the disease is not better.

Various Eruptions of the Skin.—Children, particularly those not daily bathed, or washed in water, are very subject to a great variety of eruptions on their skin, commencing sometimes the first week of their birth. Different names, as red gum and white gum are given to each kind; but it is useless, as they require nearly the same treatment. In the red gum there is a number of small, elevated red spots, scattered over the body, and sometimes on the cheek or forehead; on the feet the spots are still larger, and contain occasionally a clear fluid. In some stages it resembles the measles. Generally no medicine is requisite; but if it suddenly disappears, and the child shows symptoms of internal disease, an emetic of ipecac, as mentioned under head of Snuffles, or purgative ought to be given, and repeated, if not at first relieved. The white gum appears after the red gum, resembling itch, with white, shining little blisters, containing a little clear fluid. There are other varieties of these eruptions of the skin, but few of them require medical treatment. A vomit or purge, to clear the stomach and bowels, generally relieves. The prevention is in great cleanliness, free washing daily in soap and water, with *regularity in nursing.* When these affections of the skin are attended with fever, they require, besides the vomit and purging, applications to the parts inflamed, to lessen the action; cold lead water (two or three grains of sugar of lead dissolved in one ounce of cold water), and sweet oil, are the best for this purpose applied every three hours.

Sore Eyes.—Children are very subject, sometimes during the first month, to inflammation of their eyelids and eyes, particularly those who are exposed to a strong light soon after birth, getting soap in the eyes by the nurse when washing them, perhaps for the first time, a draft of cold air, etc. At whatever time the inflammation comes on, in slight cases, a very weak solution of sugar of lead—fifteen grains to the pint of water, should be applied every two hours to the part, by means of a piece of linen soaked in the water and laid over the sore eyes for a few minutes at a time. In many cases the warm breast-milk of the mother put into the child's eyes every time that it nurses, will cure them. (In cases where much *thick matter* is discharged from the eyes, *or they are closed up,* and the *eyelids puffed out,* a physician should be sent for at once, for the eyesight may be lost in a few hours, if not properly attended to.) If it do not speedily subside, a purge of castor oil should be given. The inflamed eyes should never be turned towards the fire, and the hand of the infant so confined as to prevent it from rubbing the part, and the room kept darkened. In cases where the inflammation of the ball of the eye is great, a leech should be applied to each temple, or cupped, after being scarified. Also five to ten drops of syrup of ipecac should be given every two hours to reduce inflammation.

This cold lead water alone is usually the proper application to the eye, and nothing should be added excepting where the eyelids adhere together.   In this case, the mildest sweet oil, mild hog's lard, or any bland grease should be applied to the edges of the eyelids before the child goes to sleep.   Avoid every *stimulating* application in inflammations of the eyes ; it has been the cause frequently of loss of vision.

EXCORIATION.—When the skin is rubbed off (termed excoriation), as is often the case between the legs, behind the ears, in the hair, between the toes or fingers, etc., you should make an application of sugar of lead (twenty grains dissolved in a pint of cold water) three or four times a day, with a soft linen cloth, or sweet oil, or fresh lard, will generally heal them up. Powdered starch is also good.

In cases of inflammation, a poultice of Indian corn meal or flax-seed meal wet with this lead water, and kept applied to the part, will expedite the cure.   When sores have been of long standing, you should, by all means, on drying them up, purge the child once or twice a week, for three or four weeks afterwards; also *diminish* its *food.*   The neglect to do this, or to make a slight issue or sore, by means of a small blister plaster, applied and kept to some part of the body, two or three hours every day, for a week or two, has often been fatal ; as the system, when the old sores are healed, not having its accustomed irritation, takes on violent disease in other parts.   Death has often resulted from healing up old sores suddenly, without taking the precaution alluded to.

WIND IN THE STOMACH AND BOWELS.

WHEN a child has wind on the stomach it may be known by wind often rising in its throat, which makes it struggle at times, as if to get its breath, and from which it is occasionally relieved by belching of wind upwards.   When it often occurs it is annoying, and interrupts rest.   It is most common with children dry-nursed.

Different articles have been given to dispel the wind; but none of them are to be compared to spirit of hartshorn : three drops in half a table spoonful of cold water, and repeated two or three times a day, as may be required.   Hartshorn, when it will answer the purpose, is to be preferred to cordials, spirits, seeds, spices, and hot things of any kind ; as, although it is fully as powerful in dispelling the wind as any of them, it will not, by a permanent heat, nor by repetition, injure the stomach as they do ; nor can any bad habit or other disadvantage arise from giving and repeating it as often and long as it may be necessary.   It is endowed with a property which makes it a

desirable medicine for children; it corrects and removes *acid-*
ity or *sourness*, a principal cause of griping with children.
The dose here mentioned is the smallest that need ever be
given, and it may be increased, as a child grows older, to five
or six drops. The child's bowels should be kept open with
mild purgatives, such as magnesia, syrup of rhubarb, etc.
Also be careful that nothing in the way of nourishment be
taken except the breast-milk, until the disease is entirely
cured. If the hartshorn can not be had, use a tea made of
aniseed, catnip, mint, or cinnamon bark, (not too strong) every
half hour.

Some children seem naturally more subject to wind in their
bowels than others; and which can be accounted for no otherwise, than as proceeding from a particular weak and tender
state of those parts. And as it is much increased by cold, a
particular attention must be paid to keeping them well covered
with flannel next the skin, and three or four times a day use
friction or rubbing with the dry hand over the stomach and
bowels.

DIARRHŒA OR LOOSENESS OF THE BOWELS.

This is generally brought on by too much, or unsuitable
food; in which case great attention should be paid to the
diet. In other cases, it may arise from disease of the
bowels, such as irritation from worms, or inflammation. In
such cases a dose of ipecac, according to the age, should be
given to produce vomiting, so as to get the stomach emptied
of its contents, (unless there is already sickness at the stomach;) then the bowels are to be cleansed by a purge of a
little rhubarb and magnesia; (four grains of each,) to be followed by small doses of chalk in some mucilage, as milk—
made more palatable by a drop or two of the essence of peppermint, cinnamon or aniseed. Flannel, soaked in whisky,
should be applied to the bowels, and the child made to *lie*
down in bed and keep as quiet as possible. If the stools continue more frequent than they ought to be, and are either
slimy or tinged with blood, the purge of rhubarb and magnesia should be repeated. Cold drinks of all kinds should be
allowed *only in very small quantities at a time:* small pieces
of ice taken into the mouth and allowed to melt, is better for
allaying thirst. Sponging the body also once or twice a day,
with a mixture of water and whisky, is often of signal benefit.
The diet must be of the simplest kind, avoiding all kinds of
solid food. Warm applications are to be made to the bowels,
and the skin gently rubbed. Sometimes the application of a

small blister to the pit of the stomach is of great service. **A** mustard plaster applied for a few minutes at a time to the stomach and bowels, two or three times a day, is often of great service. and is always to be tried in preference to the blister, especially in *small children*. In cases where the strength is fast sinking, injections of thin starch, with a few drops of laudanum or paregoric, with a teaspoonful of wine or whisky every two or three hours, should be given; and laudanum may be rubbed on the stomach and bowels, with sweet oil.

FALLING OF THE FUNDAMENT.—In children of lax habits, the lower portion of the bowel is very apt to protrude after a stool It is a source sometimes of great pain, and often of great uneasiness. In general it may be replaced by the application of a rag wet with cold water, using very moderate compression. More obstinate cases require that the child should be laid on its belly, the sides separated, and then the fingers of the hand are to be applied, so as equally to cover the protruded part; then gradually and firmly, in one continued pressure, the part may be caused to draw up. Whenever the bowel protruded is *inflamed* or painful, it should be bathed in cool water, cold green tea, olive oil, or hog's lard. Sometimes fomentations of mild articles, as flax-seed poultice, hops, with hot water, &c., are of service.

Those children much subject to this complaint, should never be allowed to strain in evacuating the bowels. The discharge had best be made in an *erect posture*. The strength of the bowel may be restored by injections of tea made of oak bark or nut galls: when the irritation is great, a drop or two of laudanum will lessen it. Pouring cold water occasionally on the parts, and always after a discharge washing in cold water will be found serviceable.

COLIC.—Colic, which in some children is of very common occurrence, is easily discovered by sudden fits of crying or screaming, which nothing can appease; the child bends back the body, spurs with the feet, and then has an abatement of the pain for a few minutes, obtained sometimes by the escape of wind from the stomach or bowels. An attack may consist of one uninterrupted fit, or of repeated screaming, with intervening moments of ease. It may be induced by costiveness, by cold, by damp clothes, by the too liberal use of panada, particularly if made of *sour bread;* by passion, or some state of the nurse affecting the milk, by collection of wind in the bowels; or it may accompany thin and slimy purging, which is sometimes produced by the injudicious use of purges.

In ordinary cases, nurses give gin and water, which is a

most injurious practice, and may in some instances kill the
child. Laudanum gives speedy relief, but it weakens the
stomach and nervous system, and produces costiveness. A
few drops of tincture of asafœtida, mixed with oil of anise, is
generally effectual, and is always safe. Two drachms of tincture of asafœtida, twenty drops of oil of anise, and an ounce
of mucilage of gum arabic, may be rubbed up together: and
of this mixture, from ten to twenty drops, in a little water, will
be a proper dose, as often as occasion may require. The warm
bath is useful, and if these means do not give relief, rubbing
the stomach and bowels with laudanum will be safer than
giving it internally. An injection of gruel and a little oil is
proper, and cloths dipped in hot water and applied to the bowels is also good; and if the child has been costive, it will be
right to give a tea spoonful of castor oil, after these remedies
have relieved, in order to prevent a return.

When children are subject to colic, we may suspect that
there is something wrong in the diet. Common panado, especially if it contain much sugar, is very apt to have this effect.
The nurse's milk may also be flatulent, and this bad property
is sometimes increased by the use of porter or ale, intended to
increase the quantity. The state of the child's bowels must be
attended to, and it should not be allowed to load the stomach
by taking *too much at a time.* If it belches up wind after
sucking, it should be gently dandled, as that promotes expulsion.

SPASMS OR CONVULSIONS.

This unpleasant and often dangerous disease may take place
at any age, and may occur either in the course of some other
disease, under which the child has been laboring for some
time, or suddenly, in apparent good health. In one case they
are highly dangerous, and often indicate a fatal result; in the
other, they are frequently attended with little hazard. Convulsions vary in degree, from a slight movement of the muscles
of the face, to a rigid, or convulsed state of almost the whole
body. In general, whatever be the degree of the movement,
the countenance is altered, both in color and expression: the
patient is insensible, and cannot follow an object with the eye
In some instances, the motion is so slight, that the child may
rather be said to be in a state of *fainting,* or *stupor,* than of
convulsion. In very young infants, there is sometimes only a
smile about the mouth; the eye, which is half closed, turns
slowly round, the breathing seems occasionally to flutter, and
the child starts, and throws out the arms on the least noise.

These motions, called inward fits, frequently proceed from wind in the bowels.

Convulsions sometimes go off in a few seconds; in other instances they continue for several minutes. The child may have only one short attack, and become well immediately afterwards, or it may remain in a languid, sleepy state; or it may have repeated attacks in a very short time, and continue insensible during the whole of the intervening period, which is always an unfavorable symptom. They may be produced by wind, or irritation in the bowels, dependent on worms, costiveness, indigestible food, griping, stools, &c. ; or by teething; or by breathing bad or confined air ; or by the striking in of some eruption ; or during the coming out of others, such as small-pox ; or by affections of the brain itself; or by other spasmodic diseases, such as hooping-cough, &c.

When the child has been ill for some time before convulsions come on, especially if the pulse has been quick, the skin warm, and the head affected, whilst these symptoms could not be traced to the effect of teething, there is ground to believe that the convulsions proceed from a diseased state of the brain.

With very young infants, if there have been no preceding disease, there is great reason to attribute the convulsion to the state of the bowels : and we shall be confirmed in our opinion by finding that the stools are not of a good appearance ; that there is much wind in the bowels ; that the child has not been nursed or fed properly ; that the nurse has been agitated by passion, or committed some irregularity in diet; or lastly, in infants a few days old, that the meconium (or contents of the bowels,) is not expelled.

When young infants have convulsions from the state of the bowels, we generally find that the face is pale and the motions slight ; but if they proceed from the state of the brain, which is still more alarming, the motions are stronger, and more deserving of the name of convulsion.

After the child is two months old, irritation of the bowels, proceeding from bad stools, worms, or indigestible food, does not produce those gentle motions, or that apparently languid state, observable at an earlier period, but generally excites pretty strong and well marked convulsions.

At the period when children are teething, convulsions may be produced by irritation of the gums, more likely than by other causes ; and, therefore, we should in every case which occurs at that time, examine the gums carefully and cut them if there be the slightest swelling or sign of teething.

When a child is seized with convulsions, great consterna‑
tion immediately prevails, and without some common sense
rules, either *nothing* will be done, or very *contradictory* plans
may be adopted.

The first general rule in such cases is, if the child seems to
to sick, or oppressed in its breathing, or has a fulness of the
stomach, or has been known to have had something which has
disordered the stomach, vomiting should be excited, by tick‑
ling the throat with a feather, during the fit, or by giving
ipecac, (five to ten grains, in warm water, or one or two tea
spoonsful of the syrup,) as soon as the child can swallow. Rub‑
bing the spine, or along the back-bone, with some stimulating
liniment, or a mixture of one tea spoonful of ground mustard,
one of salt, one gill of vinegar, and half pint of water, with a
little laudanum, will be beneficial. Cold water and vinegar
mixed in equal parts, in which a cloth has been soaked, and
applied to the head, is also good in cases where there is flush‑
ed face, fever, and insensibility between the convulsions. At
the same time, give injections of warm soap-suds, or warm
water and castor oil up the bowels, and as soon as the child
can swallow, give a good brisk purge of castor oil, Rochelle
salts, (one table spoonful in water,) or rhubard and magnesia.
Rubbing chloroform on the temples and back-bone is also
some times beneficial in severe cases, as also is pounded ice,
wrapped 'n a bladder or piece of cloth, and applied to the
head.

When there is a tendency to frequent returns, it will be
proper, besides keeping the bowels open, to give repeatedly a
few drops of tincture of asafœtida, mixed with oil of anise. In
all cases of weakness, the strength is to be supported by suit‑
able nourishment, even by injections of beef tea.

SORE HEAD.

SOME children, are subject to sore head. It often be‑
gins on the fore part of the head, in large white scabs,
which, if neglected, spread over the head, forehead and face,
in large patches. In the beginning, generally, it is dry;
at other times, it is moist and has a thin discharge. Medical
writers have named this complaint *crusta lactea,* or milky
crust, from its appearance. The children of the lower order
of country persons, who are gross in feeding, are most subject
to it; and it seems to be occasioned by a want of cleanliness
and exercise, which children, who have a bountiful supply of
nourishment, require; but to which parents, in this situation,
are not often disposed, or seldomer have opportunity to afford

them. A cabbage leaf is a very common application, as it
promotes a discharge from the head, which is supposed necessary, previous to the cure ; but as such a discharge is in no
way necessary, and as it makes the head uncommonly offen
sive, it is better not to encourage it, and the sooner the com
plaint is cured the better. For that purpose, take of brandy
(or whisky) and water, each equal parts ; mix them together,
and bathe the parts of the head and face where the complaint
is, once a day, and immediately afterwards lay on a plaster of
basilicon ointment, (made of lard, eight ounces ; resin, five
ounces ; yellow wax, two ounces ; melted together,) spread
upon a linen rag, which is also to be renewed every day, after
each washing with the brandy and water. Two or three doses
of purgative medicine must be given during the cure. *Bath
ing* in the *sea*, or salt and water, will be of great use.

SCALD HEAD.

This is different from the preceding, as the soreness is
confined altogether to the head, but will extend to the
neck if neglected. It begins in distinct brownish spots,
that form a scab and discharge a thick, gluey matter, that
sticks amongst the hair. The spots increase and enlarge so as
to cover a great part of the head. When these spots are discovered, the hair upon and about them must be cut as close as
possible, and they must be washed well, every day once or
twice, with soap and water. Should that not prove sufficient
to remove them, they may be daily anointed with a little tar
ointment, (a mixture of tar and fresh lard, in equal proportions,) or Barbadoes tar mixed with sweet oil, in equal proportions, with the point of the finger, which rarely fails of a
cure. The scald head, which is either this complaint in the
extreme or nearly allied to it, may be treated in the same
manner, and which will be going as far as can with propriety
be attempted before consulting a physician.

MEASLES.

The symptoms of the measles are, a sickness, a heaviness, a
thirst, a short, dry, husky cough, with hoarseness, a sneezing,
a running at the nose, and a running and thin discharge from
the eyes, which appear red and much inflamed, particularly
the eyelids, with sometimes cold shiverings. These symptoms
are commonly slight at first, and increase till the measles come
out, which generally happens on the fourth day from the first
attack, although children will frequently be much indisposed
for a week before they come out. At the first appearance of

the measles, they look like flea bites upon the face and neck, in distant spots; but soon after, the face, neck and breast are covered in patches, resembling a thick rash, that does not seem to rise above the skin, although it may be discovered by the touch and feel of the hand, to be a little prominent or raised upon the face and breast, but not upon the other parts of the body. The measles, like the small-pox, come out first upon the upper part of the body, and last of all upon the feet; and they observe the same progressive regularity in going off.

This disease is attended with much depression and dejection, and sickness at the stomach. It is very common for the most lively children to lie in a stupor, or state of heaviness and seeming insensibility, from the second day of the attack, during the whole of the complaint, which continues three days after the first coming out; on the third day the eruption begins to look paler, and, on the fourth, goes off with a mealy appearance upon the skin. During the whole of the complaint there is considerable fever, which often, with the cough and a difficulty of breathing, increases in proportion as the disorder advances, and will sometimes be the most violent and severe at the height, or turn, of the measles; sometimes the fever, cough, and other symptoms abate, and the child recovers, in part, his spirits soon after the measles come out, but not generally.

The patient must not be kept either very warm or very cold; he ought not to be kept near the fire, nor yet suffered to breathe the *cold* air; it will be best to confine him to one room that is *moderately and temperately warm.* Cold air will add to his hoarseness, and make the cough worse. His drink may be water, barley water, milk and water, balm tea, saffron tea, or anything of the kind; but water, or milk and water, seems most agreeable to children at this time. What he drinks ought to be a little *warmed,* but not *hot.* Wine, cordials, and all stimulating drinks are improper and injurious.

These precautions are always to be observed on the first attack of the measles. It will always be proper to give something at the beginning, to procure two or three loose stools, as the infusion of senna, salts, castor oil, prunes, or manna, &c.

Mustard plasters, applied between the shoulders or to the sides, have been found of great use in abating the cough and relieving the breathing, and may safely be applied at any period of the disease, if the cough and breathing be bad Cupping the sides and back is also of value.

A fever always accompanies the measles, and is the cause of the drowsiness and stupor which children have in the be-

ginning, and often during the whole of the complaint. Noth
ing will so sensibly check and abate this fever, remove the
drowsiness, and restore a child's spirits, as repeated doses of
ipecac and spirits nitre. Mix five grains of powdered ipecac,
(or one tablespoonful of the syrup,) and one tablespoonful of
spirits nitre, and two ounces of cold water, together. Of this
give one teaspoonful every three or four hours, unless there is
much sickness at the stomach. It may be begun with on
the second or third day ; and after the stools have been procured, as above directed, while the fever and heaviness continue, it will be particularly proper to give it in the evening, at
which time the fever is most severe, and if it operates, as it
generally does, both by vomit and stool, it will give most
sensible relief—the fever, heat, and oppression will be considerably abated, and the child will be much more easy and
cheerful, and more tranquil and composed, than before
taking it.

The fever and cough will frequently continue, without much
abatement, for a few days, or a week, after the measles are
entirely gone, but which may be greatly relieved, or entirely
removed, by giving a gentle purgative every second day. It
may also be known that the fever continues while the dullness,
thirst, and want of appetite remain, and during which time
the purgatives ought to be given, at proper intervals, if no
other cause forbids it. It may also as certainly be known that
the fever is gone off when the child's spirits and appetite return. The danger from the measles is much increased when
they happen to be connected with the small-pox or hooping-
cough ; and, therefore, so circumstanced, they require more
medical attention than is generally bestowed upon them.

The eyes, and particularly the eye-lids, will sometimes remain sore, swelled, and inflamed after the measels. The cough
also, will oftentimes continue for sometime after the fever and
every other remains of the measels are gone. While either
the sore eyes, or the cough remain, the child ought not to be
suffered to go out of doors, or to be exposed to the cold ; as
the air, in cold weather particularly, is very apt to add to and
greatly aggravate these complaints, and may make them very
troublesome and tedious. Too much caution, therefore, in
avoiding cold, cannot be observed during the disease or while
there are any remains of sore eyes, or cough. The measels
sometimes leave these symptoms for the remainder of life—
which most frequently may be attributed to a *too early venturing out,* which of course would have been prevented by *sea-
sonable confinement within doors.*

WORMS.

WHEN a child gets sick, and the mother can find no other solution of the difficulty, she is almost sure to attribute it to worms. She is oftener wrong than right. Worms of different kinds are often found in the bowels; but there are chiefly two met with in children, the lumbricus, or long worm, having a great resemblance to the common earth worm, and the ascaris, or small white worm, like a bit of thread. These two kinds inhabit different parts of the bowels, the small worms being confined to the lower part, whilst the other is found much higher. It is extremely difficult to account for the production of worms. It is observable, that few infants have worms till after they are *weaned*, which is to be accounted for on the principle that the bowels are in better order during suckling than afterwards, when the diet is more varied and indigestible.

Worms may exist without producing any symptoms, until they either accumulate in considerable quantity, when they cause more or less irritation in the bowels, or some slight indisposition takes place, and they, by their irritation, increase it. All the injury they produce, is that of *irritation :* but the *degree* of this, and the *effects* of it, must vary, not merely according to the number of worms, and their movements, but also according to the state of the bowels themselves. It is also to be remembered, that as a weakened state of the bowels is favorable for the accumulation of worms, many of the symptoms may proceed from that state alone, independent of the new irritation from worms.

The long worms may be suspected to exist, when the child complains of frequent griping or pain in the belly, has repeated and unexpected attacks of looseness, variable appetite, being sometimes seized suddenly with *extreme hunger*, has swelling of the belly, especially at night, disturbed sleep, frightful dreams, and grinding of the teeth. Always give children something to eat *at once* when they thus cry out with hunger. A failure to do this has caused the worms to *pierce the bowels through*, and cause death, when a slice of bread and butter would have saved its life. Besides these symptoms, we also observe that the countenance is alternately *pale* and *flushed ;* the child *picks its nose*, has *bad breath*, *dry cough*, and sometimes *slow fever*, or convulsive affections. These symptoms may exist in different degrees, and are ultimately attended with the expulsion of worms, either by vomiting or stool. It has been supposed that a very obstinate and protracted fever, called "*worm fever*," might also be produced : but this generally depends more upon costiveness, or a deranged state of the bowels, than simply upon worms. It resembles a most formidable disease, " water on the brain."

A variety of worm medicines have been employed, such as tin powder, tansey, sulphur, hellebore, worm seed, cowage, Indian pink root, &c., besides the thousand and one nostrums sold by druggists. In general, however, we find that with children, the most successful plan is to give *frequent and repeated purgatives, to expel both the worms and morbid stools,* and also to excite and support the due and vigorous action of the bowels. Castor oil, in tea spoonful doses, in which a few drops of oil of lemon are put, given about three times a week, is better than all the nostrums you can buy. The extent to which this plan is to be carried, and the period for which it must be continued, will depend upon the effects produced. As long as the stools are unnatural, the purging should be continued.

In cases of a third species of worms, called tænia, or tape worm, it is sometimes difficult to cause the expulsion. It is most common to *adults.* Large doses of the spirit of turpentine have been recommended, (taken in milk,) on an empty stomach in the morning. The dose to be from two to three table spoonfuls for a robust grown person.

An infusion of tobacco, applied to the stomach, has often caused the expulsion of worms, when other remedies failed.— But I never knew a case of failure when the patient was freely purged with calomel, and then given either the worm-seed oil, or the pink root in tea. The oil should be given on an empty stomach in the morning, (ten to twenty drops,) or the tea of pink root taken occasionally throughout the day, in doses to suit the age of the patient. About ten grains of the *powder* may be given to a child of eight or ten years old, two or three times a day. When in over doses, it is apt to affect the head, and the quantity is to be lessened.

The generation of worms may be prevented by whatever will strengthen the bowels. A good, healthy diet, exercise in the open air, and an infusion of tea made from Peruvian barks, (a wine glass before each meal) are advisable.

HOOPING COUGH.

Tнis disease is generally treated improperly by parents In the beginning it is always an inflammatory complaint, requiring evacuations and determination of blood to the surface of the body, by giving warm teas or ipecac in small doses every two hours to produce sweating. Instead of the variety

of prescriptions in daily use, give the child an emetic (or vomit),
of ipecac in the usual dose, to be repeated every day or other
day for four or five days, unless the symptoms lessen. For a
violent fit of coughing, the best remedy is, to pour in the back
of the mouth a teaspoonful of melted hog's lard or sweet oil,
which sheathes the part, and lessens the irritation. The tincture of asafœtida, twenty to thirty drops every four hours, is
highly recommended. A child grown enough for the purpose,
will find some relief in holding warm water in the back of the
throat. It is of great importance to children in this complaint,
to keep the skin in good condition. A coarse flannel shirt
around the breast, has been of great service by keeping up
friction on the surface. With the flesh brush or a ball of wool,
the surface of the body should be rubbed every night. Exercise in the open air, while the body is kept comfortable, is advisable, as well as change of residence for a few weeks, which
scarcely ever fails to afford relief. The juice of garlic sweetened, lessens the cough. A solution of soda, also of alum, in
doses of three or four grains, and sweetened with liquorice,
given night and morning, is a valuable remedy. A mixture
of twenty grains of tartar emetic and an ounce of tincture of
Spanish flies, nightly rubbed on the stomach, is a remedy
highly extolled.

COSTIVENESS.

THIS complaint is sometimes hereditary, or natural to
the child; when this is the case, and it does not exceed
proper bounds, it may not require the use of any remedy;
but should the infant's health begin to suffer from frequent
attacks of colic, flatulence, etc., it should be attended to, as it
may produce convulsions or fits, inflammation of the bowels,
or other diseases of a difficult and lingering nature, or establish
a costive habit for life.

If the predisposition has descended from the mother of the
same habit, or in other words, if the mother herself is subject
to costiveness, the child may be relieved for a short time, but
't will again return. When this is the case, the mother, if possible, should change the quality of the milk, by being attentive to her diet, and take occasionally some mild purgative,
which will alter the quality of her milk; for this purpose there
is no medicine superior, or more innocent than magnesia and
Epsom salts, of equal quantities, mixed and ground very fine
in a mortar. Of this, take a teaspoonful or two in a tumbler
of water every morning on an empty stomach. When the costiveness originates from the child's food, it must be changed

and simple medicines given occasionally, to act as a mild purge, such as five or ten grains of magnesia, rhubarb or manna, a tea spoonful of sweet oil, or castor oil. But the best plan in such cases is to allow the mother, if the child is nursing, or the child itself if it has been weaned, a plentiful supply of syrup, molasses, and stewed fruit, at meals, and ripe fruit uncooked, between meals, and *teaching the child to go to stool at regular hours. Children often have a disposition to go to stool but put it off till the effort of nature passes.* The mother should be attentive in these matters if she wants her children to be healthy in body, happy in mind, and *sound sleepers at night.*

DISEASES OF THE EAR.

Acute inflammation of the ear is known by the swelling, acute pain and noise in the head, and pain in swallowing or moving the lower jaw. *Chronic* inflammation is attended by some degree of deafness and discharge of matter. The disease is caused by colds, foreign bodies in the ear, measles, scarlet fever and scrofula. *Nervous earache* occurs in paroxysms of severe pain in the ear, and shooting over the face, head, neck and shoulder. It is caused by sudden cold, decayed teeth, and sometimes by fullness of blood. The warm foot bath, with some ground mustard and salt in the water, and hot applications to the ear and face, usually give relief. A drop or two, each, of laudanum and sweet oil put into the ear on a piece of warm wool, is an old and valuable remedy, to be repeated every hour or two if necessary. Holding hot coffee or tea in the mouth is also good, or gargling the throat well with tea or coffee, and then spirting the liquid out through the nose by stooping forward with the *mouth closed.* A bag of hops steamed over boiling water, then allowed to cool sufficiently to be agreeable, and applied to the ear, often affords relief in a short time. If an insect is in the ear pour warm sweet oil into it; this will generally cause it to come out to get air. *Foreign bodies,* such as beans, coffee, dust, etc., sometimes get into the ear and cause intense pain : they may be removed by syringing the ear with water, or by a small probe or blount wire, doubled, or bent into the proper shape. *Great care must be exercised not to hurt the drum of the ear.*

RUNNING FROM THE EARS.—When the discharge after an abscess does not disappear, or when running from the ears shows itself after acute diseases, such as measles, scarlet fever, etc., the symptom must not be neglected, and should be examined into by a medical man. It is most common in children of a weak or scrofulous constitution, and may be with or

without disease of the bone ; in the latter case the discharge is
extremely offensive, and often stains the linen black. These
discharges must not be *too quickly stopped*, neither can they
be allowed to go on without risk ; in the former case, the sudden stoppage may throw back the disease upon the brain ; in
the latter, this organ or its membranes may become affected
by its gradual extension to them through the bones. Counter-
irritation, by blisters, or tartar emetic ointment, (ten grains of
the powder to one teaspoonful of lard, well mixed, and applied
once a day till little pimples appear) behind the ears ; keep
the bowels open regularly. The general tonic treatment as recommended in some other diseases of children already treated
of, and syringing with slightly astringent washes, such as one
grain of lunar caustic, or two of white vitriol, to the ounce of
water, or a wash of strong green tea will constitute the most
appropriate treatment.

MALIGNANT SORE THROAT.

This species of sore throat differs from that which attends malignant scarlet fever. It is usually limited to the
upper part of the throat. It begins with redness, swelling of the tonsils, bloated face, flow of tears, chills and
flashes of fever : the redness of the throat soon changes to a
dull ash color, and then to brown or black--there is thirst,
hoarseness, difficulty in swallowing, nausea, sometimes vomiting and diarrhœa. In the more severe cases there is a bloody
or watery discharge from the nose, and an offensive discharge
from the throat—the tongue becomes brown, dry, and coated,
there is often an eruption on the skin, sinking of the powers
of life, and finally death in severe cases.

Most medical writers consider this disease contagious—it is
caused also by cold, wet, insufficient clothing and food, bad
air, and want of personal neatness : it is a very dangerous disease, and requires prompt and efficient treatment. A physician should be immediately called in ; in the meantime, or
in cases where one can not be had, give the patient an emetic
of ipecac ; then give a purgative of one teaspoonful or two of
Rochelle salts, in half a gill of luke-warm water. Apply a
mustard plaster to the throat, and give every hour or two a
gargle of sweet oil or melted lard, mixed with a few drops of
spirits camphor and a little sulphate of iron (green vitriol) dis
solved in water.

A gargle made of Cayenne pepper and oak bark tea, not too
strong, with a little salt in it, is also good. Also a gargle
made of yeast and finely powdered charcoal has been used

with great benefit. When the patient is weak the strength must be supported by tonics, such as wine and infusion (or tea) of Peruvian barks, or one grain of quinine, three times a day Where there is a feeling of suffocation or choking, it is advisable to produce vomiting, either by tickling the throat with a feather, or giving a dose of ipecac, to clear the throat.

C H O K I N G.

CHILDREN sometimes get choked by bits of food or stones of fruit, which produce cough, blueness of the face, gagging, sometimes nose-bleed and convulsions—and if relief is not given, death ensues.

When a child is choked, he should he held with the head downwards and receive two or three smart blows on the back between the shoulders: if this does not give relief, the mouth should be thrown wide open, and some person should endeavo.· to dislodge the substance, either bringing it out of the mouth or gently pushing it downwards; a few swallows of water may enable it to pass into the stomach.

CHOLERA INFANTUM. (*Summer Complaint.*)

THIS is one of the most fatal diseases to which the period of infancy is subject: it occurs mostly among children under the age of two years, and during the warmer part of the season. This is said by authors to be a disease peculiar to the United States.

It usually commences with a profuse discharge from the bowels of a light-colored fluid; after a short time the extreme irritability of the stomach is manifested by the constant vomiting of everything swallowed. The discharges from the bowels sometimes contain flakes of mucus: the passages are often involuntary, and attended by much irritability and debility: the tongue is coated with a white slimy matter, the skin dry, pulse quick and small, much thirst, bowels hot, sometimes bloated and tender: there is, at times, moaning or sudden screeching, indicating acute pain.

Digestion is so far suspended that whatever is eaten passes unchanged. In some cases delirium comes on early, and the little sufferer dies in one or two days from the attack: at other times the disease continues until extreme emaciation is produced; the skin has a wrinkled, dirty appearance, bathed in cold perspiration, the features sharp, eyes large and glaring, the whole countenance has the appearance of old age. The cholera of infants is mostly a disease of the mucus coat and glands of the bowels—often accompanied by enlargement of the livei. It is caused by impure, stagnant, or confined air, coming in contact with the sensitive surface of the air passages, skin and digestive organs, and improper food.

The disease may be produced by all the causes which produce diarrhœa: it seems mostly to prevail in low, damp situations, in towns and cities. Perfect cleanliness, pure air, good diet, and change of location when it depends on that, are indispensable in addition to medical skill.

When the child can not be taken to the country, take it often into the open air, in the cool of the day in good weather It should be confined entirely to the breast-milk when nursing or if weaned, let its food be arrow-root, tapioca, corn starch rice flour, and milk. Put it in the warm bath once a day ; keep flannel next the skin, and the bowels moistened outwardly frequently by vinegar and water, and the arms, hands, feet and legs, frequently rubbed with whisky or any kind of spirits. Give small pieces of ice to melt in the mouth in preference to water. For the vomiting, give a few drops of essence of peppermint in water frequently, or essence of cinnamon, or a tea made of allspice (cold), in which there is a little gum arabic and a teaspoonful or two of prepared chalk to the cupful of tea. Of this give a teaspoonful every hour. Acidity or sourness of the stomach seems to be the great difficulty in this disease, and, as a consequence, suddenly checking the bowels will not do until the preparation of chalk has to some extent corrected this sourness. If the discharges become very offensive, a mixture of charcoal, finely powdered, with chalk and white sugar, equal parts, and thick mucilage of gum arabic given three or four times a day will be advisable.

Mucilage of gum arabic or thick slippery elm water, made by putting the slippery elm in cold water, to which add one teaspoonful of spirits of nitre to the half pint, given in teaspoonful doses, is also a good remedy. Sometimes nothing will afford relief as quick as one half grain of calomel given every four hours, and continued until the passages are more natural.

DYSENTERY.

This disease consists of inflammation, which is confined mostly to the large bowels. In some cases, however, the inflammation extends to the small bowels and even the stomach.

The symptoms are griping, frequent and small discharges of slimy matter (mucus,) mixed with blood : the first discharges are usually, however, thin and watery, the bowels are tender, dry and hot, there is some fever, furred tongue, and sometimes vomiting. When these symptoms are not abated by timely remedies they are apt to increase in intensity until terminated by death.

Dysentery is caused by changes of weather, improper diet, worms, hot wet weather, impure air, want of sufficient food, unhealthy milk, etc. When the attack is attended by profuse discharge of blood the case is more favorable than if no blood appears.

The diet should be restricted to animal broths, boiled rice, and the like ; the clothes kept clean, the child put into the warm bath once or twice a day, fomentations of hops and vinegar, stimulating liniment applied to the bowels ; the drinks should be barley water, gum arabic water and flax-seed or slippery elm tea : injections of tea of oak bark, or starch and laudanum, four to ten drops, according to the age, and repeated three times a day if needed, are also of much value. The allspice tea, etc., recommended under the head of Summer Complaint, will be found advisable also in this disease. Also the flannel next the skin, bathing with vinegar, etc.

INCONTINENCE OF URINE.

This is a common disease among young children—and is often the result of a careless and filthy habit of neglecting the calls of nature, and not endeavoring to restrain their desires. It usually occurs at night, the child allowing the urine to pass even while awake, rather than to rise and evacuate the bladder. It is also caused by palsy of the bladder or some of its appendages, or by an irritable state of that organ. The discharge of urine is most apt to take place when the child is lying on his back : the urine sometimes scalds and irritates the legs and produces sores. Incontinence of urine is caused by the improper use of irritating medicines, certain articles of food, and by diseases of other parts of the body. The habit of incontinence, although an unpleasant one, demands indulgence and pity, rather than blame and punishment in most cases. Children thus afflicted should not be allowed much drink or fluid food ; they should be made to urinate immediately before retiring, and also to rise at *stated hours of the night* for the same purpose.

The best position in bed, in order to prevent involuntary discharge, is on the side. The diet should be digestible and nutricious, and the bowels regular—the cold hip bath at night will be of service. A tea made of uva ursi, or buchu leaves, a table spoonful of which may be given three times a day often cures the irritability of the bladder, on which the habit depends, or alters the quality of the urine, which causes its involuntary flow.

RICKETS.

RICKETS depend upon disordered nutrition, and some alter ation of the blood from its healthy standard. It has usually however, been supposed to depend upon a deficiency of phosphate and carbonate of lime in the food, to furnish the necessary earthy matter to the bones—and therefore has been considered peculiarly a disease of the bones. But recent investigations show that it does sometimes occur when there is no deficiency of lime—and that the whole system, particularly the muscles, brain and nerves, are equally implicated with the bones. It is an affection peculiar to childhood, and supposed to depend upon the action of the causes which favor the development of scrofula. The signs of rickets are, a softened gristly state of the bones, large joints, large head, prominent forehead, straightness of the ribs and flatness of the sides of the chest, prominent breast bone, looseness of texture in the bones, crooked legs and distorted spine : many other symptoms of scrofula are sometimes also present. This, like scrofula, disposes the system to other diseases : the treatment of rickets is nearly the same as that of scrofula, (which you will find in its proper place in another part of this work,)—rickets, however, is a more curable disease, and less apt to continue after adult age.

FOREIGN BODIES IN THE NOSE.

THE nose, like the ear, is very liable to be made by children the receptacle for any thing that will pass into it; beans, buttons, stones, &c. Sometimes they have been in the nose, unnoticed, for days or weeks, and are not discovered until inflammation of, and perhaps discharge of matter from, the lining membrane attracts attention ; a reason, when such symptoms occur in a child, for always examining the nose for the presence of foreign bodies. The extraction of a foreign body from the nostril is always best done by a surgeon. If, however, circumstances render it desirable to attempt the extraction without waiting, it must be done by means of the flat end of a probe, or of a bodkin, bent about the eighth of an inch, nearly at right angles with the rest of the instrument, which bent end being carefully passed beyond the body, must be used as a scoop to take it out. The flat end of a pair of tweezers also answers for this purpose, or use them as a pair of forceps, if the foreign substance can be got hold of. Sometimes, when the foreign body is not very far in the one nostril, if that on the opposite side be closed, and the child can be made to blow forcibly through the other, the obstruction will be shot out.

The lining membrane of the nose is liable to become inflam
ed and ulcerated. In a mild case, washing with warm water
—if necessary, by means of a syringe—containing a little car
bonate of soda in solution, will be of service: soap and water
is also good. It is a common popular error to suppose that the
nose communicates with the brain: it is sufficient to remark
that it does not.

WEAK ANKLES.

IF children are put on their feet when too young, before hav
ing requisite strength, or who are allowed or taught to turn
their toes outward too much, or those who are of a weak and
relaxed muscular system, are apt to have weak and crooked
ankles, or bowed legs. The soles of the feet are flat, the ankles turn inward so that the child walks almost on the ankle
joint, and with lameness and difficulty. The general health
should be improved by a good diet, cold bathing, and exercise
in the open air: the child should also wear high boots made
of leather sufficiently stiff to support the ankles in the proper
form and position. Also use friction with the hand or rough
towel to the legs and feet every day.

BRONCHITIS.

THIS is an inflammation of the bronchial, or air tubes of the
lungs, and is common to childhood. It commences with chills,
flashes of heat, slight cough, oppression and tightness in the
chest, breathing difficult, wheezing and rattling, and hoarseness of the voice. Breathing is more distressing when the
patient is lying down—the cough is at first dry, but a copious
discharge of stringy phlegm, resembling white of eggs, soon
appears, with some relief to the cough: the skin is dry, and
the tongue is covered with a white mucus. In more severe
cases, these symptoms may all be augmented and attended
with much danger. The disease is caused by cold, wet, sudden changes of weather, insufficient clothing, loud speaking,
crying, dust, and noxious vapors. When a physician can be
procured, in cases of this kind, as well as in inflammation of
the lungs, by all means do so; but as delays are always dangerous, while you may be endeavoring to procure a physician,
or if you can not obtain one, the following course of treatment
is advisable:—In the first place, give the child a purge of
say, one or two tea spoonsful of Rochelle salts in a wine glass
(or half gill) of cold water; or castor oil will answer as well,
made more palatable by a little essence of lemon, peppermint,
or cinnamon. Also, of the following mixture, let the child
take one tea spoonful every four hours, unless vomiting is pro-

Cultivation and Carelessness.

A form representing a full-chested woman.—Such a person would naturally have a strong constitution, and could endure a great amount of labor, either mentally or physically. The European ladies are more generally of the above form than the American, because they take more interest in cultivating a full chest and fine form. In future let it be truthfully said that the American ladies not only have "pretty faces" but healthy forms.

FORMS THAT CAN BE CULTIVATED.

This is a fac-simile in form of a great many women that are daily met with. Such persons are usually troubled with that sinking sensation, or "goneness" at the pit of the stomach, which is always produced by the pressure upon it in stooping, and might be prevented by care in keeping back the shoulders, expanding the chest, and taking that kind of exercise so much needed, but so much neglected, called "House-work!"

FORMS CONTRACT-ED BY CARELESS-NESS OR HABIT.

We here see represented a full-chested and erect man, one so rarely seen, although it is no more than can be obtained in nearly every person by cultivation. A person with such a chest would usually be free from disease of the Lungs or the Heart, and would have all the indications of being a robust and long-lived person. It is as easy to have this form as au improper one, by a little timely training.

This represents a man of stooping form, with small Lungs and Chest. Such a person would be almost sure to have some disease of the Lungs, Heart, or Stomach, and would naturally be Consumptive and short-lived, because the vital powers are small. Care should be taken to avoid contracting such a form. It is simply the result of carelessness and habit.

duced ; if so, reduce the dose :—Take one table spoonful of syrup of ipecac, (or five grains of the powder,) half a gill of cold water, one table spoonful of spirits of nitre, 20 grains of chlorate of potash, and a few drops of essence of lemon or cin- namon, mix thoroughly together and keep in a cold place, to be used while the active symptoms continue. Also, wrap the chest and neck in a flannel cloth, saturated (soaked) with this mixture :—Sweet oil, spirits of turpentine, spirits camphor, of each one ounce, to which add a table spoonful of laudanum, and shake well before using. Renew this twice a day.

The child should be allowed to drink freely of cold water, in which put plenty of gum arabic, or slippery elm. The bow- els should be moved every second day, to remove the phlegm which is usually swallowed by the child. In robust children, if there is much fever and oppressed breathing, a few leeches applied to the chest will be advisable ; or the application of cupping may be tried once a day for two or three days. When the inflammation is somewhat reduced, applying a more stimu- lating liniment all over the chest will expedite the cure. A table spoonful of tincture of cayenne pepper added to the lin- iment of turpentine, &c., will be about as good as any—appli- ed twice or thrice a day.

INFLAMMATION OF THE LUNGS.

AMONG children, as well as adults, this is a frequent and dangerous disease. It begins with symptoms similar to those of bronchitis, and is produced by nearly the same causes.— The treatment is the same as in *Bronchitis.*

STAMMERING.

STAMMERING in the speech cannot be said to be a disease, being rather a functional disorder. This is evident from the fact, that, under certain circumstances, an habitual stammerer does not stammer, and that cases have occurred in which most inveterate stammering has been completely cured by the exer- tion of the will. Moreover, stammering is often caused either by imitation in children or by nervousness in both children and adults. This nervousness is often the result of *debility*, and of *weak constitution*—a fact which should not be lost sight of, for, if such be the case, every means of strengthening should be used. At the same time, while the general health is sustained, much may be done by *checking* children, and making them speak at all times *slowly*. Much pains with children, and much perseverance and self-command in adults, is required in the efforts to overcome this defect. The one

great matter of importance in curing children or others of
stammering is, as soon as the least hesitancy in speech, or
stammering, is observed, make the patient *stop at once, then
draw in a full breath,* filling the lungs thoroughly, when they
can begin again. Let this rule be observed only a short time
and you will be surprised at the rapid improvement. The
difficulty, in almost every case I have observed, has been that
*the person attempts to talk when he has already exhausted the
air from the lungs.* Overcome this, and the case is cured.

SCARLET FEVER.

In this disease, usually, the first symptom complained
of, in the incipient stage is sore throat, either accompanied or quickly succeeded by the usual symptoms of a
feverish attack, shivering, headache, loss of appetite, perhaps
vomiting, followed by heat of skin, quick pulse, and thirst.
The eruption appears early, on the second day after the first
symptoms of indisposition. It first shows itself in the form
of minute red points on the chest and arms, especially about
the elbows, the points becoming more numerous, till they form
one diffused surface of a tolerably bright scarlet eruption,
which extends to the neck, face, and abdomen, and body gen
erally. On the second day, when the eruption is appearing,
the symptoms of general fever, and especially the heat of skin,
continue unabated, the throat is more inflamed, and the
tongue assumes the appearance characteristic of this disease.
It is probably covered with a white, creamy-looking fur,
through which, on its forepart, about the tip especially, pro-
ject red points. This appearance may continue, but in many
cases the fur comes off, as it were, in patches at a time, and
ultimately leaves the tongue preternaturally clean and red.
The eruption in scarlet fever generally looks more patchy upon
the *extremities* than it does upon the body. In a moderately
favorable case of scarlet fever, the eruption begins to fade between the third and fourth day from its appearance, and with
it the feverish symptoms, and other general symptoms of the
disease, such as sore throat, &c. The chief care is required
until the peeling off of the skin is completed. *During this
period also the power of communicating the disease by contagion appears to be retained.*

Favorable cases of scarlet fever pass through the course
nearly as described above, but there are much severer forms
of the disease. The feverish symptoms from the first may
have a high inflammatory form ; or the reverse may give evi
dence of an extreme condition of bodily weakness, with a ten

dency to malignant or severe disease. In such cases the eruption is slow, and, when it does appear, patchy, and dusky in color, the swelling of the throat is great, and, if they can be seen, the tonsils are evidently ulcerated, the breath offensive, the tongue swollen, and swallowing difficult, if not impossible. Offensive discharges take place from the nose, and at the same time there is extreme weakness, with delirium.

Scarlet fever requires confinement to bed, in a well ventilated room; the diet should be kept low, and consist of milk, corn-starch, farina, &c., and the patient may be freely indulged with drinks, such as flax-seed tea, with a slice of lemon in it, &c. The patient must not be so warmly covered with bedclothes as to keep up feverish heat. The first thing to be done, in all cases of scarlet fever, is, *in the very outset, to give a good vomit,* of ipecac, so as to clear the stomach. In a great majority of cases this will render the disease less dangerous and more easily managed, and very often will almost cut short, as it were, or break up the disease. If the heat of skin is great, sponging the surface of the body with tepid water, with or without the addition of a little vinegar, is at once most beneficial and grateful to the patient. A gentle purge should be repeated once or twice in the course of the disease, a tablespoonful of castor-oil, a dose of magnesia and rhubarb, or from half to a whole seidlitz powder, may be required.

From five to ten grains (according to age) of *chlorate of potash,* given every six or eight hours, dissolved in a little sugar and water, is one of the most appropriate cooling medicines in this disease.

If the feverish symptoms run high, of course the lowering and cooling remedies must be more actively enforced. In most cases much relief is afforded to the throat by the frequent use of warm gargles, made either with simple gruel, or with gruel with one or two tablespoonfuls of vinegar to each half-pint. Externally, hot bran or bread poultices, frequently renewed, are also of much service to the throat. Great enlargement of the glands around the jaw and in the neck must always be regarded seriously. When a case of scarlet fever presents symptoms of great severity, every method of supporting the strength by wine, broths, &c., must be used, and the preparations of chlorate of potash employed both internally and as washes and gargles to the nose, mouth, tonsils, &c. The chlorate of potash, in from five to ten grain doses, must be given every three or four hours; or muriatic acid, in five drop doses, in sweetened water. Two teaspoonsful of the solution of table salt, in the half-pint of water, will make a convenient wash, to be used with a syringe, if the child or person is unable to gargle.

After the eruption has faded, the person may sit up, and gradually return to fuller diet, such as pudding, broth, fish, &c., the bowels being kept free, but not purged. At this stage, too, much comfort and benefit will accrue from the use of two or three warm baths. These relieve greatly the discomfort arising from the harsh and dry state of the peeling skin, and, what is more important, encourage and keep active the perspiration, which is apt to be impaired or impeded, and thus to give rise to one of the most serious incidents connected with the disease in question, that is, to a dropsical condition connected with a disordered state of the kidneys. The occurrence of dropsy after scarlet fever is always a serious matter. It is observed that the attacks of dropsy after scarlet fever are by no means in accordance with the *severity* of this attack itself, and this is supposed to be because those who have had only a mild attack are more careless as to after exposure than those who have suffered a severe one. However this may be, it is certain that many, who have passed safely through the disease itself, fall victims to the subsequent dropsy, purely as the result of *carelessness* on their own part, or on that of others. The attacks of dropsy are most likely to occur from the end of the first fortnight to the end of the fourth week after the decline of the eruption. Its symptoms are generally those of languor and oppression, with headache, and it may be vomiting, the swelling coming on simultaneously. Usually, the face (especially the eyelids) is first affected, and the dropsical swelling may go no further, but generally the feet and legs, the hands, arms, chest, &c., become filled; the urine is scanty, high-colored, or "smoky" in tinge.

Should dropsy occur, warm baths ought to be used to restore, if possible, the functions of the skin, hot bran poultices applied to the body, and if there is pain about the kidneys, blood taken by leeches or cupping. The bowels should be well purged. At the same time a draught, consisting of ipecac. spirits nitre, &c., the same as ordered in measles, may be given every four or five hours. Besides dropsy, scarlet fever is liable to be followed by other affections, particularly in those of weak or scrofulous constitution. If the affection of the throat has extended to the ears by the Eustachian tubes, which lead from the throat to the ear, the structure of the organs of hearing may be materially damaged, and deafness, total or partial, be the result. Frequently, running from the ears, from the nose, or eyes, continue long after the subsidence of scarlet fever; and if the attack has been a severe one, a permanent state of impaired health may be the consequence. Of course, if a patient, after an attack of scarlet fever, remains weak, tonic medicines (one grain of quinine three times a day, or a wine glass full of tea, made of Peruvian bark, cold, before each meal,) and good nourishment will be required—also warm clothing.

Few diseases are more contagious than this, and few retain the power of propagation longer; indeed, it is difficult to say when this totally ceases, at least for some weeks. Probably, when the *peeling stage* is complete, the risk of contagion is gone, or nearly so. The contagion from scarlet fever is very persistent, and unless the rooms which have been occupied by patients, and indeed everything which has been about them, are very freely cleansed, aired, or fumigated, there is always some risk for a considerable time. The power of belladonna, in protecting individuals against the contagion of scarlet fever, has been much discussed. It has been used extensively, and with apparent success; at all events, the evidence is sufficient to make it worth a trial during the prevalence of a very severe or malignant form of scarlet fever. Eight grains of the extract are to be rubbed up with a fluid ounce of water, and of this, from five to twenty drops, according to age, given twice a day. It would be right to try the remedy during the prevalence of this disease, as well as measels, hooping cough, &c. Scarlet fever is generally a disease of childhood, and is usually passed through once in a lifetime; but adults who have escaped it early in life, are liable to be affected. Second attacks are rare. Although those around persons suffering from scarlet fever may not have the disease, they are very liable to suffer from *sore-throat*, often in a severe form. It is a serious thing for women to be exposed to the contagion of scarlet fever *soon after child-birth*, and it should be avoided if possible. Where there are several children in a family, let those that are well be kept in different rooms from those that are sick with this disease, and the belladonna be used as directed, and let them be kept out in the open air, if the weather will permit.

PRICKLY HEAT.

THE sensations arising from prickly heat are perfectly indescribable, being compounded of prickling, itching, tingling, and many other feelings for which there is no appropriate name It is usually, but not invariably, accompanied by an eruption of vivid red pimples, not larger in general than a pin's head which spread over the breast, arms, thighs, neck, and occa sionally along the forehead. This eruption often disappears in great measure when the patient is sitting quiet, and the skin is cool ; but any exercise that brings out a perspiration, or any warm or stimulating fluid, such as tea, soup, or wine, brings out the pimples, so as to be distinctly seen, and but too distinctly felt.

In reference to the imagined dangers of repelling this eruption, Dr. Johnson says—"I never saw it even repelled by the cold bath, and in my own case, as well as in many others, it seemed rather to aggravate the eruption and disagreeable sensation, especially during the glow which succeeded immersion. It certainly disappears suddenly, sometimes on the accession of other diseases, but I never had reason to suppose that its disappearance occasioned them."

An application every half hour, by means of soft linen, of a mixture of vinegar and cold water, in equal parts, to which add a few drops of laudanum to each cup full of the mixture, and afterwards applying finely powdered starch or common wheat flour, is about as good as any. Strong hop tea may be tried. Always give a brisk purge of Rochelle salts, (one tea spoonful in a gill of water,) or Epsom salts will do. Low diet and quietude are very necessary.

RING WORM.

This unsightly and unpleasant disease is too often treated *prematurely* by stimulating applications : its symptoms are generally well marked. It consists of minute water blisters, arranged somewhat in rings: it begins with slight redness— small blisters form and are filled with a colorless fluid—these break in four or five days, and are covered by a thin brownish scab, which falls off about the eighth or ninth day, leaving a red surface which gradually disappears. The eruption seldom lasts more than ten days, but it sometimes appears a second time, and continues for several weeks: it is always attended with itching, smarting, and burning. It often appears on the face, neck and arms of children—and may be communicated by contact. A wash of white or blue vitriol (one teaspoonful to half pint of water,) or nitrate of silver, (lunar caustic) a stick half an inch long dissolved in a gill of water, and applied once a day, or an ointment made of yellow dock root, boiling two or three of the roots in half pound of lard, for one hour, will usually effect a cure. After the first and most inflammatory stage is passed, the application once in forty-eight hours, of *tincture of iodine*, or iodine ointment, applied two or three times, will thoroughly cure ring worm.

ITCH.

THIS is a very unpleasant affection. It usually comes first between the fingers and on the wrists, in small pimples filled with colorless fluid, attended by intense itching, which is always increased by heat. The friction and scratching, used to allay the itching, ruptures the pimples, and they are by this means extended to the surrounding skin. In cases of long continuance, some of the vesicles (or sacks) become filled with matter and covered with brown scabs, which extend over a great part of the body. This disease is contagious and is communicated by contact—and probably sometimes produced by want of cleanliness, and other causes. It seldom gets well without treatment: it is not dangerous, but may continue during the life-time of the patient, with varying degrees of tormenting nights and uneasy days. In neglected cases, an *insect* is to be seen in or near the vesicles. It may be seen burrowing under the skin, and when removed by the point of a needle, resembles the "cheese skipper." Whether the disease is produced by this insect is not yet determined. Itch usually occurs in four or five days after exposure to its contagious matter. The best remedy in most cases, is sulphur, mixed with lard, (equal proportions, to which add a few drops of oil of lemon,) and applied night and morning to the parts affected: sulphur and cream tartar, mixed with molasses, may also be given in teaspoonful doses every night.

The diet should be simple and digestible, the bed and clothing of the patient kept perfectly clean and well ventilated. Two or three times a week, the entire surface of the body and limbs should be washed with lukewarm water and soap, and afterwards rub well with a towel, just before going to bed. Sponging the skin all over once a week, (after bathing, as before recommended,) with a mixture of water and cologne water, or bay rum and water (equal parts,) is of service in such cases.

CHICKENPOX.

A DISEASE which is preceded by feverish symptoms, such as chilliness, quick pulse, hot skin, restlessness, diminished appetite, thirst and headache. In some cases the fever is severe, and attended with distressing retching, great agitation during sleep, and even delirium. In others it is scarcely perceptible. On the third day, the eruption appears, first on the body, and then on the face, and lastly on the extremities; when the eruption appears the fever declines. The pustules, which are very itchy, contain a yellow matter, and by the fifth day are covered with scabs, which leave no pits. There are different varieties of this disease, for in some the pustules are larger than in others, or go off sooner. This is scarcely ever dangerous, and is seldom even troublesome ; nor is it generally necessary to confine the patient, or do more than give one or two doses of some gentle purgative, such as previously recommended in treating diseases of children. The fever and uneasy feelings may be greatly mitigated, and the eruption rendered lighter by washing the surface with cold water in the commencement of the disease. The itching may be abated afterwards, by sponging the skin occasionally with cold vinegar and water. In some cases, especially if the bowels be neglected, and the child be allowed to eat freely, the fever will be greater and the pustules become much inflamed. Some of them may even end in sloughs, which leave deep marks, worse than those of the small-pox, and as in that disease, so also in this, very troublesome boils may harass the patient for a long time.

VACCINATION.

To Dr. Jenner, of England, belongs the credit of this great discovery, which, in the order of the providence of a merciful God, has been such a blessing to mankind. It is a well-known fact that vaccination or " cow-pox," is almost a certain preventive of the contagious effects of small-pox. It is true, however, that in some *few cases* it fails, but this should by no means prevent the vaccination of every child : it is thought by some authors, to lose its efficacy in a few years, so that a second or third vaccination may be necessary : this is, however, doubtful. Children may be vaccinated at any age from three months upwards ; but there is some difficulty in securing a thorough operation of the virus in very young children, or in those affected with any disease of the skin. However, it may be performed at any time should circumstances call for it, owing to exposure to the contagion of small-pox. The vaccine matter may be taken in the fluid state from the arm of another person, or a bit of the scab which has been preserved, may be used. Almost any mode of vaccination, which will secure the formation of a pustule (pock), will answer. The usual mode is to raise a small piece of the skin on the arm, with the point of a lancet, and insert a little of the fluid or scab, and cover it with a piece of "court plaster :" or simply taking a blunt-pointed needle and gradually scratching the arm till it begins to show signs of bleeding : then dipping the point of the needle into the vaccine matter a few times, and working it into the skin, will answer. If a scab is used, it a.ust be softened by a drop of water a few minutes, till it gets the consistency of paste.

Should the vaccination not take effect in four or five days, it should be repeated until it does, as there is no safety without it. It is always best to make one or two pocks on each arm. About the third day after the vaccination, there is a red, elevated pimple, which, on the fourth, is surrounded by a faint red circle : on the fifth day there is a pearl-colored pimple filled with transparent fluid, on the eighth day the pock is at its height of development, at which time there is usually some fever, chills, lassitude, and more or less pain and swelling of the arm and glands of the arm-pit : by the tenth day the pock is red and painful, on the eleventh it begins to shrink and assumes a darker color, so that by the fourteenth day it is covered by a thick, brown scab, which falls off about the eighteenth day, leaving a white scar. All the care necessary is to see that the vaccination passes through its course without getting the part *injured;* and should there be much fever give a gentle purgative.

To preserve a " scab," keep it in a dry bottle or vial, wrapped in paper, and the mouth well corked, with some beeswax over it.

NETTLE-RASH.

It cannot be better described than as an eruption which closely resembles nettle-stings, both in appearance and in the sensations it gives rise to. When acute, it is generally accompanied with more or less fever. The nettle-rash, in almost all cases, arises from disorder of the digestive organs, caused either by indigestible food, or, in some persons, by particular kinds of food. Kernels or seeds, such as almond, peach, &c., which contain prussic acid, seem especially apt to cause nettle-rash, and in some individuals even the pips of an apple have been known to produce the disorder. Fish, particularly shellfish, or mushrooms, also bring it on ; also certain medicines, such as turpentine ; teething in children, hurry and agitation of mind in adults, and other irritations, also give rise to nettle-rash. The generally known causes of this affection indicate the remedy—the removal from the stomach and bowels of offending matters. If there is a tendency to sickness, and if the eruption appears soon after a meal, an emetic is the appropriate remedy ; but, whether this is administered or not a purgative should be given. As acid in the bowels often accompanies the disease, a dose of magnesia with rhubarb is very suitable, and, afterward, a dose of castor oil. External remedies are comparatively of little service in the acute form of nettle-rash. A lotion of sugar of lead, one drachm, in half a pint of water, will give relief, or you may try the effect of flour dusted over the surface.

Rose-Rash.—Occurs both in children and adults, in the form of rose-red *patches* of various sizes, somewhat resembling measles in many cases, but of a redder hue. The disease is generally accompanied with slight fever, but the symptoms differ from those which accompany measles. It is devoid of danger, and generally subsides after the administration of a simple purgative. If either rose-rash or nettle-rash are thought to be connected with teething, the gums should be scarified. Wheat flour, applied to the affected parts, or powdered starch, is beneficial.

INDIGESTION.

This is manifested by the food being imperfectly, or not at all, digested, but is discharged by vomiting or stool without being changed; there are often no symptoms of inflammation. Indigestion in infants is almost invariably caused by too much or improper food.

The most common symptoms of indigestion are, nausea, vomiting, sour odor of the breath, the milk discharged is sometimes curdled, and at others unaltered, and there are more or less griping colic pains. Children who are weaned early, or reared without the breast, are liable to attacks of indigestion, and often become pale, weak, emaciated, the tongue furred, bowels bloated and tender, mouth sore, thirst, fretfulness, moaning, eyes glassy, and finally, in some cases, death ends the suffering. In older children, indigestion is caused by *unripe fruits, too much food, confectionery, pastry, improperly cooked or hard food, eating too often* and at *irregular* and unsuitable hours. Affections of the brain, convulsions, spasm of the windpipe and inflammation, sometimes result from this disease.

The child should be carried or permitted to go into the open air; the tepid bath and flesh brush should also be used every morning. Also give from four to six grains of rhubarb and magnesia, or a teaspoonful of castor oil, once every second day, and let the child's diet be so regulated as to avoid those things which have caused the disease.

WATER ON THE BRAIN.

Children of scrofulous constitution are most liable to this disease, and should be closely watched, especially from the

second to the sixth or seventh year of life, the most genera.
period of attack, and particularly after the child has suffered
from any of the diseases incidental to childhood. At first, the
patient is languid, looks heavy, is subject to irregular heats
and chills; the appetite is variable, the bowels irregular, and
the discharges from them unnatural in color. The sleep is
disturbed; there is frequent starting, moaning, perhaps screaming; the teeth are grated, and the thumbs folded across the
palm of the hand. When awake the brow is contracted; the
nose is continually picked, and the child, if able to speak, complains of the head, which is hot. As the disease advances beyond the first stage, all these symptoms become more marked,
and probably obstinate vomiting, and when the stomach is
empty, retching occurs. Toward the termination of the disease, insensibility, dilated pupils, convulsions, etc., come on;
but long before the latter stages, the case should be under
proper medical treatment. The object here is to put parents
on their guard as to the advances of an insidious and very fatal
malady—not to induce them to incur the responsibility of its
treatment. Many of the symptoms above detailed undoubtedly occur, in less alarming combination, in many of the diseases of children; but come as they may, and when they may,
they should not be neglected. Some amount of treatment
ought however to be employed to save time, and the most important and safest indication is to give a brisk purgative, say
one teaspoonful or two of Rochelle salts in a teacupful of water. The head is to be kept cold, and quiet strictly to be observed. One or two leeches may be applied to the temples,
and then obtain proper medical advice soon as possible. Its
causes are numerous, but sometimes the disease arises without
any being distinctly traceable; the irritation of teething, long-
continued disorder of the digestive organs, falls or blows on
the head, exposure of the child's head to the heat of the sun,
and fevers, may any of them give the first impetus to the diseased tendencies. Those children who are most liable to its attack are often the most endowed intellectually; and there is a
morbid tendency to excitement in the brain, which gives it
power beyond what is natural to its age. If permitted or encouraged, the child will give up the sports and exercises of its
time of life, for the sake of mental employment, and sometimes
a parent's pride permits the erroneous system, which, in all
probability leads either to early death from active disease of
the brain, or to the possession in after life of a sickly body and
morbid mind. In no children is it so necessary to insist upon
strict observance of all the laws of *physical* health, previous ▼

treated of in this work, as in those who exhibit precocious development of mind.

DIET FOR SICK CHILDREN.

UNDER this head will be found many valuable recipes for preparing suitable diet for the sick room. They can be relied on, as they have been prepared with a view to their adaptation to the delicate stomachs of children during sickness, besides being "not bad to take," many of them, by well children or grown people.

Panada.—Pour boiling water on toasted bread, and season with butter, white sugar, lemon and nutmeg.

Boiled Custard.—Beat one egg in one pint of milk, add salt and sugar to the taste, and boil two minutes.

Starch Pudding.—To one pint of boiling milk, add two tablespoonfuls of starch, and one egg, beaten together; season with sugar, salt, wine and nutmeg, and boil one minute.

Rice Caudle.—Make a paste of two tablespoonfuls of rice flour in a little cold water, boil in one pint of water, and season with salt and nutmeg.

Dyspepsia Bread.—Mix together three quarts of *unbolted wheat flour*, one quart of warm water, one gill of fresh yeast, one gill of molasses, and two teaspoonfuls of salt; let it rise, and bake.

Lemonade.—To one pint of water add the juice of one lemon, and the beaten whites of two eggs; sweeten with white sugar.

Orange Jelly.—Squeeze the juice from six oranges and half a lemon, add half a pound of white sugar, half a pint of water, boil, and strain through flannel; then add one ounce of isinglass, and, when this is well dissolved, put it into a mould or dish to cool.

Biscuit Jelly.—Soak one biscuit or Boston cracker in one pint of water, boil, and add white sugar, wine and nutmeg or lemon to the taste.

Sago Jelly.—Soak two tablespoonfuls of sago in water one hour; pour off the water and boil the sago in half a pint of water, until it is transparent; then season with salt, lemon, wine and sugar to the taste.

Tapioca Jelly.—Soak the tapioca eight hours, and then prepare like sago jelly.

Isinglass Jelly.—Boil two ounces of isinglass in one quart of water down to one pint, and add one ounce of white lemon candy.

Rice Jelly.—Boil three tablespoonfuls of rice and three of white sugar, in just sufficient water to cover it, until it bo comes a jelly, and season to the taste.

Calves' Feet Jelly.—Boil one calf's foot in two quarts of water till reduced to one pint, strain, and, when cold, skim carefully, and add one teaspoonful of salt, the whites of three eggs, beaten with four ounces of white sugar, one gill of wine, and the juice of two lemons; boil the whole, stirring constantly, for four minutes, then strain through flannel.

Moss Jelly.—Soak half an ounce of Irish moss a few minutes in cold water, then drain it off, and boil it in one quart of water until it becomes a jelly; strain, and season with cinnamon, wine and white sugar.

Rice Pudding.—Boil one teacupful of soaked rice in one quart of milk, then add two tablespoonfuls of white sugar and one egg, beaten together, and one teaspoonful of salt; bake one hour.

Milk Toast.—Toast a thin slice of wheaten bread slightly brown, pour on to it some boiling milk, and season with nutmeg and salt.

Boston Cracker Toast.—Split Boston crackers, toast them brown, pour on boiling water, and drain it off; then season with butter, sugar, lemon juice, and nutmeg or orange peel.

Broiled Meat.—Broil the lean round or sirloin of beef or mutton, on the coals, until tender, and season with salt or tomato catsup.

Boiled Eggs.—Boil eggs until the white is partly cooked, and the yolk slightly turned; remove from the shell, and season with salt.

Roast Potatoes.—Roast pink-eyed potatoes in the fire until well done, remove the outside crust, mash, and season with salt and cream.

Oaten Gruel.—Boil two tablespoonfuls of sifted oat meal in one quart of water for ten minutes, then add a teaspoonful of salt, one of wine, one of lemon juice, and a little nutmeg.

Indian Gruel. Boil two tablespoonfuls of Indian meal in one quart of water for twenty minutes, add salt, sugar and nutmeg, or lemon to suit the taste.

Sweet Corn Gruel.—Boil three tablespoonfuls of dried sweet corn in one quart of water for half an hour, season with salt and strain through linen.

Sago Milk.—Soak a teaspoonful of sago in a pint of cold water one hour; pour off the water and boil the sago in a pint and a half of milk fifteen minutes, stirring constantly. Season with salt, sugar, ginger or nutmeg, and sometimes wine.

Sago Mucilage.—Soak a teaspoonful of sago in a pint of warm water two hours, then boil the same fifteen minutes stirring constantly; season with salt, lemon juice, sugar, nutmeg and wine. Arrow root and tapioca may be prepared in the same way as sago.

Beef Tea.—Cut one pound of lean fresh beef into shreds, and boil in one quart of water for twenty minutes; add one teaspoonful of salt and strain through linen.

Mutton Broth.—Boil the same quantity of lean fresh mutton and water as above, for one hour; add a few crusts of bread—season with salt and parsley, and strain.

Milk Porridge.—Boil one pint of water and one of milk; add one tablespoonful of wheat flour made into a thin paste, season with salt and boil five minutes.

Oyster Soup.—Boil four oysters in one pint of water for five minutes, add one small cracker and a little salt.

Barley Water.—Boil two ounces of pearl barley in one quart of water down to one pint—season with salt, lemon and sugar, and strain through linen.

Apple Tea.—Boil a middle-sized sour apple in one pint of water, strain and sweeten with white sugar. Peach tea may be made in the same way after removing the stone.

Wine Whey.—Boil one pint of new milk, and while boiling, add a large wine glass full of sherry or madeira wine; let it boil a few minutes, remove it from the fire, let it cool a few minutes, then strain from the curd and sweeten with white sugar.

Tamarind Water.—Boil six tamarinds in one pint of water for ten minutes, and strain through linen. This is a gentle purgative.

Currant Water.—Boil equal quantities of currant juice and water a few minutes; strain through flannel and season with orange peel and loaf sugar. Cherry water, may be made in the same way.

Chicken Water.—Take half a chicken, remove the fat, break the bones, and boil in two quarts of water for half an hour; add two teaspoonfuls of salt, and strain through linen

DISEASES OF FEMALES.

MONTHLY TURNS, OR MENSTRUATION.

This important function should be well understood by both male and female, as by so doing many of the ills of life could be avoided, and things which otherwise would be mysterious are rendered intelligible.

From the womb of every healthy woman who is not pregnant, or who does not give suck, there is a discharge of a fluid having the appearance of blood, at certain periods, from the time of puberty to old age, called menstruation, or courses.

Some few menstruate while they continue to give suck, more frequently after having suckled over six months. Some are said to menstruate during pregnancy, but which latter is very doubtful, for it will be borne in mind that the menstrual discharge is not blood, but a secretion resembling blood, and that every such discharge from the womb is not *menstrual*, but may be blood, dependent on morbid action. Although the term *unwell*, is by common acceptance used among women, yet, a woman during menstruation cannot from that cause alone be said to be unwell, for that is a process of health, and which, when *regular*, requires very seldom more than to be let alone.

This is a very important process to females, and ought to be particularly so to mothers who have daughters coming to maturity : that *mother* is very remiss in her duty who does not inform her child that menstruation is expected, and point out to her what it is, so that the child may not be taken by surprise, and through fear and alarm do things, which may lay the foundation of disease and unhappiness during life. The delicacy attendant on the subject too frequently prevents the afflicted from obtaining the necessary information, and gives rise to groping in the dark, and administering medicine at random.

It should be borne in mind that the time of life at which menstruation commences, depends some upon the climate, much upon the constitution, and delicacy of living; in this country, girls begin to menstruate from the fourteenth to the eighteenth year of their age, but seldom later without inconvenience ; but if they are luxuriously educated, menstruation usually commences at a more early period. About the time that the constitution is establishing menstruation, a variety of important changes show themselves: the complexion is improved; the countenance is more expressive and animated; the attitudes more graceful; the tone of the voice more harmonious; the whole frame expands; the breasts are enlarged; the nipples protrude, &c.

Generally there are symptoms which indicate the change that is about to take place; these are usually more severe at the first than in the succeeding periods, such as a sense of fullness at the lower region of the belly; pains in the back and inferior extremities; a slight head-ache, ringing in the ears; a sensation of choking or a lump in the throat; palpitation of the heart; easily affrighted by slight and unexpected noise; irregular appetite, twitching of the limbs, sometimes convulsions, all of which cease soon after the flow commences. The first discharge is sometimes very small and not colored; for several times it is apt to be irregular, both as to the quantity discharged and period of its return, but after these it usually observes stated times, and nearly the same quantity at each visitation. The time occupied and quantity discharged, vary much in different women, from two to five days, and from one to five ounces may be stated as the average.

Frequently it occurs when the time for menstruation has arrived and it does not appear, or when a girl begins to menstruate in small quantities, or when it wants color, that instead of being let alone, she is compelled to swallow one nauseous portion after another, until that process by which nature was about gradually establishing an important and necessary change, is interrupted, and a train of morbid actions laid, which entail upon her a great amount of suffering.

When about to commence, or having commenced, being small in quantity, the girl should be directed, during the time it is upon her to avoid the extremes of either heat or cold; if cold weather, should add some clothing, keep dry and warm feet, abstain from laborious exercise, such as violent running, jumping, dancing, lifting or carrying heavy burdens, or any thing else by which the body may be strained, or the system over-heated; avoid sudden exposures to currents of cold air when heated, or fatigued; she should also be taught to have a command over her temper, so as to avoid violent outbreakings of anger, and paroxysms of excitement from terror or fright; and she should wear flannel next the skin. Observing these directions, together with a moderate, rather low diet, avoiding all high seasoned victuals, hot aromatic teas, spiced stews, all and every intoxicating liquor, she should continue much her usual indoor employments; and thus, the principle of letting well enough alone being observed, a few periods will generally

establish the regularity of their return, on a healthy and per
manent basis. In general, no medicine is required, except she
be costive; it should be removed by purgatives, such as small
portions of Epsom salts, or castor oil, but the more violent and
drastic purgatives should be strictly avoided. But when it
occurs that the courses are either retarded in their progress, or
do not appear, and symptoms of ill health be present, great
care and attention will be required, that proper remedies be
administered.

RETENTION OF THE MENSES OR COURSES.

This is a condition in which the courses do not appear at the
proper age, in consequence of which the health suffers. The
general rules for the treatment of which are, that when robust,
florid girls, about the age of fifteen or sixteen, begin to complain of flushings, headache, and general uneasiness, they
should observe a spare diet, consisting chiefly of vegetables;
use moderate exercise, carefully avoid all that is violent, particularly in crowded and heated rooms; should carefully
attend to the state of their bowels, and keep them freely open
by saline purgatives; such as cream of tartar, Epsom salts,
Rochelle salts, one tablespoonful, of either, in a glass of water
before breakfast, every second or third morning, or a seidlitz
powder. If the symptoms continue or increase, and the discharge of the menses does not take place, take a tablespoonful
of Epsom salts every two hours, beginning in the morning, until
freely purged; bathe the feet and legs, or rather sit awhile in
warm water for several evenings; after the bathing remove all
moisture, and rub the feet and legs freely with a coarse cloth.
In obstinate cases this proceeding will require to be repeated
for two or three times, at the end of every four weeks.

On the other hand, relaxed and feeble young women, with
pale complexions, when they are subject to delay in the appearance of the menstrual evacuation, and are suffering in
health in consequence thereof, should make use of such remedies as strengthen the system in general. One grain of
quinine, taken night and morning, on an empty stomach, is an
excellent tonic, or mix together the following: Gentian root,
half an ounce; Columbo root, half an ounce; orange peel and
wild cherry bark, each, half an ounce; all to be beat or
ground fine, and put with half a pint of whisky and same
quantity of water. Let stand for ten days, then strain, and
take a tablespoonful half an hour before each meal, during the
use of which the bowels must be kept open, and a more nutritious yet easily digested diet allowed. Take sufficient exercise

in the open air, such as riding on horseback, jumping the rope in moderation, or walking with agreeable persons, so that cheerfulness is blended with exercise. For the same reason, a journey, a short residence at watering-places of public resort, independent of the quality of their springs, contribute greatly to their relief; and when the impregnation of such springs is chalybeate, (iron,) they may be drank with moderation, remembering to precede their use by an active purge, by which chalybeates and all tonics are rendered not only more safe, but more beneficial.

When the young girl thinks, from her feeling, that nature is making an effort to bring forth the discharge, which is known by an increase of uneasy feeling in the back, hips, or lower part of the belly, she is to use the warm bath as before directed.

Owing to previous debility, or other diseases, the courses are sometimes retained or obstructed, and no medicine or treatment will avail until the disease under which the person labors be removed.

Retention of the menses for a length of time soon undermines the general health, even in the best constitutions, and degenerates into what is called *green sickness*, a very dangerous disease, and difficult to cure. Every symptom of feebleness prevails—a pale skin, and even a greenish complexion, succeeds to the rosy hue of health; the lips and gums become almost white, the breath offensive, the skin under the eyes puffy, and of a leaden color, the whole body lax, swollen and doughy; the judgment, memory and natural cheerfulness impaired; the pulse is generally slow and feeble, but easily excited, and it is then accompanied by shortness of breath, a palpitation of the heart, and an almost unconquerable disinclination to motion; the appetite is destroyed, and the stomach so deranged that the food, instead of being digested. sours on the stomach. Hence the patient finds gratification in chalk, lime, pieces of old wall, and other improper substances; there is also costiveness.

The treatment should be more energetic under such circumstances; it should be commenced by removing the costiveness by repeated doses of active purgatives, such as ten grains of jalap with four grains of aloes well mixed, and made into a bolus, or pills, with some syrup, and followed, in six hours after, by one tablespoonful of castor oil, or a gill of senna tea, which should be repeated every four hours, until free evacuations take place; to be repeated every two or three days, according to the effect of the preceding, until the feverish or inflammatory symptoms have been removed; after which the tonic mixture may be used, as formerly prescribed. During the use of either of these, the bowels must be prevented from becoming costive; but, should sourness of the stomach be present, which is known by a burning sensation at the stomach, sour belchings, sour taste in the mouth, soreness or tenderness of the stomach, particularly on external pressure, the following preparation should be used occasionally with the above: Loaf sugar, four teaspoonsful; essence of cinnamon or peppermint, two teaspoonsful; powdered rhubarb, two teaspoonsful; carbonate of soda, one teaspoonful; carbonate of magnesia, four teaspoonsful; mix well in a bowl, adding half a pint of cold water. The dose is, one tablespoonful night and morning. Wearing flannel drawers, using moderate exercise, never going to the length of fatigue, and the aversion to motion to be overcome, by proposing such exercise as may be most agreeable, a nourishing diet, such as may agree best with the stomach, is the proper course to pursue ; and at such times as the menstrual efforts are felt, to assist them by the hip bath, and friction of the feet and legs.

SUPPRESSION OF THE MENSES.

AFTER being fully established, if the courses are arrested, or do not return at their usual period, when not caused by pregnancy or suckling, it is called a suppression. The most fruitful sources of these derangements are exposure to cold, in some form or other, violent exercise, great mental agitation during their flow or immediately before their appearance. As soon as it is discovered that they are arrested, remedies should be immediately employed ; if their suppression be not complicated with general disease, it is not difficult to induce their return. The feet and legs should be bathed, or the person should sit in warm water; an anodyne may be given, such as a teaspoonful of elixir paregoric, or anodyne cordial; promoting its operation by catmint, penneroyal, or spruce pine tea.

Should these remedies fail, and there is pain in the head, back, and lower extremities, and the circulation excited, you should purge the bowels freely. After the free operation, give fifteen drops of antimonial wine, with five of laudanum, or ten of paregoric, in a spoonful of water, repeated every two hours, until nausea be produced ; then reduce the dose to one-half. If the feverish symptoms still continue the purging must be repeated on the third day following, and either of the medicines again used as before directed; during which time a very low diet is necessary. This treatment must be pursued until the fever shall be abated and the pain relieved.

You must not expect the courses immediately to return, but probably they will at the next period be restored; and in order to facilitate their re-appearance, an active purge should be administered about four days before they are expected. Also bathe in warm water, etc., as before.

DEFICIENT AND PAINFUL MENSTRUATION.

FEW persons have an idea of the amount of suffering among females from this disease. It seldom attacks any until they have menstruated some time with considerable regularity, and little or no pain; afterwards, they begin to suffer more or less pain, which increases until it becomes grinding and severe as those in labor.

It soon affects the general health; the patient loses her complexion, and becomes very irritable and fretful. At the approach of each menstrual period, the pain generally begins in the back, extends to the loins and hips, to which soon ensues an alternate and pressing down pain resembling in severity and suffering those of labor. At first a slight discharge takes place, but which suddenly ceases, after some time is renewed and becomes more plentiful, which, together with the pain, gradually ceases. The appearance of the discharge differs from that of a healthy menstruation, being mixed with lumps, and clots of flaky matter, having the appearance of membrane or skin. The breasts sympathizing with the womb, frequently swell and become painful. Women are mostly barren who have this disease in a severe form.

Painful menstruation must be treated by having the bowel well opened a few days before the anticipated attack, in being confined to a very light vegetable diet, strictly avoiding the use of all spiritous liquors. The patient should be kept in bed, drink freely of tea made either of penneroyal, catmint, sage, or the leaves of spruce pine, until the discharge be fully established; after which the pain seldom returns for that period. Bateman's drops, or tea made of the bark of the root of tulip poplar tree, lovage, tansy, hops, or black snake root, may be used. What I have found the best in my own practice is this: Get at a drug store four ounces of tincture of ergot; commence about a day or two before the expected return of the monthly sickness, and take a teaspoonful every four hours, until the discharge is *fully established*. Take a purgative before commencing with the ergot. Do the same way for two or three periods of the courses. The directions heretofore given as regards diet, clothing, and exercise, keeping the feet warm and dry, and the bowels open, together with an occasional use of

the warm bath should be strictly adhered to, and persevered
in for a considerable length of time. Sometimes one or two
grains of powdered ipecac, or half a teaspoonful of the syrup
taken every two hours, will bring on the flow freely, when
other means fail. Keep warm in bed while using the ipecac.

PROFUSE MENSTRUATION.

There is a great difference in different women as to the
amount of the discharge during their courses. When a
scanty evacuation is followed by a general uneasiness, a sense
of fullness, flushing and headache, it is to be considered a
suppression; and when a considerable flow is followed by languor, paleness and general weakness, it is to be considered
as profuse, and should be checked.

If feverish symptoms, such as headache, oppressed breathing, increased heat, and a full, firm pulse, precede or accompany a sudden and profuse flow of menses, the evacuation frequently becomes its own cure; and if the woman be careful to
keep her bowels open by moderate purgatives, to observe a
spare diet, to drink only cold water, to keep her person cool
by thin clothing, sleep on a hard bed, and have free exposure
to the open air, she may not only moderate the evacuation in
future, but probably will derive considerable advantage from
its present excess. But if, notwithstanding these precautions,
the flow continue or return, still accompanied with the above
febrile symptoms, she must take a brisk purgative of Epsom
salts, or senna and salts, to be repeated until *full* and *free
evacuations* from the bowels take place. Also if necessary,
take one or two grains of ipecac, or a half teaspoonful of the
syrup of ipecac every two hours until sickness at the stomach
is produced, *but not vomiting.*

Should there be such a profuse flow as to cause great prostration, faintness, vomiting, the lips becoming pale, nails blue,
extremities cold, with convulsive twitchings, the danger is
great, and it is then no longer profuse menstruation, but must
be considered a true *uterine hemorrhage;* she must lie down on
a hard bed, and be kept perfectly quiet. Motion of every kind
must be forbidden, not even permit her to turn herself; she
must be freely supplied with fresh air, or the use of the fan
drink cold, or even iced water; bladders half filled with cold
water, or cloths wrung out of cold water, applied to the belly,
and frequently renewed, provided there is no chill on her at
the time. These, together with oak bark tea, or alum whey
every hour or two, with the addition of from ten to fifteen
drops of laudanum to each dose; all to be given cold. These

proceedings generally give relief, or at least will control the disease, and probably preserve life until medical aid can be procured, which should always be had if possible. After the hemorrhage has been moderated, she must for many days avoid exertion, remaining in bed, be confined to a strictly vegetable diet, and avoid every kind of spices, and also all spiritous and fermented liquors. Keep her bowels open by the use of purgatives, such as Rochelle salts—one teaspoonful in a glass of water—castor oil or magnesia, taking three times a day some light tonic bitters, such as cold watery infusion or tea made of wild cherry bark, thorough wort (boneset) or dog-wood bark, and while the body is to be kept cool, the feet are to be kept warm and dry ; if they are cold, they must be frequently rubbed with a woolen cloth or flesh-brush.

CESSATION OF THE MENSES.

As a general rule, a woman ceases to have the menstrual flow between the age of forty and fifty ; in some women gradually, in others more suddenly. This is an important and critical period of a woman's life, and great care is to be exercised that the health of the patient be not injured by improper treatment. The greater number of cases require only to be let alone ; many, particularly the weakly, will probably be benefited by the cessation, and will enjoy better health. When, therefore, this discharge shall decline or altogether cease, and not be succeeded by other disease, it will require no other attention than a strict regard to temperance, so as not to interrupt nature in effecting an important change. But in constitutions in which there is a predisposition to some disease, the cessation, more particularly the sudden stoppage of the courses will expose the woman to an attack of that disease. Therefore, if upon the decline of menstruation there shall occur general feverish uneasiness, such as flushings, restlessness, headache, throbbing, either in the head or under the ears, singing or ringing in the ears, dizziness, darting pain through the head, palpitation of the heart, piles, hard or painful swelling of the legs, it will then be necessary in addition to strict temperance, both as regards body and mind, to keep the bowels freely open by the occasional use of salts, seidlitz powders, castor oil, senna, or if necessary, more active purgatives, such as cream of tartar and jalap, etc. These remedies and precautions will have to be persisted in for some time, at such intervals as the urgency of the symptoms may require, and should be so timed as to anticipate the attacks, until the system becomes used to the change.

At this time of life, some women, instead of the menses ceasing, become subject to repeated and excessive discharges; in such cases, before the patient becomes weakened by exhaustion, and if the discharge be accompanied by fever, it will be necessary to use gentle laxatives and anodynes, such as a teaspoonful, each, of cream of tartar and sulphur in a glass of water before breakfast. Also put one tablespoonful of spirits nitre in a tumbler of cold water, and take one tablespoonful every three hours. As an anodyne, take half a teaspoonful of paregoric in a little water at bed time. But if such discharges become so excessive as to come under the denomination of uterine hemorrhage as formerly described, the case should be treated the same as advised in that disease.

If it occurs that a woman, after menstruation has ceased for a few periods, becomes again subject to discharges either of blood, or matter resembling that from a boil, especially if it be accompanied with pain at a particular spot, darting from thence across the abdomen, through the hips or down the thighs, there is reason to apprehend cancer, or other serious ulcers in the womb or adjacent parts. It is advisable to have such cases placed under medical advice. Nothing has a greater tendency to retard the progress of these cases, than a strict regard to temperance, and nothing will more surely hasten their advancement into incurable disease than the opposite course.

UTERINE OR WOMB DISEASES.

FLUOR ALBUS, OR WHITES.—Female weakness, (as this disease is often called,) is among the most prevalent ailments among women: even young girls before marriage, are often troubled with it without applying for medical aid, owing to false notions of modesty. It is a discharge from the privates, not colored with blood; in general, if the disease be not the consequence of falling down of the womb, or other organic derangement of that organ, the discharge is easily arrested, when taken in the first stage. At first, it is in most cases strictly local, but if suffered to run on, its necessary consequence is to undermine the constitution, and ruin the general health. The discharge is at first mild and semi-transparent, resembling in appearance that of the white of eggs, or thin starch made by boiling; but becomes in its progress opaque or milky, yellow, greenish, and scalding—so much so as to irritate and inflame the parts over which it passes. In the first stage it is accompanied by little or no pain, but as it progresses towards the second, and succeeding stages, the person suffers from loss of appetite, pain in the back and loins, weariness

and fever. When persons of a robust and full habit are sub
ject to this disease, or when it occurs in more delicate constitutions, but accompanied by a feverish state of the system,
the treatment should be commenced by a purgative of ten
grains of rhubarb, a dose of senna, or castor oil. A cooling
and spare diet is absolutely necessary, and cleanliness must be
strictly enforced. In order to secure this latter requisi.e,
(which is all important to the speedy termination of the dis
ease,) injections of luke-warm water, or milk and water of the
same temperature, should be thrown up the privates three or
four times a day : this may be accomplished by the use of the
female syringe, to be obtained at most drug-stores. When the
feverish state of the system is thus in some degree subdued,
gentle astringent injections will be proper ; such as weak tea
of white oak bark, with or without a small portion of alum ;
or 20 grains of white vitriol, 100 drops of laudanum, and half
a pint of water, to be used three times daily. Injections of
green or black tea, very strong, or tea made from sassafras,
sage, or dogwood bark, (cold) thrown up the private parts three
or four times a day ; also, at the same time, washing the outer
parts in cold water, and always keep the bowels open, is advisable.

When the disease has already progressed until it has passed
the inflammatory stage, or is not originally attended by a
feverish state of the system, but by general weakness ; an inability or disinclination to exercise ; pain in the back and loins,
want of appetite, &c.,—after the purging, as above directed,
and during the use of the astringent injections, the tonic bitters previously mentioned, (of wild cherry bark, &c.) together
with bathing the outer parts in cold water thoroughly three
or four times a day, should be used. In those cases in
which the discharge is offensive and of a greenish color, or re
sembling matter discharged from a boil and streaked with
blood, it is advisable for the purpose of cleansing out the vagina, (or canal,) first, immediately preceding each astringent
injection, to throw up a few syringes full of weak soapsuds.
Too great attention cannot be paid to cleanliness, and indeed
all other directions are useless if that be neglected. The outer
parts should be well washed with soap and water two or three
times a week. If this course of treatment does not effect a
cure, make no delay in seeking advice of a physician.

FALLING OF THE WOMB.

WHEN a prolapsus takes place, an uneasy dragging sensation
is felt in the loins while standing or walking ; a mucous dis

charge is perceived, sometimes bloody, accompanied with a kind of pressing or bearing down; an inclination to go to stool, frequently a light slimy purging and a sense of numbness shooting down the thighs; when first rising from a lying to a standing position, a sensation of falling from above into the passage below, which prevents the free evacuation of urine; these symptoms all subside, or are much mitigated by lying down. Whatever weakens the parts concerned, has a tendency to produce this disease: such as, frequent miscarriages; improper treatment during labor; severe and protracted labor; the use of instruments in delivery; too early rising, and too violent exercise after delivery; improper treatment of profuse menstruation; long continued whites; violent exertion during menstruation, such as jumping, dancing, lifting heavy weights; blows on the abdomen, &c., are the most frequent causes. The means of cure, in those cases in which the womb will not return to its place on lying down, is to lower the head and shoulders, whilst the hips are somewhat elevated, then with the finger oiled, gently press the prolapsed part into its proper situation, and lie down for many days, or in severe cases, weeks, and two or three times a day make use of astringent injections as mentioned in treating whites. If the bowels are costive, some gentle cathartic, such as castor oil, seidlitz powders, magnesia, &c., should be administered, but all irritating purgatives, as well as stimulating diet and drinks, strictly avoided. If these means fail, recourse must be had to the use of a pessary, which can be obtained at most any drug-store. The " Ring Pessary" we consider preferable. It should be taken out once or twice a week, and washed in soap and water; and at the same time, inject a pint of cold green tea up the privates.

PREGNANCY—Its Signs and Diseases.

DURING pregnancy, the natural irritability of the womb being increased, manifests itself by a variety of symptoms and sympathies. Although such symptoms may, and frequently do, arise from other causes than pregnancy, still, when a healthy married woman finds that the menstrual discharge does not return at its usual period, finds her breasts enlarge, and the circle which surrounds the nipple change from a light pink to a dark brown color, and that she soon after becomes subject to languor, nausea, and vomiting in the morning, heart-burn during the day, and some degree of restlessness and want of sleep during the night, she may with confidence attribute these symptoms to pregnancy. Women who have borne children will, in consequence of their peculiar feelings formerly experienced, seldom be mistaken in their judgment whilst those who wish to hide their pregnancy, and others from an overweening anxiety to have children, will be led into many ridiculous and frequently dangerous errors. As all, or at least most, of the symptoms above mentioned may be present, yet no pregnancy exist, therefore, in cases of doubt, it would be most expedient to act as though it did, for, about the end of the third and beginning of the fourth month, the rising of the womb, and the feeling of a tumor or fullness below the stomach—and, between the fourth and fifth month, the motion of the child, termed "quickening"—will, generally, put the matter beyond doubt; no risk can be encountered by thus delaying any measures which might have an injurious tendency in case of pregnancy.

DISEASES OF PREGNANCY.—A common attendant on early pregnancy is a slight degree of feverishness, but which, unless excessive or accompanied with other diseases, will seldom require any other remedy than to open the bowels, and a low diet, of which fruit and vegetables should form the principal part; there is generally a dislike to animal food of every kind, and it should be avoided as much as possible, for, if indulged in, it gives rise to much inconvenience. Another means of keeping up a healthy action, and thereby obviating the effects of fever, is moderate exercise in the open air. It is a mistake to suppose that pregnant women should be encouraged in living more luxuriously and indolently than what is habitual to them; they should, therefore, not be confined to close or heated apartments, but be allowed a full share of out-door exercise, yet be cautioned to guard against extremes, such as carrying or lifting heavy burthens, running, jumping, dancing, &c. The irritation of the fever will cause them to be fretful, peevish, and desponding, which is often by others mistaken for ill-temper, which erroneous opinion only leads to farther inconvenience and unhappiness. Women, whose happiness always is a matter of deep interest to the civilized man and Christian, are, during the state of *pregnancy*, more than at any other time, entitled to the tender regard and affectionate consolation of their friends, particularly of those who claim the title of *husband*. The desires and dislikes of pregnant women should not be overlooked, and the effects of despondency prevented by kind words and by everything calculated to encourage them.

Frequently the feverish heat, full pulse, headache, uneasiness and restlessness continue to increase as pregnancy advances, becoming sometimes very distressing and alarming; recourse should then be had to purgatives, after which two or three doses of soda powders, during the day, or cooling drink of water, acidulated with lemon juice, vinegar, cream of tartar, tamarinds, cherries, plums, &c. The *mild* purging may be repeated as occasion requires, but should not be excessive, or such as produces griping.

Morning sickness is one of the early symptoms of pregnancy, which generally ceases after quickening, and returns towards the conclusion, sometimes at intervals during the whole period, which, when moderate and confined to the early part of the day, should be left to nature, not being of a dangerous tendency, but generally found serviceable, as a woman will generally find she enjoys more ease during the remainder of the day after having vomited in the morning than when she has not.

When sickness and vomiting prove more severe, and the stomach continues to reject the food taken during the day, recourse should be had to medicine. When the vomiting is accompanied by a costive state of the bowels, a tablespoonful of carbonate of magnesia should be given several times during the day, until they are moved, and continued once or twice a day for some time ; or the following : water, one gill ; loaf sugar, two teaspoonsful ; essence of cinnamon, one teaspoonful ; rhubarb, one teaspoonful ; calcined magnesia, one teaspoonful—mix together.

Of this mixture one tablespoonful may be given every four hours. A seidlitz powder taken occasionally will be found useful. If the vomiting be accompanied by a diarrhœa, a teaspoonful of prepared chalk, diffused in cold water, three or four times a day, soda powders, lime water and sweet milk, or a piece of flannel moistened with laudanum and camphor applied to the pit of the stomach, will all prove very serviceable.

Heartburn is a painful sensation of heat in the throat and stomach, attended with a sudden coming up of thin, sour froth into the mouth. There is often reason to think that it is occasioned by food highly seasoned, or not easily digested, and by fermented liquors—and sometimes by *sleeping in a sitting posture after a meal*—but it most frequently depends upon some sympathy of the stomach with the womb, and although more disagreeable and difficult to remove than dangerous, and always removed by delivery, yet there are some cases which are extremely distressing, and cannot be considered as entirely void of danger.

Even if we cannot remove this unpleasant attendant on pregnancy until after delivery, it may be much mitigated by giving, two or three times a day, a large teaspoonful of carbonate of magnesia, or prepared chalk, in cold water ; a tablespoonful of lime water, or ten grains of carbonate of soda in a gill of fresh water. In obstinate cases, the following will be found a very beneficial prescription : Water, half a gill ; essence of cinnamon, one teaspoonful ; calcined magnesia, one teaspoonful ; hartshorn, ten drops ; mix and put into a bottle. Two or three teaspoonsful may be given occasionally, particularly after meals, each dose to be taken in a large tablespoonful of water. If costiveness prevails, it should be removed by gentle purgatives. Magnesia is among the best in this case.

Costiveness is one of the most prevalent, as well as the most obstinate, diseases of pregnancy, and is, generally, *most neglected*. When excessive, it gives rise to many other affections, such as colicy pains, headache, flushing of the face, frequent desire to make water, unavailing straining at stool, piles, palpitation of the heart. Costiveness, if not the *cause* of vomiting, heartburn, and flatulence, is known much to *increase them ;* and there is much reason to believe it is frequently the cause of abortion, and therefore of much importance that it should be removed. But, in accomplishing this, care should be used to select proper purgatives, strictly avoiding all such as are *griping*, as profuse purging is hurtful in pregnancy. Small doses of Epsom salts, seidlitz powders, manna, senna and manna, castor oil, rhubarb, sulphur and cream of tartar, are the most proper purgatives. These ought to be used in rather small and repeated doses than given largely.

Much may be done to remove this unpleasant state of the bowels by a suitable diet. Bread made of unbolted ground flour, (called sometimes "Graham flour,") and bran tea, sweetened with molasses, or manna, deservedly rank high; also such other articles as are known to have a laxative effect upon the bowels should not be neglected. Stewed fruits at meals, or ripe fruit uncooked, once or twice a day, or a roasted apple at bed time, assist in keeping the bowels open. If the bowels have been in a costive state for many days, it will always be advisable to assist the operation of the medicines by an injection, composed of a pint of milk and water, with the addition of a tablespoonful of melted lard, or sweet oil, which will facilitate the evacuation, and cause the medicines to operate with less pain and sickness.

Piles.—They are a cause of much suffering to pregnant women, and, in consequence of the delicacy attendant, they

will generally suffer a long time in silence rather than apply
for aid.

The treatment proper for this disease will be found unde.
the head of PILES, in another part of this work.

STRANGURY, OR PAIN AND DIFFICULTY IN URINATING.—This
is a frequent desire to pass the urine, and painful discharge in
small quantities. It is sometimes caused by not complying
with the calls of nature, and retaining the urine too long, but
most frequently arises, during the early period of pregnancy,
from the sympathy which exists between the womb and bladder; subsequently, from the pressure of the enlarged womb,
a retroversion or falling down of the womb, from its pressure
against the neck of the bladder, may also give rise to it; in
such cases, a tumor in the passage will be evident. For the
relief of strangury, the bowels should be moved by laxatives,
assisted by injections up the bowels of lukewarm water, in
the meantime drinking frequently of tea made either of elm
bark, marsh mallows, flax-seed, parsley root, or lovage, adding
three or four times a day, to some of the tea, twenty drops of
spirits of nitre, and five drops of laudanum, or, in the place
of the laudanum, twenty of paregoric. Should the strangury
be caused by the falling down of the womb, it must be removed by replacing it according to the directions formerly
given, and, while so attempting to replace the womb, let her
make effort to void urine; and if the bladder be thus emptied
the tumor will probably return by further gentle pressure upwards, and, if so, she should keep her bed for some days.

ITCHING OF THE OUTER PARTS, with an irresistible desire to
scratch, is, in some cases, a very distressing complaint, more
so, as women will suffer intense misery in silence, and, even
when forced to consult medical men, will not make it known,
unless drawn from them by close and repeated questioning.
Women are liable to this affection at any time of life; it is
frequently dependent on some disease of the bladder or womb,
and when so no treatment will avail as a cure, unless the
primary disease be first removed; it is also sometimes an
original disease, and sometimes appears to be caused by the
pregnant state, and in all cases where having previously existed is much increased during pregnancy. Use the following
mixture: Water, half a pint; borax, three tablespoonsful
tincture of myrrh, one tablespoonful; laudanum, one teaspoonful. To be injected into the passage, two, three, or four
times a day, and during the intervals to keep pledgets of old
linen, or a soft sponge, moistened with the same, in close contact with the outer parts, where there is the most itching

Avoid scratching as much as possible. The bowels should be kept well opened by proper purgatives. Wet tea leaves or hops applied to the parts are often beneficial.

CRAMPS, in the legs, which are very troublesome sometimes may generally be relieved by some stimulating liniment or mixture of vinegar and mustard rubbed on the parts. ' More exercise should be taken.

ABORTION, OR MISCARRIAGE,

Is the expulsion of the contents of the pregnan* womb, at a period of pregnancy so early as to render it impossible for the life of the child to continue.

Although in many cases no evident cause can be assigned for the production of this unpleasant occurrence, it may be mostly traced to some of the following: violent exercise; severe fatigue; sudden exertion; contusions or shocks on the body; heating and stimulating food; indulgence in spiritous and other intoxicating liquors; violent operations of emetics and purgatives; fear, grief, and excessive joy; a full gross habit and feverish state of the system; exciting and debilitating diseases. A delicate frame, and weakly constitution, will predispose to it, yet, there are some in whom there exists a predisposition to abortion, and who nevertheless appear otherwise healthy; in such, the slightest causes will excite the womb to cast off the contents; and having once done so, are apt to miscarry again, and if the habit be once acquired, it becomes very difficult to remove—ruining the health of women, and disappointing the fondest hopes of parents.

The first symptoms are, the absence of the usual morning sickness, a subsidence of the breasts, discharge of water or of blood from the womb, commonly known by the name of flooding; pains in the back, loins, and lower parts of the belly coming on in paroxysms with intervals of ease, resembling those of labor.

The hemorrhage being the surest symptom of abortion, demands our first regard, for if it continues, abortion must ensue. If then, this alarming symptom be present, our only hope lies in restraining it. Yet it must be borne in mind that frequently the hemorrhage is the effect of the aborting process already going on, and therefore the discharge cannot be stopped until the contents of the womb have been cast off; but still we can generally, by correct conduct, so moderate it that the woman will be sustained, and her life be preserved.

The bowels must be attended to; if costive, to be opened by some gentle purgative, (Rochelle salts or castor oil,) and if diarrhœa be present, it must be relieved by a few drops of laudanum or of paregoric. *She must lie down and keep quiet,* on a hard bed with light covering ; every thing that will heat the body and quicken the circulation must be carefully avoided, cold air freely admitted into the room, and she be kept perfectly still, every kind of exertion strictly forbidden even to conversing with her friends ; the diet must consist wholly of vegetables, fruits, butter milk, cold water, lemonade ; all cordials, spiritous liquors, spices, and stimulating food must be rejected. Cloths wrung out of cold water, or vinegar and water, should be applied to the back, bowels, thighs, and external parts ; and when the heat of the body is considerable, and the hemorrhage profuse, the coldness of these applications may be increased by ice or snow ; but these cold applications are limited to the stage of *excitement,* and to be discontinued on their producing pain or a continued chill. If the fever is reduced, and the system brought rather below the natural heat, then, if there be a continued pain, thirty or fifty drops of laudanum, given in a spoonful of vinegar, or one-fourth of a grain of opium, with half a grain of ipecac made into a bolus, two, three, or four times a day, will abate it. But if the pains be in paroxysms, with ease between them, and bearing down with expulsive effort, and more particularly if this kind of pains have preceded the flooding, then opium and laudanum are improper, as they will prolong the suffering. Very little expectation can be had of saving the child, our attention must then be directed chiefly to the saving of the mother ; for if the contents of the womb must be parted with, the sooner the better—which laudanum and opium retard. The aforesaid treatment to be pursued steadily from day to day, until all appearance of abortion shall have vanished, and then to try, by rising slowly and without exertion, whether she is safe in so doing ; but on the least appearance of the return of flooding or pain, again resume the bed as before. During all this time the bowels must be kept gently open.

If the case be one of exhaustion, of which if the reader will judge by the symptoms before mentioned as indicative of that state, the treatment must be varied accordingly. Before she has been brought into this state, she will probably have flooded much, and the hemorrhage will be considerably diminished at least for the time. Fainting, or disposition to it, will then be the most prominent and alarming symptom, which is not only a consequence of the loss of much blood, but is the remedy which nature makes use of to check the further effusion. This, although very alarming to those unacquainted with its good effects, should not in recent flooding be interfered with; no efforts should be made to rouse her, or prevent a recurrence of fainting by administering cordials or other stimulants; but she should be left in that languid state which always accompanies fainting: during which the blood moves slowly through the vessels, and an opportunity is afforded for the mouths of the bleeding vessels to contract, the blood to coagulate, and the bleeding to be stopped. When some time has been allowed for the contraction of the blood vessels, and coagulation of the blood, and the fainting should still continue to an alarming degree, dash cold water on the face, give a tablespoonful of wine, or a teaspoonful of brandy, or camphorated spirits, or twenty drops of ether, in fresh water, which repeat every ten, twenty, or thirty minutes, as the recovery may be quick or slow, having regard to the hemorrhage; if it show a disposition to return, desist from the brandy, &c., which, at all events, must be used no longer than absolutely necessary to call back the powers of life, which, when recovered, must be left to themselves; keep her perfectly quiet, not permitting her to speak or move hand or foot. After being somewhat restored, having still hopes of averting the abortion, and there be pain, opiates may be given, as before mentioned, a soda powder, in the state of effervescence, given a few times at intervals of a few hours, and if there be still a slight discharge, alum whey may be given.

Occasionally cases of pregnancy occur, accompanied by a slow or chronic hemorrhage, continuing for many days in a small degree, sometimes being scarcely perceptible, at other times more profuse, but not sufficient to excite much alarm, until some new excitement or exertion, suddenly brings on formidable hemorrhage, and abortion with great risk of life to the woman, takes place. A dull, heavy, aching pain in the back, at all times easily excited into a sharp, pungent pain, darting through the womb, in different directions, and down the thighs is commonly attendant.

In such cases the most perfect quietude in bed must be observed, and the mind kept tranquil, and free from every excitement or care. Also injections of *cold* green tea two or three times a day up the privates, may be employed, and cloths wrung out of the same laid across the bowels, renewing the application frequently.

ABORTION FROM EXTERNAL VIOLENCE.—External violence is very frequent cause of miscarriage. A pregnant woman having received an injury from which abortion might be apprehended, should immediately be put to bed, and kept perfectly quiet; if chilly, some *warm*, not *hot* tea, be given her for drink, and moderately warm covering, but no longer than the chill may continue; as soon as that is off, or if no chill has taken place, tnen to be kept cool, admitting the fresh air freely, and cold water for drink, and when fever shall succeed such injury, give a dose of Epsom salts, or Rochelle salts. When the bowels have been freely moved, put one teaspoonful of spirits nitre, one of paregoric, and twenty drops of essence of peppermont into a tumbler of cold water, and of this mixture let a tablespoonful be taken every hour or two.

Of course, rest in bed must be enjoined, and perfect quiet, etc., as before mentioned. Also the cold green tea to the powels.

In all cases where there is reason to fear abortion, a state of *absolute rest* in bed is to be enforced with great perseverance, as the first rule of practice. By rest alone, without any other assistance, hemorrhages may be restrained and abortion prevented; but without it, no woman can be safe. All other means will be unavailing unless assisted by rest. Even after the immediate alarm of the attack is over, and she be in a prosperous state of recovery, she must still recollect her danger. She should be confined to a hard bed, for several days after, and keep her room for a much longer period.

If an abortion does take place even after all that can be done, the case must be treated the same as after an ordinary confinement.

SWELLING OF THE LOWER LIMBS.—This is sometimes very troublesome; it can, to a certain extent, be relieved by keeping the bowels open, and giving fifteen drops of spirits of nitre in half a glass of cold water three times a day.

PALPITATION OF THE HEART—When it occurs in pregnant women, may be relieved by taking fifteen drops every four hours of a mixture composed of equal parts of tincture of valerian and spirits of lavender in a little water. Take more exercise and avoid costiveness of the bowels.

Bandaging the Bowels, for a length of time before confinement, is of very great benefit, with delicate females, or in cases of unusual enlargement.

THE NIPPLES.—During the last months of pregnancy they should be daily bathed in a tea made of oak bark, borax, or alum water, or strong green or black tea (cold). This will harden or toughen them so that they are not near as likely to become sore, soon after confinement. And women who are subject to having sore nipples, should draw them out with a breast-pump, or get a friend to do it for her once every day during the last four or five weeks before confinement. **They** thus become hardened beforehand.

Too much on the Feet.—It is not good for either the mother or her child, for her to be *too much in the erect posture*, more especially those of a delicate constitution. Sitting down or lying down for a few minutes at a time, several times during the day, will be of signal benefit, as it keeps the womb from settling down too low.

Cramps in the Stomach, may be relieved by drinking a few spoonfuls of a tea made of cloves and cinnamon, repeated as occasion may require.

MIDWIFERY.

LABOR.

This takes place at the completion of the term of pregnancy, a period of about forty weeks, or nine months. At this period, the child being able to live without its connection with the mother, the womb begins to contract itself so as to lessen its cavity, and thereby expel or thrust off its contents, which are propelled downwards, towards the mouth of the womb, which opens and dilates so as to give them a ready exit, at the same time those parts through which they are required to pass, assume a disposition to dilate or open.

Natural Labor.—All such as come on at the full period of nine months, in which the head of the child presents or comes first, and which are completed by the unaided efforts of nature; are so denominated from the frequency of their occurrence, and the regularity with which they proceed.

The first stage of labor commences with the true labor pains, and ends when the mouth of the womb is completely opened, about the time the membranes usually burst and the waters are discharged. The second stage is occupied in the passage of the child's head so low as to begin to press upon, and to dilate the external parts. These two stages frequently go on together, although the mouth of the womb is usually dilated before the head has descended low down; yet it sometimes happens that it is unyielding and not disposed to dilate, and therefore descends before the head, which is thus covered by the neck of the womb, presents at the external opening. The third stage of labor commences with the distention of the external parts into the form of a large protuberant tumor, and continues until the external orifice shall be so far dilated as to

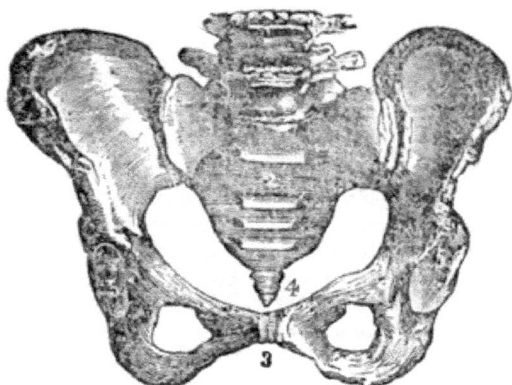

The Female Pelvis.

We have here a front view of the Pelvis. 1. 1. The *ossa in-nominata*, or hip bones.   2. The *sacrum*, perforated with two rows of holes, for the transmission of nerves.   3. The *symphysis pubis*, or os pubis. 4. The *coccyx*. 5. 5. Articulations for the thigh bones.   The *antero-posterior diameter* of the brim, from symphysis pubis to middle of sacrum, measures four inches and a half.   The *transverse*, from the middle of the brim on one side, to the same point on the opposite, five inches and a quarter   The *oblique* diameter measures about five inches.   The *antero-posterior* and *transverse* diameters of the *outlet* measure, each, four inches.   The cavity of the pelvis measures, in depth, one inch and a half in front, and four inches and a half pos-teriorly.

suffer the child to pass through it. The last stage is taken up in the care of the infant, in tying and cutting the navel string, and in receiving or gently aiding the delivery of the after birth.

Midwives should keep this division of labor into four stages constantly in their minds, and perfectly understand what is going on during each, by which they will avoid all unnecessary hurry and confusion, and they will expect no more in any one period than is intended by nature to be then performed.

EXAMINATION OF THE PASSAGE.—To perform this properly, and to draw from it certain conclusions in intricate cases, can be acquired only by attentive practice and experience, aided by previous anatomical knowledge of the parts. But we hope to be able to give a few plain directions, which may teach others how to judge of their progress, and to direct them in the conduct of a natural labor; and also to discover those which are unnatural, and are likely to be difficult. It is a rule among physicians never to perform the operation except in the presence of the nurse, or some other married woman; but with female midwives, this delicacy is not absolutely necessary; still it is advisable that some other woman be present. The patient should lie on her side or back on the edge of the bed, with her knees drawn up, and a light covering should be thrown over her. The midwife sitting at the side of the bed, the forefinger anointed with lard or sweet oil, is to be carried up to the outer parts into which the finger is to be cautiously introduced. The introduction should be made with all possible tenderness, carefully avoiding all hurry, force or rudeness, by which the parts may be irritated or wounded; and, above all things, take care not to break or burst the membranes.

The finger will probably first reach the neck of the womb, covering the head of the child, and pressing down into the vagina, or birth passage; passing the finger toward the backbone, and upward, the mouth of the womb will generally be found (in the beginning of labor) far back and high up, very different in different women. In some, hard and irregular; in others, thick, soft, and smooth, a little open, and beginning to discharge a thick mucus; whilst in some few it is worn quite away, although still close shut. The examination is to be commenced a little before the time that a pain is expected, and should be continued during the pain, and until it ceases, so that the effect of the pains upon the internal orifice or opening may be noticed; and having the finger introduced, continue it until satisfied of all that is desired, or can then be discovered.

If the mouth of the womb be pressed down tight, and begins to open during the pain; if this general tightness relax during the intermission, and especially if those parts remain soft and slippery, and a thick mucus, with or without some tinge of blood, begin to ooze from them, we conclude the labor to be actually begun. But if, on the contrary, we discover no extraordinary pressure, and the mouth of the womb be neither opened during the pain nor relaxed again as the pain goes off, we may conclude the present pains to be false—that labor has not yet begun.

FALSE PAINS—Frequently resemble true labor-pains so exactly as to be mistaken for them, particularly by young women with their first child. But they are carefully to be distinguished from true labor, or the mistake may lead to error and mismanagement. If it is ascertained that the pains are false, the woman, if she be feverish, with a full pulse, and hot skin, should be put to rest in bed; if costive, the bowels should be moved by a gentle purgative, assisted by an injection u.° the bowels, of warm water, and perspiration to be promoted by drinking frequent draughts of weak tea. By such means false pains will generally be removed; but if they still continue after the fever has been moderated, and the bowels opened, a teaspoonful of paregoric, or twenty drops of laudanum given (and repeated if necessary, in two hours), assisted by rest and quiet, will seldom fail in suppressing them.

FIRST SIGNS OF LABOR.—Some days before the time, a woman begins to feel the symptoms of her approaching labor, she moves with difficulty, and frequently complains of restlessness and pain in her back and loins. As the period approaches she becomes smaller around the abdomen. Sometimes a diarrhœa comes on, but generally she is rather costive; she perceives some enlargement, relaxation, and a degree of forcing down of the external parts, and frequently a glairy mucus tinged with blood, is discharged; but this latter symptom more frequently comes on after labor has actually begun.

It is often the case that the anxiety, restlessness and uneasiness of this period prompt many women to wish it over; and some are so imprudent as to attempt to shorten it by rough exercise, with a view to bring on their labor. But no conduct can be more faulty or absurd; at any rate, they increase the present uneasiness, and should they succeed in their attempts to precipitate their labor, before nature is properly prepared for it, they will unquestionably render it more tedious, more painful, and more difficult. On the contrary, let them, according to the dictates of nature, give themselves more rest than usual, attend carefully to the state of their bowels, keeping them freely open.

The subsidence of the belly, which denotes the approach of

labor, is caused by the womb beginning to contract at the upper part, and proves, not only that the womb has begun to act, but, that it is prepared to act in a favorable manner. In like manner, the discharge of mucus, and the relaxation and distension of the external parts, show that they are prepared to dilate. The difficulty in urinating is owing to the pressure of the child's head upon the neck of the bladder; the constant desire to pass the urine, to the same pressure upon the body of the bladder; both are favorable symptoms, and indicate a natural presentation of the child.

Diarrhea is sometimes a sign of labor, which, if moderate, is always favorable and should not be interfered with; if profuse, a few drops of laudanum will generally check it.

Costiveness, when it exists, is not only distressing for the present, by increasing heat, restlessness and pain, but may become very inconvenient during labor. If, therefore, a pregnant woman has neglected to pay attention to this circumstance before, she must now take care to remove it, by mild laxatives, or rather by repeated injections of warm water, which is the best mode at this late period, when all active medicines are improper.

FIRST STAGE OF LABOR.—The first stage of natural labor, which is occupied in opening the internal orifice of the womb, frequently commences with a slight shivering, which, when connected with regular pains, is rather a favorable symptom; but, if succeeded by fever, is unfavorable. But most commonly, labor begins with pain in the back and loins, stretching from thence across the belly, and ending at the upper part of the thighs. It soon leaves the woman free, and returns again periodically, at longer or shorter intervals. These pains, at first, are slight, and return at long intervals; but soon the intervals grow shorter, and the pains become exceedingly sharp and cutting, and, at the same time, the mucus discharge from the outer parts is generally discolored by some blood; after which, the very sharp and cutting sensation commonly abates; and, although the pains grow stronger, return at shorter intervals, and the pressing down increases, they become less distressing, and are borne with less impatience.

On making an examination at this period, (which should be very seldom, and always with the greatest caution, for fear of breaking the membranes,) this variety of the severity and sharpness of the pains, will be found connected with the situation, and to depend on the state of the mouth of the womb. At first, it is found far back, and high up, with edges more thick and hard, and the opening small and hardly perceptible;

as the pains continue, the internal orifice or mouth of the womb descends, and comes forward; the edges become thin and soft, the opening enlarges, and after some time, will admit the finger—a small bag is then felt within, which, during the pain, tightens, and is distended, but, as the pain goes off, becomes loose. As the internal orifice enlarges, this bag passes through, and assists in dilating it, until the thick edges of the orifice, or mouth, being entirely obliterated, the membraneous bag, no longer supported by them, gives way, and the waters are discharged. Sickness and vomiting are frequent and salutary symptoms of this stage of labor, the nausea contributes to the entire relaxation of the whole system, and the retching adds somewhat to the dilating effects of the pains.

The duration of this first stage of labor is very different in different women, and in the same women at different labors— but, in general, it requires more time with the first child than with those which follow—and in all women, it commonly takes up more time than any other stage of labor. If the membranes burst early, before the labor begins, or very soon after, then, the mouth of the womb, requiring to be dilated by the child's head, the pains in the back are more grinding, and the dilation or opening, in all respects, more tedious and more painful, but still requires only more time, patience, and caution. And, if the neck of the womb descends very low, before the mouth begins to dilate, as it sometimes does, it likewise tends to protract this stage of labor.

Bear in mind, that in *this stage*, no skill or art of the midwife, no exertion of the woman, can in the least contribute either to lessen the severity of the pains or shorten their duration. They are intended by nature to accomplish a necessary and important object, that is, the complete opening of the internal orifice of the womb, which, from a hard ring of some considerable thickness, and generally close shut, is to be softened, relaxed, and worn away, until it is entirely obliterated, and so astonishingly enlarged, as to permit the child to pass through.

Nevertheless, if we can during this stage neither lessen the woman's pain, nor shorten its duration; and, although we are forbid interfering in any manner with the progress of natural labor, yet the presence of a midwife, or such other woman as may be selected to officiate, is far from being useless, but is very necessary. They should inquire into the state of the patient's bowels, and, unless they be in a laxative state, administer an injection; indeed, whenever there is time for it, it is a good rule always to do this, as by its effects the injection has in all respects a tendency to promote and render labor easy, and is particularly useful in the case of a first child. She must likewise pay attention to the evacuation of the urine—direct her patient to discharge it frequently, and if she fail in one posture to try another; sitting over warm water, or lying on either side, on her back with her hips raised; on her knees with the head low. Should all these efforts fail, the midwife should, while the patient is lying on her back, her head and shoulders lower than her hips, and during the remission of pain, introduce a finger under the pubes, or front part, and endeavor to raise the child's head a little from its pressure on the neck of the bladder; and while so raised let the patient make effort to void urine; and if notwithstanding these efforts, a total suppression should take place, she must be relieved early in labor by the catheter. For, as labor advances, the difficulty of discharging the urine will increase—the pain of the distended bladder may become so great, as to intercept and suspend those of labor—and the bladder being over distended, may lose its power of contracting ever after; and may inflame, and bring on fever, convulsions, mortification and death. Such are the evils which may follow NEGLECT—but which can very generally be avoided by proper care. But, we would here remark, that the above suppression must not be mistaken and confounded with that suppression, which takes place towards the conclusion of the labor, when the child shall have descended *low down*, the pains being forcing with considerable effort, and there is reason to hope that the labor will soon be completed—this latter suppression will soon be relieved by delivery.

After such attentions as before mentioned, the labor is to be suffered to go on without any interference—the pains continuing gradually to open the internal orifice of the womb, and force the membranes through it, in the form of a purse, which, acting as a soft wedge, contributes in the easiest way to its further dilation. Of this, the midwife may now and then assure herself by examination. Although the finger may be introduced on the accession of a pain, no accurate examination must be made until it remits, lest the membranes should be burst, and the waters let out before the internal orifice be fully dilated, which accident always protracts labor, and renders it more painful and more difficult. Do not make too frequent examination; it is injurious, by removing the mucus from the soft parts, intended by nature for *lubricating* them, and causes irritation.

The abominable practice of stretching the soft parts of the

mother by the midwife, under the idea of making room for the child to pass, is preposterous and cruel. It is impossible to censure this idle, indecent, and dangerous practice too severely; it is always wrong, nor can there be any one period in any labor, the most easy and natural, the most tedious and difficult, in which it can be of the least use, and in which it will not unavoidably do great mischief.

Therefore, leaving nature to her own unassisted and undisturbed efforts, the midwife is to encourage her patient, by appearing perfectly calm and easy herself, without hurry or assumed importance—by assuring her, that as far as can now be discovered, all matters are natural; by encouraging cheerful conversation with those around, permitting her to walk about the chamber, or from room to room, to sit or lie down, as she finds most agreeable, and if she can, to sleep between the pains—but although inclined to it, she should not lie constant, until the mouth of the womb shall be completely opened. The midwife should excite the hopes and confidence of the patient, by manifesting her own ease and firm belief in the natural progress and happy termination of the labor; hope and confidence will very much tend to give regularity and strength to the pains, whilst on the contrary, fear and despondency will tend to disturb and protract the labor. Let there be no "hobgoblin" stories, "ghost tales," nor superstitious nonsense in the room of a woman in confinement.

Partaking of food at the proper time during labor, is allowable, but it should always be light and sparing.

There is great impropriety in directing the patient, *at this period*, to assist her pains, as it is called, by holding her breath, and exerting her strength; by forcing, straining, and bearing down—which inevitably will exhaust and waste her strength, now in the beginning of labor, which may be necessary for her support at the conclusion. Young women, with their first labor, are most apt, from impatience, to be guilty of this error, by which they necessarily overheat themselves, and may bring on a fever; it may likewise occasion a premature bursting of the membranes.

A still more dangerous practice is that of giving strong aromatic teas, cordial, and spiritous liquors, with a view to strengthen the pains; but which can only increase the resistance to their proper effect, by heating the patient, bringing on fever, and checking the natural secretions; on the contrary, let the patient's food, if she take any, consist of cooling fruits, thin gruel, and weak broths; and her drink, lemonade, apple water, weak tea, or what is still better, fresh water. In sum mer, let her chamber be kept cool by open doors and windows : and in winter, comfortable but not too warm.

The patient should be strongly impressed with the fact, that the best state of mind she can be in at the time of labor, is that of submission to the necessities of her situation ; that those who are *most patient*, actually *suffer the least*—that, if they are resigned to their pains, it is impossible for them to do wrong—and that attention is far more frequently required to *prevent* hurry, than to forward a labor. Instead, therefore, of despairing, and thinking they are abandoned in the hour of their distress, all women should believe, and find comfort in the reflection, that they are at those times under the peculiar care of Providence—and that their safety in child-birth is ensured by more numerous and powerful resources than under any other circumstances, though to appearances less dangerous.

The practice of gathering many attendants, is certainly very improper, and is one from which serious inconvenience has resulted ; yet, the presence of a few elderly women is of very great service. If their attendance is requested, it becomes them to consider for what purpose—most assuredly, to assist in an important business, in which the welfare of one of their sex is at stake ; and not by indulging in idle gossip, and magnified reports and misrepresentations of unfortunate cases, and perhaps such as never occurred, to overwhelm the sick women in gloom, fear, and despondency. All questioning and unnecessary inquiry, all smothered and mysterious conversations, should be carefully abstained from, as highly injurious. On the other hand, all levity of conduct, and unfeeling mirth, should be as strictly guarded against, as having an equally injurious tendency, by creating excitement in the mind of the patient, owing to its annoyance.

The first stage of labor is thus to be passed, now and then cautiously examining its progress ; in doing which, when the internal orifice is sufficiently open to admit the finger, the head of the child may frequently be felt and distinguished by its regular shape, smoothness, and hardness, through the membranes ; and may be made another source of consolation and encouragement to the patient, by assuring her of it. But we must be cautious how we predict a speedy termination of the labor, for many circumstances which we can not at the time discover, may concur to deceive us, and nothing will tend more to render the patient anxious, and robs us of her confidence, than disappointment in this respect.

It will now be proper to arrange the patient's dress and bed

A flannel petticoat, and short-gown, with the shift turned up over
the hips, so as to preserve it dry ; but any dress will do if it be
not too cumbersome, and the under side of which can be doubled up under her hips as she lies on the bed.   On the middle
of the bed lay a quilt or blanket, four double ; over this the
sheet, doubled back toward the head of the bed.   Near the
foot of the bed, on that place where the patient will lie when
she is being delivered, put down upon the under bed a folded
sheet with one end hanging over, so that the midwife may take
the edge of it on her lap ; over this latter sheet, a blanket or
quilt doubled and redoubled, but not hanging over.   A piece
of thin oil-cloth under all these, is also advisable, to keep all
moisture, etc., from getting on the bed.   This arrangement
will be found very convenient during labor, and by means of
which the patient, after she has been delivered, may be made
dry and comfortable with very little fatigue—it being a matter
of very considerable importance to have the woman put to bed
dry.   If another bed is intended for her lying-in, it should be
prepared as was first above directed, but without the sheet
being reflected upward ; and after delivery, she is to be *lifted*
from one to the other, *without being suffered to rise up.*

SECOND STAGE OF LABOR.—This stage of labor commences
with a full and complete dilatation or opening of the internal
orifice of the womb, and is ended when the child's head has
sunk through the brim of the pelvis (or bones) so low as to begin
to rest upon and distend the soft parts of the mother.   These circumstances can be certainly known only by examination ; but
there is, likewise, a remarkable change in the patient's manner
of expressing them.   Whilst the internal orifice of the womb
is opening, the pains are cutting, sharp, and grinding—the
patient is restless, bears them with impatience, and expresses
her sense of them by sharp and shrill cries ; but when this is
accomplished, or nearly so, the pains become more supportable,
and the patient finds herself instinctively called upon to make
some voluntary exertion.   She lies quiet, holds her breath, and
expresses her sense of pain in a grave tone of voice, or frequently bears them in silence.

About the commencement of this stage of labor, as previously stated, the membranes frequently break, and the
water is discharged.   This in well-formed women, especially
such as have borne several children, is generally a period of
some little alarm ; as, when the child is small, the head falls
almost by its own gravity through the pelvis, and delivery
succeeds immediately ; for this circumstance the midwife
should always be prepared ; and for some time previous at

least, the patient should be *laid on her bed*, that at all events, the necessary assistance may be afforded, and that no accident may happen from hurry, confusion, or mismanagement. But more frequently this stage of labor takes up a longer time; and although in a well-formed woman and a small child, it may end in a few minutes after the perfect opening of the internal orifice, in others it may sometimes require many hours.

Of course the duration of this stage of labor depends on the proportion which exists between the size of the child's head and the openings of the bones, or upon some irregularity in its shape, or some awkwardness in the presentation of the head—circumstances which nature, *when left to herself*, most frequently will vary, so as wonderfully to adapt one to the other in every stage and progress of the labor. The imperfect hardening of the bones in the child, and the loose manner in which they are connected by membranes, is the provision which nature has made for overcoming these difficulties. Where the head is large and the pelvis or bony passage narrow, the bones ride over one another as the head is forced through the brim, and the shape becomes more oval and pointed, entering the brim of the pelvis or outer bony passage, with one ear towards the lowest portion of the back-bone, and the other toward the pubes or front part, that is, with the narrowest part of the head to the narrowest part of the mother; it turns as it descends, where it finds most room, until the face is brought into the hollow of the sacrum, or lower part of the back-bone, and the vertex, or smallest and most pointed part of the head, to the external orifice.

Our great care in this stage of labor, especially when it proves tedious, is to regulate our patient's conduct, to soothe her sufferings, to calm her fears, and above all things, make her to avoid *fatigue* by over exertion. Although the woman feels some disposition to voluntary efforts, she is not to be encouraged to exert herself during her pains, more than she can well avoid.

The bursting of the membranes, likewise, is a circumstance of great uncertainty : it most frequently happens at the end of the first stage, or during the second stage ; but it sometimes occurs with the first pain, sometimes many days, or even weeks before the commencement of labor—at other times, after having, in the form of a distended sack, contributed to dilate the internal orifice of the womb, they continue in the same manner to dilate the vagina, or canal leading to the outer parts, and external orifice ; and now and then are expelled entire, covering the child's head with the placenta (afterbirth), and water but this is a circumstance by no means to be wished, and ought not to be permitted, as it may be followed by a dangerous flooding, or by an inversion of the womb. Whenever therefore, the bag appears at the external parts, it ought to be ruptured, and the water let out, which is now easily done by keeping the finger hard against them, or by placing the end of the finger firmly on the presenting part of the child covered by the membranes, in the absence of the pain, and keeping it in contact during the succeeding pain, which will seldom fail to rupture the membranes; for that point on which the finger is kept will have to bear the whole of the pressure of the water urged upon it by the effort of the womb. During this stage of labor women are less inclined to move than during the first stage; still they are not to be confined to one posture, but indulged, and permitted to rise from the bed, to walk about, and endure some pains by leaning over the back of a chair, supported by their friends, or kneeling at the side of the bed.

It is just at this time that the impatience and apprehensions of the patient are frequently much excited, the pains return at short intervals, and are strong and bearing; she longs and hopes, and strives for a speedy termination, and it requires much prudence and no little management to check her impatience, at the same time that we support her hopes. She may be assured of her safety, but must be informed that much of that will depend upon her proper conduct, and longer time, and above all things, the midwife must not appear too busy in any apparent or real efforts to shorten it.

THIRD STAGE OF LABOR.—This begins at the time when the head of the child, having sunk through the pelvis, or bones, begins to rest on, and distends the soft parts of the mother, at which time the vertex, or central top of the head, presents at the external parts, and the forehead and face occupy the hollow of the sacrum, or curved lower part of the back-bone, and continues until the perineum, or space between the privates and the fundament, being stretched and distended into the form of a large protuberant tumor, the external parts is so far dilated as to suffer the head and body of the child to pass through without injury. The pains during this period, whilst the perineum and soft parts are undergoing so great distension, become more severe, and at last, when the child's head is passing the external parts, are most exquisite. But they always are least when the labor has been suffered to go on from the first with little or no interference, and much more excruciating and dangerous, when these tender parts have been fretted and inflamed by improper conduct at the beginning.

The part which is most apt to suffer during this period, is the perineum, or space between the privates and the fundament, which, from the extent of one inch, or an inch and a half, and thickness of the hand, is stretched to that of four or five inches, and reduced to the thinness of paper so that in the most natural and well-conducted labor, it will sometimes give way at this extremely thin edge. The perineum and adjoining parts are relaxed, and prepared for so great a change by the secretion of a large quantity of mucus, by which the parts are softened and a disposition to yield and stretch, is given to them, at the same time that they are lubricated by it, so as to suffer the child to slide through them ; and whenever there happens to be a deficiency of this mucus, or when, by improper handling, it has been rubbed off, and its secretion checked, or when a violent and sudden labor does not allow sufficient time for this secretion to take place and give to those parts a proper disposition to dilate, the perineum is apt to be torn, always an unfortunate accident, and one which, if extensive, subjects the woman to great misery and inconvenience during the rest of her life.

It should be your great care in a natural labor, to prevent this, and to which attention should be directed, from the very commencement of labor to the complete delivery of the patient. With this view, avoid irritating these extremely tender parts by too frequent and unnecessary examination, or rude and preposterous attempts to stretch and extend them, as well as avoid heating the patient by improper diet, cordials, and spiritous liquors *in the commencement.*

During this stage of labor, a woman becomes less inclined and less able to move, and the delivery may be expected to be accomplished in a short time. She is, therefore, now, or rather before, if she was not there already, to be laid on her bed in a proper posture for delivery ; that is, on the folded blanket or quilt, at the foot of the bed, on her back, her clothes being turned up under her, and she laid near the edge of the bed, and her knees moderately drawn up and her feet supported against the bed post, or against some person sitting on the bed, taking care not to draw her legs too much up toward the belly, nor to separate the knees very wide ; both of which put the perineum on the stretch, and increase the danger of its being torn.

The bed being prepared, and the woman laid on it as directed, the midwife is to sit herself behind on a low chair, taking the end of the folded sheet, which had been laid across the bed, on her knees ; she will then find herself most conveniently placed to afford every assistance. Still, however she has nothing to do, and it may require some time before she will perceive the perineum sufficiently distended, and the external parts so far dilated that the crown of the child's head shall begin, during each pain, to protrude. She is then to take into her left hand a soft linen or cotton cloth, which, being several times folded, and placing it over the tumor, with her fingers extended, and the palm over the perineum, make a *gentle* pressure on the tumor during each pain, so as in some measure to retard the sudden advance of the child's head, or rather, to be ready to retard it, when a violent pain shall threaten too sudden a delivery; for, let it again be recollected that in a slow labor, well managed from the beginning, where the soft parts are properly prepared to yield, the perineum never is torn, and that all the danger of this unfortunate accident arises from a sudden and violent labor, or one that has been mismanaged in the beginning, or the patient throwing herself beyond the reach and support of the midwife.

Any considerable resistance is seldom necessary; but as the child's head passes through the external orifice, it is always proper, whilst one hand is kept in the position just now described, to place the fingers and thumb of the other hand collected together, upon the protruding part of the child's head; in this position, the midwife has it in her power to make such resistance with her hand as the rapidity with which it advances may require; and to make it on the head, rather than on the perineum, the dilatation of which by too great pressure may be prevented, and the perineum itself bruised. Experience alone can teach the degree of resistance required, and until the midwife has acquired this experience, she must exercise her judgment, and be cautious not to make more than is necessary, for by too much, mischief may also be done, and where the labor has been well conducted from the beginning much is seldom required.

The direction in which the pressure on the perineum is made, is likewise of some consequence; it must not be directed upward and backward, but it must be directed upward and forward toward the pubes, or front; by which, as the centre of the head advances under the arch of that bone, the nape of the child's neck will be pressed up against it, and immediately relieve the perineum. When it is perceived towards the last, that from early mismanagement, or from any other cause, these parts are not properly prepared to dilate, and there is a dryness and rigidity about them, we are directed to anoint the parts with hog's lard or sweet oil, once or twice, and over that, to apply a soft flannel wrung out of warm water; this, we have reason to believe, is of advantage, as it promotes the secretion of the mucus, and otherwise relaxes the parts.

At this moment the sufferings are at the highest—extremely severe, and sometimes almost beyond endurance; and, in hopes of shortening their continuance, she is often inclined, and too frequently called on to exert her utmost strength. Still, however, her present safety and future comfort may very much depend upon submission, patience, and *gaining a little more time;* and all extraordinary exertion beyond what she is in some measure compelled to make, is hazardous. Therefore, do not discourage her crying out, but encourage her to speak, by asking questions, which will check the bearing down efforts, and gain a little more time, which may be necessary, not only to the safety of the perineum, but to the preservation of her life.

Just before the birth, the head is often found to advance during the pains, and to retire again as they remit; and this alternate advancing and retiring is frequently of much consequence to the perineum, by the strength and elasticity of which it is generally occasioned, and which, after some little time, will thereby become perfectly relaxed, and easily distended. If, however, it should lodge upon the chin of the child, by introducing a finger within, between the pains, it may easily be slipped over it, and the next pain the head will generally be delivered; but this must not be attempted whilst the perineum is tense, nor during a pain, but only when it is relaxed and easily distended. As the head emerges from the external parts, it should be received by the midwife on the extended fingers and palm of the hand, stretching the fingers round, so as to examine whether the mouth and nose be covered by any of the membranes, or any part of the clothing, by which its breathing might be obstructed, still keeping the hand in contact with the perineum, and under the child's chin.

A short respite ensues when the head is delivered; but the pains soon returning, the shoulders of the child are perceived as they descend, to make the same turns as the head had done before; and after a pain or two, they are delivered, the one toward the abdomen, and the other toward the back of the mother. While they are passing the external orifice, the same attention is to be paid to the perineum which the passage of the head rendered necessary, as the distension is rather greater, and the danger of laceration not diminished, which, in fact, has frequently taken place at this period, through neglect of the midwife, induced by the supposition that the danger was all over as soon as the head was delivered. The next pain advances it to the hips, so that the arms of the child are delivered with little or no assistance. Another short respite again takes place, whilst the hips of the child advance, of which one or two pains effect the delivery, and the birth of the child is accomplished.

In this gradual way, time is allowed for the regular contractions of the womb, from the top, around the afterbirth, pressing it down before it; whereas, in a more sudden delivery, when the head, shoulders, and body of the child are ejected by a single pain, the womb may, and frequently does, contract from its sides, expelling the child, but retaining the afterbirth high up in the top. The midwife should, therefore, never, as it is too frequently done, take hold of the child's head, and drag it forth : a most dangerous practice—generally the cause of severe after-pains, and frequently the cause of much worse consequences, as a ruptured perineum, retained afterbirth, flooding, inversion (turning inside out), or falling of the womb, by which many women have *lost their lives.*

At the time of the child's emerging from the external parts, the extreme suffering of women will impel them to implore anxiously for relief; her friends also will call upon the midwife and expect her to assist in the delivery, and not unfrequently medical aid is sent for at this stage when it is utterly impossible for the physician to arrive before the case has terminated, unless he be very convenient; but no entreaties should cause the midwife to deviate from correct principles, and make her endeavor to hasten the expulsion of the head; after that event there is not so much inducement. Should there, however, be a considerable interval betwixt the expulsion of the head and of new pains, she may press gently on the belly, or cause gentle pressure to be made by some assistant. Or she may gently insinuate the finger into the arm pit, and slightly pull or agitate the child, so as to excite the womb to contract—even this assistance is rarely required. But on no account is she to attempt the delivery by pulling the head.

The birth of the child is always followed by the discharge of what water had been retained in the womb, frequently by some clots of blood, and generally by some fresh blood flowing from those parts of the womb from which the afterbirth has been wholly, or in part detached.

Fourth Stage of Labor.—After the child has been delivered it should be laid in an easy posture on its side, a little inclined towards the back, and close to the mother; its head and body should be covered, with its face and mouth exposed to the air

The mother at this time is generally in great heat, and wet with perspiration ; and in consequence of her exertions suddenly ceasing, and being much fatigued, she is very apt to suffer from a chill, often followed by disease, which, if not dangerous, will yet delay her recovery; some additional covering should, therefore, be carefully spread over her, without causing a current of air. The doors and windows should now be shut, if the weather be cool, and some dry clothes slipped under, should the part of the bed on which she is lying, be wet: whilst, however, these directions are being executed by the nurse and other assistants, our attention should be directed to the child.

DUTIES RELATIVE TO THE CHILD.—Our first object is to ascertain whether breathing be established. Should this be fully established, and the pulsation or beating in the cord have ceased, the midwife may at once separate it from the mother., This is done by applying two strings, one about two inches from the umbilicus or navel of the child, the other about one inch further and cut between them ; this should not be done under the bedclothes, but uncovered, that she may see distinctly what she is doing, for fear of cutting some part of the child, which has happened with careless midwives. Sometimes a finger or a toe has been thus cut off. The child should not be separated from the mother, until the pulsation in the cord has ceased. Any strong piece of string, tape or ribbon, will answer, and it should be tied sufficiently tight to prevent the blood from escaping. But if the child fail to cry, or breathe freely soon after delivery, our attention should be directed to ascertain the cause—and to the use of proper means for inducing breathing. In determining the treatment necessary to be pursued, the first thing to be ascertained, is, whether circulation is still going on in the cord; which is done by squeezing it between the thumb and finger, to see whether it continues to beat or pulsate. In cases in which this is going on, we need in general apprehend no great danger, as long as it continues; some cold water or spirits should be dashed on the child's breast; if this does not succeed, the little finger, surrounded by a piece of fine rag, should be made use of, to remove any mucus that may have collected in the mouth or throat of the infant. If we have reason to infer that the impediment is situated in the wind-pipe, the hips and the body of the child should be elevated higher than the head with the mouth downward. at the same time gently shaking, so as to assist the mucus in flowing out of the mouth. As soon as this takes place, the child generally cries lustily. But, should the child be still-born, and the pulsation in the cord have ceased, it should immediately be seperated from the mother, as above directed, as no good can arise from suffering it to remain any longer in connection with her. In these cases, after having carefully removed any mucus that may have collected in the mouth or throat as formerly directed, we should endeavor to imitate natural breathing by placing the mouth immediately in contact with that of the child, and forcibly breathing into it, whilst, at the same time, the nostrils are held, to prevent the air from passing through them. Should we be successful in forcing air into the chest, which will be known by the elevation of it, and should it not be immediately returned, gentle pressure must be made upon the breast of the child, so as to cause its expulsion—occasionally elevating the hips and body to permit any mucus which may have collected in the throat or wind-pipe to flow out. In addition to this, warm applications should be made, by applying heated clothes to the child, with gentle friction upon the chest. Should breathing not be immediately established by this method of proceeding, it should not be at once relinquished, for in many cases, perseverance in properly directed efforts will prove successful.

After the child has been separated from the mother, according to the above directions, it should at once be delivered over to an assistant, who should proceed to wash it immediately, but if no such assistance be present, it may be wrapped up in a warm, soft cloth, (flannel is preferable,) leaving an opening, only sufficient to admit the air necessary for its breathing, and laid on its *side* in a warm place, until the mother be safely put to bed.

THE DELIVERY OF THE AFTER-BIRTH.—This should also be left to the efforts of nature, unless some deviation call for our interference. If the delivery has been properly conducted, the womb will have contracted successively, upon the body, hips and lower extremities of the child—so that by the time they are delivered, it will be only sufficient to contain the after birth; and the succeeding contractions not only contribute to separate it from the womb, but press it out into the passage and from thence it is delivered, and thus prevents any serious hemorrhage. Some blood is generally discharged from the womb after the birth of the child, and always after the delivery of the after-birth. A small quantity, therefore, to the amount of even a pound, is no reason for alarm. A short interval, however, generally happens after the birth of the child, before the contractions of the womb are manifested by actual pains, which time is required in the care of the child.

No interference is necessary, if strong pains come on in half an hour. The contractions of the womb will throw the after-birth out of the passage, which is the very best security against either flooding or after-pains. But, if within that time, there be no contractile force exerted by the womb, the mid-wife should take the cord in the left hand, and pass a finger of the right hand up along the passage ; if she can, with the end of the finger, reach that part of the after-birth to which the cord is attached, she may rest satisfied that all is safe ; and that the womb has begun to contract and throw it off; but if, on the contrary, she cannot reach the root of the string, the after-birth is probably still attached to the womb. Under these circumstances, she should place her open hand upon the patient's belly, and if she find the womb soft and resting on the lower side, she must take it in the hollow of her hand, and raise it towards the middle of the belly, press it moderately and rub the surface of the belly over gently, and change the patient's posture from side to back or from back to side ; she will then very probably soon perceive the womb to contract, by its assuming the form of a ball of considerable firmness. She may now again take the cord in her left hand, putting it just so much on the stretch, as to prevent the after-birth (which descends a little during inspiration) from ascending again during expiration, still holding the end in her hand, pass the fingers of the other hand as high as possible toward the root of the cord, and pull gently while at the same time, an assistant should make gentle pressure upon the belly with the open hand, by which the womb is frequently excited to make vigorous contractions. When the after-birth shall have descended into the passage, and is somewhat protruding from the external parts, it is best then to leave it entirely to the expulsive efforts of the womb.

The most common cause of delay in the delivery of the after-birh is this want of expulsive pains; and this inaction of the womb is a very common consequence of fatigue after a severe or tedious labor, especially if mismanaged. But this weakness, so far from being a reason for haste and precipitancy, is a most powerful argument for waiting, and making no attempts to separate and extract the after-birth; a hasty delivery of which, before the womb has begun to contract with some degree of vigor, will expose the patient to great danger of a flooding or inversion of the womb. Under such circumstances, therefore, our efforts must be directed to compose the patient's mind, and attempt to excite the action of the womb as previously directed.

Great care should be taken by the mid-wife how she exerts any considerable force on the cord, which in some instances is small; in others, inserted by several branches into the after-birth, and easily torn from it—at all times an inconvenience, and on some occasions a very serious accident. Or, if the string should be so strong as to endure much force, more terrible accidents may follow; the after-birth may be torn from its attachment to the womb, of which a violent flooding will be the consequence, or the womb may be in part or wholly inverted, turned in side out and actually brought out of the body, which has been the unhappy consequence of imprudent force applied to the cord. Let it, therefore, be an invariable rule, never to tighten the cord, and put it on the stretch, until the womb can be felt like a hard lump or ball under the hand applied to the woman's belly; on the same principle, coughing, sneezing, or blowing into the hands, and every such exertion of the woman, are likewise improper as they tend to cause flooding. Should the after-birth however be retained or flooding ensue, directions for the treatment of them will be found under those different heads when treating of the dangers of difficult or tedious labor. The after-birth being delivered, carry the finger into the passage up along the membranes, which are continued from the edge of the after-birth, and slowly and cautiously assist them in coming away; for if they are left, they cause after-pains, and in a few days a very offensive smell.

PUTTING TO BED.—The external parts should be annointed with lard or sweet oil, and a soft cloth applied; and lift her to the upper part of the bed, having previously brought down the reflected part of the sheet; while she is being lifted let some assistant bring down her clothes which had been for security tucked up over her hips; if she has lain in a petticoat, it ought now to be removed, and the clothes from above supply its place. When laid in her place she should be covered with bed-clothes, more or less, as she may be inclined to be chilly. A towel, four or five double, is then to be applied to the belly, and over that a broad bandage round the waist, is pinned so low, as to take in the bottom of the belly, and afford some support to its loose and relaxed sides, but not so tight as to give the least pain or uneasiness. The use of such a bandage is evident, but a *twisted hand-kerchief*, applied in the form of a cord, and drawn down tight, as is frequently done, (to keep down the mother, as the women express it,) is the very extreme of absurdity and must do harm, yet, when a *broad* bandage is skillfully applied, as here directed, it will be found very agreeable and very comfortable by the woman, and has a tendency to prevent and relieve that faintness, which is sometimes very alarming to women, recently delivered. We would observe, that with regard to putting to bed, if the patient be much exhausted and fainty, it had better not be attempted to move her before some hours, only change her posture so as to extend her limbs, putting dry clothes under her; give her some nourishment until she is somewhat recruited. For immediate nourishment we would advise as the best, a cup of coffee, with plenty of milk, and as much sugar as may be agreeable, or chocolate, either with a few mouthfuls of bread and butter, or rusk or boiled milk with some stale bread in it, yet, if the exhaustion be *very great*, a small portion of wine and water, or a tea-spoonful of brandy, with some fresh water, may be allowed, or panada, with the same quantity of wine or brandy in it; but unless the feebleness be *very great*, we would advise the abstinence from all and every intoxicating liquor.

AFTER-PAINS.—Very few women but who will be troubled with after-pains, harassing them, and disturbing that repose so necessary to their comfort. Those pains, however, very much depend for their severity, upon the manner in which the after-birth has been delivered, yet it must be considered that they sometimes follow the best conducted and most natural deliveries. They come on soon after delivery, resembling in some manner those of labor, returning in paroxysms, though with longer intervals between them, throwing off during their action whatever lumps of blood may remain in the cavity of the womb. If these are in a moderate degree, and not of very frequent recurrence, they demand no remedy—but, more frequently, they are very excruciating, and therefore call for remedies.

Opiates, are the surest means to be relied on; two tea-spoonfulls of Paregoric, or twenty drops of Laudanum, with thirty of Sp. Camphor, or thirty drops of Laudanum by itself: either of these, given in fresh water, repeated in half doses for a few times at intervals of an hour or so, will seldom fail to give relief.— There will be some who cannot use Laudanum ; for such, use fomentations of Hops and Vinegar or Camphor to the bowels, occasionally, just warm enough to be agreeable.

A horizontal position, laying down in bed, should be observed for nine days or more, and a low diet is always necessary. Every indication of pain and soreness, should be early attended to; directions for which will be hereafter prescribed. If the woman should have no evacuation from her bowels within three days, one should be procured by the use of Castor Oil, Sedlitz Powders, or injections of warm water up the bowels And, if within ten or twelve hours, she passes no urine, she should be solicited to do so, and, if necessary, be aided by fomentation of clothes wrung out of warm water. The following will be found very beneficial: Spirits of Nitre, four table-spoonfulls ; Laudanum, half a tea-spoonfull, forty drops of which, may be given every hour in water until relief be obtained: but should these also fail after a fair trial, recourse must be had to the catheter.

WASHING THE CHILD.—Children when born are frequently

coated with a tenacious unctious substance, wnich is somewhat difficult to remove. It has been found by experience, that rubbing the child over with hog's lard, until it becomes completely incorporated and mixed with this substance, and then making use of soft dry flannel to remove it, is the most simple, expeditious and perfect manner of getting rid of it. Afterwards, warm soap-suds may be used. The process of washing should be performed, so as not to unnecessarily expose the child to the influence of cold ; if in the winter it should be done near the fire. It may also be well enough here, to insist upon the person having charge of this office, to use as *much despatch* as is compatible with the proper execution of her duty, as it is to be feared, much injury is frequently done the child by too long exposure. After washing, it should be carefully dried. The highly injudicious and culpable practice which some nurses have, of using spirits to bathe the body and head, especially the latter, should be strictly prohibited, as it tends to carry off the heat by evaporation. Be careful not to expose the child to a strong light or to get any soap in its eyes. After having performed the necessary cleansing and washing of the child, the next thing to be attended to, is, dressing the naval. Nothing more is necessary, than after examining if it be properly secured, to pass the remainder of the cord through a hole made in the centre of a piece of linen or cotton rag, with the extremity of the cord towards the breast, fold the cloth over it so as to envelope it, and secure it by a bandage about two inches broad, pinned round the belly. Be careful *not* to pin the bandage *too tight.*

DRESSING THE CHILD.—Let its clothes be put on loose, as every thing like lacing, or tight clothing, prevents the proper exercise of the lungs, impedes the circulation,—and sometimes entails impaired *health*, and a bad *shape for* life. Nothing is needed on the head except the weather is very cold. The rest of the child's clothing should consist of a shift, and a wrapper of fine flannel, with a diaper. All children cry when shifted and dressed, therefore the more short and simple the process can be, the better. Also, be careful of pins, always using those with elastic and protected points.

FIRST OPERATIONS FROM THE BOWELS.—The bowels of all newly born infants are loaded with a mattter, technically called the *meconium*. It has been found, that it conduces much to the health and comfort of the child to carry off this substance, and its retention has even been known to give rise to fatal diseases. But dangerous as its presence may be, it should not lead to the administration of *active* and *violent* purges to remove it , for, there is no practice fraught with worse and more dangerous consequences than this. The mother's milk is the best purge for an infant, or a little molasses and water. If this, however, should not be sufficient to produce the desired effect, the child generally shows symptoms of being ill at ease ; will become sleepy ; frequently starting up ; moan and cry loudly. As soon as we find this to be the case, and especially if the stools still possess a tenacious and greenish appearance, resort must be had to a teaspoonful of Caster oil, administered warm, and repeated in four hours, should the first not produce the necessary evacuations. A mild injection of warm milk and water, with some molasses dissolved in it, will also assist much; as soon as the evacuations assume their proper color and consistence, which might be compared to tolerably thick mustard, all purgative medicines should be immediately relinquished. We have now described the process of an ordinary natural labor, and any woman endowed with common understanding, may, by a careful study of these remarks, easily qualify herself to give every necessary assistance in ordinary cases of natural labor, and distinguish most of those of difficulty in time to have a physician sent for.

DISEASES AFTER DELIVERY.

Flooding.-If after the woman is put to bed, she loses too much blood it will manifest itself by a faintness; it will be necessary to observe that, if she has been over heated by too much clothing, or the temperature of the room, this should be remedied by the removal of some of the clothes, and fresh air freely admitted, cold water should be freely and repeatedly sprinkled on her face, a cold hand rubbed over her belly, a towel or napkin several times doubled and wet with cold water applied to the belly confined by a broad bandage, and to be renewed if it becomes warm; frequent drinks of cold water must be given her; if she has repeated vomiting, a dose of soda powders after every motion, will have a good effect; if these should fail, and the case become alarming, a lump of alum, about the size and shape of a hen's egg, and a nick cut around the middle, so that a tape can be securely tied round it, which is then to be pushed (enclosed in the hand) into the womb, and left there for sometime; and when it has accomplished the restraining of the flooding, it should be s'owly withdrawn by pulling the tape. The applications of cold have their limits: they must be discontinued when they have accomplished the restraining of the hemorrhage ; or, if they fail, and are productive of continued chills, it may then be supposed that all the advantages to be expected from them are obtained; they must, therefore, then be omitted. If the system does not show signs of returning life, and the patient continues to become

colder, and appears in imminent danger of dying, stimulants
are then administered in small quantities and often repeated, as
the urgency of the symptoms may require. Wine or brandy,
diluted with water and made warm, will answer; a julep made
of the yolk of eggs and warm wine, or part of brandy or whisky, diluted with three parts of water, sweetened with loaf sugar, flavored with a few drops of Ess. Cinnamon, is also an excellent cordial
If there be pain, fifty drops of Laudanum, or teaspoonful of Paregoric, may occasionally be added to the stimulents; but the
exhibition of these must cease as the patient shows signs of
the return of life. Above all, she is not to be disturbed, or
raised to an erect posture, but with perfect quiet—the small portion of life is to be carefully husbanded; for there is often a
power of living in a quiescent state, or when laying down, when
the patient would be destroyed by the least exertion, or by being raised to an erect position; she ought, therefore, not to be
raised or even *moved*, before she is quite revived; and then,
only with the utmost care. Persons have suddenly and unexpectedly died through want of attention to this matter. And
when immediate danger is no longer appr-hended, the flooding
ceased, it will still not be prudent to replenish the emptied vessels *too hastily* by high living, for, by so doing, the blood vessels may be easily again stimulated to immoderate action, and
the hemorrhage renewed.

FAINTING.—Sometimes this comes on immediately after delivery; but, more frequently, not before an hour after. Some
person ought, therefore, have the special observance of the patient during that time, as the fainting may come on suddenly
and unexpectedly; if it be caused by flooding, the method to be
pursued has been considered in the preceding paragraph: but if
it proceed from other causes which we are perhaps unable satisfactorily to assign, a tea-spoonfull of Paregoric, two tea spoonfuls of Bateman's Drops, twenty drops of Ether or Spirits of
Camphor, or either of these in some fresh water, together with
fresh air, and forcibly sprinkling fresh water on the face, will
generally soon be found to afford relief; frequently the fit will
be terminated by vomiting. Fainting appears in some manner
connected with the sudden evacuation of the contents of the
abdomen; as a preventive from this cause, the bandage should
be applied; this should, in all cases of fainting after delivery,
be examined, and if it has moved, so as not to give the necessary
support, it should be rectified and tightened.

INFLAMMATION AND SORENESS OF THE EXTERNAL PARTS —
After delivery there is sometimes much inflammation and swelling of the external parts: by washing and bathing with warm
milk and water, and annointing with fresh lard, it will in general

go off in a few days; but if it continue to get worse, so as to require attention, the parts are to be several times a day anointed with the following cerate: Beeswax 4 ounces, Lard 6 ounces, add to this, Sugar of Lead 20 grains, dissolved in two table spoonfuls of vinegar, melted over a fire, and simmer for half an hour; strain, and stir until cool. If the parts be hot and tender the following poultice will be found very useful:—one handfull of Hops with three of Wheat Bran, pour over them boiling water sufficient to make them into a poultice, which wrap in a piece of fine muslin, and apply it warm over the parts after putting on the above cerate.

MILK FEVER.—The secretion of milk is usually accompanied with a slight fever, often amounting to a considerable degree of inflammatory action, preceded by shivering, and going off with perspiration; it is in general more severe, and of longer continuance with the first than with subsequent children. If properly managed, it will seldom continue longer than twenty-four hours; during its continuance the breasts are full, hard, and painful, which distinguishes this from fever. A tea-spoonful of Spirits of Nitre, in a gill of toast water, should be given every two hours; balm, sage, mint, or elder flower tea should be freely drank about luke-warm, and a poultice of bread and milk freely spread with lard and applied warm to the breasts.

SORE NIPPLES.—These are the immediate result of some inflammatory state of the system. The inflammation impeding the evacuation of the milk, the suction of the child peels the outer skin from the nipple,—the inner skin cracks, from which blood is discharged. As a means of *prevention*, when the important and highly interesting duties of a mother are about to devolve on her, she should, during the last two months of her pregnancy, have her nipples drawn out by some other person or by breast-pump—at first, very gently and but once a-day; and, as she approaches towards the completion of her time, more force should be used, (but at no time so much as to cause pain,) and the frequency of the operation increased to three times a day.

After each suction the nipple should be washed with cold water, and exposed for a few minutes to the air. As soon as the mother gets settled in bed and rested, the child should be put to the breasts. When the nipples become sore or tender a Poultice of Flaxseed meal anointed with sweet oil or Lard, should be kept to the nipple all the time except when necessary to remove it long enough for the child to nurse. This is better than any other application. When the nipple feels tender on the child sucking it, the attention should immediately be directed to the state of the system ; if costiveness prevail, to have the bowels moved; and if fever prevail either general, or local in the breast, then to live on low diet.   Always attend to gently drawing out the nipple for the child before putting it to the breast, and immediately after the child has finished its suction, to wash the nipple with cool Sassafras or Sage Tea.   By care and attention to this treatment on the occurrence of sore nipples, they will in general be relieved.

SORE BREASTS.—Sore nipples, and their consequent distention from milk, are perhaps the most fruitful source of inflammation of the breast, and also the most uncontrollable kind: those which arise from cold or from that feverish state called the weed, will generally be more under the control of remedies. Sometimes, a chill will precede affections of the breast; at other times, a painful swelling, without a chill.   In either case, fever is soon excited—pain and swelling increase rapidly.   There are two varieties: one is confined to the cellular or spongy substance between the skin and the breast, and is soon brought to an issue; the other is within the substance of the breast, and is more slow in its progress, and frequently renders the breast, ever after, useless—which the first variety, of itself never does.

The patient should be put to bed, and under the most rigid restrictions of diet—allowed no kind of animal food, nor any kind of spices, or stimulants; toasted bread, and water only should be allowed unless the patient has been previously much reduced; then bread and milk, mush and milk, tea or coffee, may be taken.   Her drink should be water, or cool tea of tamarinds, apples, cherries, peaches, or cream of tartar whey: she should be kept in a room not too warm, and a purge of Castor oil, Rochelle or Epsom salts, be given so as to open the bowels freely.   Local bleeding, by cupping or leaching, on the body, near the circular margin of the breast, will answer a good purpose.   For a local application to the breast, take hot vinegar, pour it over some hops, let it stand for a few minutes, strain out the hops, soak fine linen or muslin cloths in it, and apply them frequently warm to the breast.   They will be found particularly useful when the breast becomes much distended with milk, and cannot be drawn: it will diminish the secretion, and relieve the swelling.   The breast, however, should be drawn as long as it can be done, always washing the nipple clean before the child is put to it.

This treatment is to be continued through the whole course, until we have no more hopes of a resolution or scattering the hardness, and suppuration or coming to a head is considered inevitable, then the further reduction of the system, by purgatives and low diet, will be unnecessary.   At this stage a plaster of honey, lard and flour, is among the best applications. If we have reason to believe that matter is formed, and is ready for its exit, then a small poultice of bread and milk, smeared with lard, should be applied, slightly warm, to that part where we think it will burst or have to be opened. As soon as there appears a small, elevated, soft, rather dark spot, which elevation is easily indented with the finger, but which also quickly reappears upon the withdrawal of the finger and gives the sensation of having a fluid enclosed; that should be punctured with a lancet, and the matter discharged. The bread and milk poultice should then be again applied: after a few days, it may be dressed with basilicon ointment, or some other cerate, until the part be healed. If there remain a hard lump in the breast, it should be rubbed with camphorated oil, opodeldoc, or volatile liniment, keeping the parts covered with fine flannel.

EPHEMERAL **FEVER, OR** WEED.—This is a fever of common occurrence to lying-in women. It is usually of short duration, the paroxysm being completed generally within twenty-four hours, and always within forty-eight. It consists of a cold, a hot, and a sweating stage; but if care be not taken, the paroxysm is apt to return, and we either have a distinct intermitting fever established, or sometimes, from the co-operation of additional causes, a continued and a very troublesome fever is produced. It is generally caused by exposure to cold, irregularities of diet, fatigue, want of rest, &c. It is ushered in by a shivering fit, accompanied by pains in the back. When the cold stage has continued for some time, the hot one commences, and this ends in profuse perspiration, which either carries off the disease completely, or procures great remission of the symptoms. The head is usually pained, often intensely, especially over the eyes, in the first two stages, and in some instances accompanied by a slight delirium. The thirst is considerable, the stomach generally oppressed with wind, and the bowels bound. The pulse, until the third stage has somewhat advanced, is extremely rapid. In the cold stage, we give frequent small quantities of warm tea, such as Virginia snakeroot, balm, **mint,** or sage; and apply a bladder filled with warm water, or a dry **warm** flannel, to the stomach and back, and something warm to the feet ; by this, we shorten the cold stage, and hasten on the hot stage. When the chilliness is gone off, and the hot stage is fully established, we then gradually remove the warm application and lessen the quantity of bed-clothes. We now also, in the place of warm drink, give cooler, about lukewarm, such as toast water, **lemonade or** apple water. If the heat of the body be **very** great, and the thirst distressing, soda powders will be found to be very grateful; but the water in which the powders are dissolved should be previously made about lukewarm: these may be repeated every half hour during the continuance of the hot stage. When we find the heat of the system considerably diminished, the pain and restlessness much abated, together with other symptoms of of perspiration about the breast, we then add some covering, and again resort to the free use of the teas, *fresh made*, and given *warm*, but *not hot* ; keeping the patient perfectly quiet, in a state of gentle perspiration, for the space of five or six hours. We then refrain from the use of the teas; and, when the process is over, the patient is to be cautiously shifted, the clothes being previously well dried and warmed; and, if she have an inclination, let her have something to eat. During the whole course of the paroxysm, we must carefully guard against the *sudden application of cold*—it renews the shivering and prolongs the disease; but, at the same time, we must also avoid *too much* heat. A comfortable room, with a moderate quantity of bed-clothes, is what we are to have for the patient. Do not give purgatives until the sweating stage is over, for fear of giving the patient fresh cold, but then a dose of Castor oil or Epsom Salts may be given.— By the foregoing treatment we shall frequently be able to confine the disease to one paroxysm; it however will in some cases return at irregular periods; if the intermission become longer it is favorable, but if they become shorter it is unfavorable; if it does return the same treatment must again be pursued.

LOCHIA AND ITS DERANGEMENTS.—We mean by the term *lochia* those bloody discharges from the womb after delivery, which continue for some days, becoming greenish, and lastly, pale, then decrease in quantity, and disappear altogether within a month, and often in a shorter time. The variableness of this evacuation should prevent it from becoming an object of very great solicitude; for it differs very much in different women, and in the same woman at different confinements. If the quantity discharged is small, therefore, or its entirely ceasing to flow at a very early period, need not create any degree of alarm, if the woman be, otherwise, in as good a condition as may be expected from her situation; and, no irritating or propelling medicines should be used : they cannot do any good, and may be productive of many evils. In those cases in which no very obvious cause can be assigned for the derangement, and which nevertheless appear to be producing injurious effects, some mild purgative should be administered, and, after its operation, some weak Cammomile or Sage tea : these, with occasionally sitting up, will frequently restore the discharge.

SWELLED OR MILK LEG OF LYING-IN WOMEN.

During or after confinement, women are liable to a swelling of the legs, called *milk leg;* its first symptoms are great pain and difficulty in moving the leg. The disease does not appear to be connected with any peculiarity of constitution or preceding complaint, nor the kind of preceding labor, or on the treatment, before or after child-birth. It occurs at any period from the first or second day, to two or three weeks after delivery; it is preceded by general uneasiness, lowness of spirits, slight pains about the womb, with a discharge from it peculiarly offensive. These symptoms seldom command much attention, until the patient is seized with pain on the inside of the limb, commonly about the calf of the leg, which soon extends from the heel to the groin, along the course of the vessels called absorbents. The limb soon after begins to swell : the soreness extends all over it, so that it cannot bear the slightest touch, and every attempt to move gives exquisite pain ; the skin becomes glossy and pale, the countenance is expressive of great anguish and dejection, the pulse is quick, the heat of the skin increased, the tongue white, and the urine muddy. These symptoms strongly mark the presence of some irritating matter ; and no doubt it is in the womb. The prevention must depend on cleansing the birth-place, by injecting water so as to enter and cleanse it: also injecting powdered charcoal and water (one tea-spoonfull to the pint of cold water) up the birth-place three or four times a day. The cure of this complaint is often tedious ; sometimes the other leg takes on the disease, as the first subsides. The bowels should be opened by some good purgatives once in every two days The leg may be rubbed with a mixture in equal parts of Sweet oil, Laudanum, and Spirits of Camphor, frequently through the day, and poultices of Bread or Flaxseed meal applied during the night to the groin and upper part of the leg. When the soreness has somewhat subsided, bathe the leg in a mixture of half a pint of Whiskey and one quart of cold water, night and morning, while at the same time the leg must be tightly bandaged, beginning at the toes and going up entirely to the groin, the bandaging to be renewed, night and morning, as the swelling decreases. Keep the affected leg elevated on pillows, higher than the body, and give a wine glass of Tea of Wild Cherry bark, Dog wood bark, or Culumbo root, before each meal, when the patient is getting better.

VARIOUS DISEASES.

PIMPLES OR WORMS IN THE FACE.

This is an affection usually met with in young persons of both sexes. It is characterized by small, more or less, red pimples, which penetrate the tissue of the skin to a greater or less depth, and is slow in coming to a head. There are several varieties mentioned by dermatologists; ACNE SIMPLEX, which appears on the forehead, face and shoulders, the pustule in the form of small, hard, red elevations, inflamed at the lower portion, in which pus forms, which is thin and mixed with a thick sebaceous (sticky) matter. They dry off, leaving a dark red raised mark. When the pustules are mixed with a number of black circular points, which are the orifices of follices (or bags) filled with sebaceous matter, and are often converted into pustules, it is what is known as *acne punctata* or maggot pimple—worms in the skin. These follices are the habitation of a small parasitic insect, called *Acarus folliculorum*. If not attended to, they are liable to spread and accumulate, and disfigure the face similar to marks from a mild form of small pox. If let alone two or three weeks, the tops of the pustule become yellow, break and suffer a yellowish pus to escape, and, by pressure, a kind of " core" looking substance is forced out. It mostly appears on the face, but may extend to the back, and become very severe and troublesome. A similar trouble is often produced by the continued use of intoxicating drinks, and known by the name of "copper nose," "grog nose," "grog-blossoms," &c. As this disease is usually produced by too rich food, the remedy consists in living on *less meat* and more *vegetable food*, taking once or twice a week a Seidlitz powder, and washing the face two or three times a day in Bay rum, Cologne water, or, if nothing better, common whiskey.

SHINGLES.

Shingles is usually situated near the waist, surrounding one-half of the trunk of the body, like a zone or belt : it may however, extend in other directions over the trunk ; it is always situated on one side, and that, generally, the right. The eruption of shingles is generally preceded by symptoms of general indisposition, and especially by severe darting pain in the parts where it is about to appear. At first red patches show themselves at the extremities of the site of future eruption, and gradually become more numerous till they form a line—upon these patches, shining points form, which gradually enlarge into vesicles, or blisters, a little under the size of small peas, these vesicles containing a clear fluid, which gradually becomes thick and cloudy in appearance. At length in the course of eight or ten days, the vesicles burst, discharge and **dry** off in the form of scabs, or, it may be, in very weak subjects, leaves sores or ulcerations. The belief was formerly entertained among physicians, **and** still retains its hold of the popular mind, that if the **belt of** the eruption of shingles was continued *round the body,* **so as to** meet, the disease proved fatal. This is perfectly erroneous. When the disease occurs in the young and robust, the diet must be reduced to milk and bread, or vegetables, and all sources of heat and excitement avoided. Five grains of blue pill at night, followed by a dose of senna or Seidlitz powder in the morning, may be repeated once or twice; and, in the course of **the** disease, if there is much fever, five grains each of the carbonate and nitrate of potash, may be taken twice or three times **a** day, dissolved in half a tumblerfull of water; or the proportion of carbonate of potash may be doubled, and a teaspoonful of lemon-juice used to form the effervescing draught. The painful itching of shingles often causes much distress. It may sometimes be allayed by simply keeping the eruption covered with a cloth soaked in tepid water, or by using the common lead lotion, (half a teaspoonfull of sugar of lead to a cup of water) in the same way. Pencilling the shingly eruption with a strong solution of lunar caustic (ten grains to the ounce of water,) as recommended in erysipelas, is found to relieve the severe pain.— Applying Tincture of Iodine with a Camel's **hair** brush, or soft linen mop, once a day, **is** also highly recommended by medical men.

When shingles occurs in the aged and debilitated, instead **of** the diet being reduced, it requires, perhaps, to be improved; at all events, the system must be sustained with nourishing broths, and with one grain of quinine three times a day.

SMALL POX.

The small pox attacks people of all ages; but the young of **both** sexes are more liable to it than those who are much advanced in life; and it may prevail at all seasons of the year; but in general is most prevalent in the spring and summer. It very seldom happens that a person is attacked a second time with the disease, however afterwards exposed to its infection. The disease is divided into two kinds—the *distinct* and *confluent.* In the distinct, the eruptions are *quite separate* from each other,

but in the confluent kind, they *run much into one another.*—
The distinct may be distinguished from the confluent, before
the eruption appears, by the mildness of its attack, by the inflammatory state of the fever, and by the late appearance of the
eruption.

The disease commences with shivering and languor, followed
by heat, thirst, and headache; there is usually either pain or great
oppression at the pit of the stomach, and not unfrequently
vomiting; there is severe pain in the back or loins, an l in
children not uncommonly, and more rarely in adults, convulsions
On the *third day* after the setting in of the above symptoms, usually toward evening, minute red spots, somewhat resembling flea
bites, show themselves on the forehead, the neck, the wrists,
and arms, the chest and abdomen, and finally on the extremities: this, is the course of the eruption, but it does not reach.
the lower extremities till at least the *fourth day.* If the eruption on the parts first mentioned is discovered over night, by
morning it is much more distinct, and the spots are much more
numerous than they first appeared to be; they are, too, slightly elevated—from this they continue enlarging; on the third
day after their appearance, they contain a little fluid on their
summits, which gradually increases in quantity. Towards the
fifth or sixth day, they contain pus or matter. About the seventh or eighth day of the eruption they begin to "crust," that
is, to break, allowing their contents to escape, and then to harden into a crust or scale. At this period of the disease, that
of "maturation," the eighth day of the eruption, the eleventh
of the disease, what is called the secondary fever comes on; the
fever, which had more or less abated after the eruption appeared, becomes again aggravated, and continues so for a few days.
At length, if the disease has progressed favorably, toward the
end of the third week from the first showing of the eruption,
some of the scabs begin to separate and fall off, leaving either a
pit or a stain of a deep red color.

Such are some of the most prominent characteristics of small-
pox.

This disease is to be treated, by avoiding every thing of a
heating, inflammatory nature; and by keeping the subject of it
in a cool, quiescent state. The diet should be of the vegetable
and mildest kind; and the drinks of a similar kind, made agreeable by the addition of the most palatable acids. The bowels
are to be kept open by a table spoonfull of Rochelle salts, or
Epsom salts, in a glass of cold water, once every two or three
days, or a Seidlitz powder will answer; and, above all, the pa.
tient is to have cool and pure air—never oppressed by clothing

or a heated room. The temperature of the chamber should always be such that he may experience no disagreeable degree of heat; but rather a sensation of cold: and, except he complains of being chilly, no fear need be entertained of carrying the cooling regimen too far. His bed should be a mattrass covered only with a few bed clothes.

But although the bowels should be kept open throughout the disease, *when the eruption is coming out* all attempts at purging should be dropped, cooling drinks, such as lemonade, toast water, &c. being given. If the surface is very hard and dry, sponging with tepid water is very useful and agreeable. When the secondary fever comes on, it may be requisite to act more freely on the bowels by means of the purgatives already mentioned, while at the same time the distressing restlessness requires opiates; twenty to forty drops of Laudanum or a tea-spoonfull or two of Paregoric, may be given at bed time. In some cases, if signs of sinking come on, with weak pulse, tardy eruptions, and pustules not filling, all lowering measures are to be avoided, and good broths, wine, wine whey, &c. administered, as the case may require. The principles of treatment are, in the onset of the disease to moderate the fever and through it the eruption, by cooling purgatives, (such as mentioned above,) when the eruption is coming out, to interfere but little beyond keeping the bowels easy, regulating the diet according to the strength; and, again, in the stage of secondary fever, to purge moderately.

If much swelling and distress about the throat should result in the course of the disease, leeches ought to be applied, in number proportioned to the age and constitution of the patient.— This treatment employed in a case far distant from medical aid might save life. To prevent being "Poc-marked" or scarred by the disease, keep the face, neck, hands, &c., well anointed with Sweet oil or Lard, until the inflammatory stage, and fever is passed, and guard against scratching or picking the scabs.

WARTS.

These are enlargements and thickening of the different coats of the skin, or one part growing into or through the other, causing the skin to break, and admitting this unnatural growth to protrude through. They are most common with children, and generally occur on the hands, sometimes on the face. In the latter situation they are better not interfered with. When situated on the hands, they often disappear of themselves.— When their removal is desired, strong acetic acid, or very strong vinegar, applied every two or three days, is the best remedy. Nitric acid is sometimes applied, or caustic, with the same beneficial effect. The juice of the green rind of the common black walnut, applied once a day for a week or two, will usually remove warts. When warts have a narrow neck, a horse hair or silk string, tied tight around them near the skin, will soon make them fall off. When a wart on the face, especially in those advanced in life, appears inclined to become ulcerated, cr irritated, it ought not to be interfered with, but show it to a physician.

CORNS.

Tight shoes are one of the most frequent causes of corns; they are often troublesome to females and others who are particularly attentive to appearance, and who wish to exhibit a neat and small foot, by compressing it in a shoe of too narrow dimensions.

Corns sometimes exist without giving much pain or trouble; but in other cases, they give so much uneasiness, as absolutely to incapacitate for walking. They are made more particularly intolerable, by every thing that quickens the circulation, or which heats the feet, or causes the corn to press on the neighboring parts. Tight shoes, much walking, warm weather, heating liquors, all tend to render the uneasiness of corns very great, and they are generally worse in summer than in winter; and persons are frequently obliged to sit down to take off their shoes, and rest the foot in a horizontal posture.

Corns may often be readily cured, by avoiding the above exciting causes, by wearing large soft shoes, adapted in form to the shape of the feet, and by continuing for some time at rest. It is useful to take a considerable number of folds of linen, covered with some softening ointment, cut a hole in the middle for the corn to lie in, and to apply them to the foot; and if it be on the sole of the foot, it may be useful to have an additional moveable sole, with a hole cut in it in like manner. If, along with this mechanical and palliative treatment, we use the following method, a corn will be easily and quickly eradicated : it is to be touched with lunar caustic, and wrapped round with adhesive plaster ; and generally at the end of a fortnight, the dead skin will be removed, with the corn adherent to it. If the corn does not come away, the operation is to be repeated. Several other remedies of the same kind, are recommended, of which the principal are, soap plasters or mercurial plasters, or blistering ointment. The following plan may also be tried: every night and morning the foot is to be put into warm water for half an hour, while there, the corn is to be well rubbed with soap. All the soft white out side of the corn is afterwards to be scraped off with a blunt knife, or what is better, with a piece of pumice-stone; but we must not persist in this scraping, if there is pain in any part of **it**. This treatment is to be continued without intermission till the corn is totally eradicated, which it will be in about a fortnight. Strong vinegar, applied to corns, after bathing in warm water, once a day, is also a good remedy. **It** is generally a difficult and painful operation to cut out a corn. Unless it be completely taken out, it is **apt to grow** again, and **this** it does faster than if **it** had been **let alone. In** old people, it is highly dangerous to cut **a corn, as this** fre quently **excites** an inflammation, and **consequent mortification** which carry off the patient unless carefully **treated.**

Bunions.—This is the **result of** chronic inflammation **of** the Mucous Bag **(or** *Bursa*) which is situated over the front **of** the great toe, **and is generally caused** by tight shoes. It ought to be **attended** to **at once;** one or two leeches, warm fomentations **of** hops and hot water with vinegar and a poultice used to allay irritation, and the offending shoe being at once discarded. **A** wrong position of the bones at the joint is a frequent attendant, and, perhaps, an antecedent **cause** of bunion. When **the** disease is fully formed, **the best plan is to avoid, by the make** of the shoe, &c. &c., **every source of irritant pressure.** [Bathing often in salt and water is advisable.

BARBER'S ITCH.

This is an eruption **of inflamed but not very hard tubercles** (or pimples) appearing **on** the hairy **parts of the face**—the chin, upper lip and whiskers—sometimes **in the** eyebrows **or** the neck, and on the scalp. It usually clusters together **in** irregular patches, with the hair passing through the little elevated points. The pimples **are** of a pale yellow **color, and in a** few days **they** burst ; **matter then runs** out forming a hard brownish **crust,** which **fall off** in **a few days,** leaving pimples of **a** purple color, which are **slow in healing.**

Before the appearance **of the eruption, there is heat, pain and** a tingling sensation in the parts.

In the *treatment* avoid the *use of a razor altogether,* live **on** a **low** diet, and bathe the parts in a solution **of sugar of lead** in water, [one Teaspoonful] to the half-pint **of** water] three or four times a day. Give a mild purgative of **Rochelle Salts, a** dose of Senna or a Seidlitz Powder, about **twice a** week. Also wash the parts *thoroughly* in soap and **water twice a** week. Also take one gill of Tea made from Sarsaparilla **root** before each meal.

INFLAMMATION OF THE BRAIN.

Its characteristics are violent fever, severe pain in the head, redness of the face and eyes, great intolerance of light and sound, watchfulness, and delirium. It is usually preceded by long continued watching, pains in the neck and crown of the head, defect of memory, diminution of urine, and irregular pulse. As the disease advances, the eyes sparkle greatly, there is ferocity in the countenance, restlessness, deafness, ravings, and increased pulsation in the arteries of the neck and temples. The tongue is dry and of a yellow or black color, the face of a deep red, and the pulse becomes small, quick and hard. It is always a most alarming disease, and often terminates fatally about the third or seventh day. It is produced by all causes which tend to excite apoplexy or fullness of the blood vessels of the head,—such as exposure to the hot sun, &c.

The patient should be bled from the arm freely ; and it ought to be done, if practicable, while he is sitting up. Bleeding by cupping the temples, and by the application of several leeches, should not be neglected if the symptoms are violent. Cold applications of ice, or iced water to the head are to be made and renewed frequently. Powerful purgatives are to be administered : (Jalap and Cream Tartar, of each one or two teaspoonsfull, in a gill of cold water,) and injections of ten or fifteen grains of tartar emetic in half-pint of warm water should be given daily so long as the symptoms continue violent. The patient's head should be kept as elevated as possible, to lessen the determination of blood to it : and the same effect will be produced by partially scalding the feet, or by blistering the arms and legs—but these are only to be applied after the violent symptoms are reduced. When the fever subsides, and the mind returns to reason, it will be very necessary to observe the utmost caution respecting all exciting causes : as when the inflammation has once been excited, slight causes bring it on again.

The diet and drinks are to be of the mildest kind. Light should be excluded, and indeed every thing which can excite the system, particularly a hot room and foul air. A shower bath, or cold water poured on the head, every morning, after the patient's recovery, will go far to prevent a return of the disease. The bowels should be kept open, by all means, to prevent a termination of blood to the head.

DELIRIUM TREMENS.

Usually this disease is the result of the excessive and continued use of intoxicating liquors, though it may also be produced by opium. The first symptom of this disease is a state of restless irritation, and if the exciting cause be continued, sleeplessness follows; there is no rest, and if there is any approach to sleep, it is haunted by dreams and imaginary figures that excite the greatest dread. The nature of the disease is, unhappily, in almost all cases too palpable, from its exciting cause. It *is* an exhausted condition of the brain and nervous system ; and the great effort must be to alleviate this exhaustion, which is too great even for sleep. Opium is *the* remedy among others, and must be given in full doses. A medical man, will, of course, give it more freely at once than another person ; but in a confirmed case of delirium tremens, thirty drops of laudanum, should be given at once, and ten drops every hour afterward, until sleep has been procured. Often it happens that the stomach is in so irritable a condition that it will retain neither food nor medicine ; in such a case the opium is better given solid, in the form of pill, one grain and a half at first, and half a grain repeated at hour intervals, if requisite. If the stomach is still irritable, a drop of creasote, in a little spirit and water, may be given, and a mustard-plaster applied to the pit of the stomach. In cases of delirium tremens, the liver is more or less affected and a purgative with opium had better be given. Five grains of powdered opium, ten grains of calomel and twenty of compound colocynth pill, are to be compounded together and divided into twelve pills ; of these, two or three should be given for the first dose, and one at intervals of an hour between each, till six have been given. Under this treatment, after sleep has continued for some time, the bowels are generally acted upon, with immense discharge of dark, black-looking bile, much to the relief of the patient. After this, the remaining pills may be given, two every night, and castor-oil in the morning, if required ; five, ten, or fifteen-drop doses of laudanum, or two teaspoonful doses of paregoric, being given, if the nervous irritation is unsubdued, or threatens to return. After the nervous irritation has tolerably well subsided, the next object must be to restore the tone of the stomach Eight-grain doses of the carbonates of soda or potash combined with a bitter tonic, as columbo, gentian, or chamomile tea, may be given for this purpose, every eight hours ; or one grain of quinine every three hours in water will be found useful

During the whole treatment, it has been customary to allow the unfortunate subject of the disease a *certain regulated* portion of alcoholic stimulant, such as brandy and water, in some degree proportioned to the previous habits ; but it is much the better way to give one grain of quinine every two or three hours, and, as soon as the stomach will bear it, the nourishment of strong meat-broths, yolk of raw egg, beat up with boiling water and lemonade should be given. If the tongue is very red at the tip, and if the pit of the stomach is very tender, milk should be substituted for the above ; fifteen drops of the solution of carbonate of potash, or one or two tablespoonfuls of fluid magnesia or of lime-water, may be added to the milk with advantage. In cases of persistent sickness, soda water, and ice given in small fragments, frequently repeated, are often useful. The reception of nourishment by the system is of the highest importance in this disease : so much indeed, is this the case that as long as a man continues to take food freely, he is not likely to become the subject of delirium tremens. One point never to be lost sight of in this disease, is that the stomach must not become empty—even cold water, to keep down thirst should be given if food will not agree with the stomach—mustard-plasters to the spine and stomach are often of great advantage.

APOPLEXY OR APOPLECTIC FITS.

This is characterized by a sudden diminution of all the senses, and the patient falls down. The heart and arteries—unlike in fainting, continue to perform their functions. The peculiar breathing and profound apparent sleep, distinguish it from an attack of palsy : and the *absence of convulsions* makes the difference between it and *Epileptic Fits.*

It chiefly attacks in advanced life and those of short necks and large heads—of full habit of body, free eaters, and great drinkers of ardent spirits. The immediate cause of these fits, is a compression of the brain : often occasioned by the bursting of a blood vessel within ; sometimes from the sympathy of the brain with the stomach—as in case of persons drunk or under the operation of opium or other poisons. It is sometimes preceded by giddiness, pain in the head, drowsiness, loss of memory, and faltering of the voice : though frequently it occurs suddenly, the person falling down without the least warning.

In cases of apoplexy, the person should have the head elevated : and ice applied to it, or cold water, applied by means of folded cloths and frequently renewed and free air admitted : all bandages or any thing compressing, particularly around the neck—should be removed. In persons of full habit, they should instantly be bled freely: particularly from the temples by cupping and leeches: and it is to be repeated according to circumstances. They should be cupped over the head—and indeed the more they are cupped the better. A large blister should be applied to the neck and shoulders. An injection of one tablespoonful of Epsom Salts dissolved in a pint of lukewarm water, with three grains of tartar emetic, should be given. Emetic or vomits of ipecac after other evacuations have been found serviceable. Mustard, or water nearly scalding hot, should be applied to the feet, to rouse the system. All who have reason to apprehend this dreadful disease, should live very low, on a vegetable diet, lead a very industrious life, and cause the return of any suppressed evacuation or renewal of sores which have been healed. The bowels, by all means, must be kept open, and the head bathed in cold water by means of pouring, or a shower bath, every morning, and never let the hair grow long.

PARALYSIS, OR PALSY.

This is produced by the same causes which produce apoplexy: by suppression of evacuations: by constant handling of lead and inhaling the fumes of poisonous metals: and by sedentary and luxurious living.

When it takes place in persons of full habits, as in apoplexy, free evacuations by local bleeding, with cupping and leeches to the temples, and purges of castor oil, salts, seidlitz powders, or some good purgative pill are necessary. Electricity and galvanism have proved serviceable in chronic cases. When the disease affects several different parts of the body, it is customary to use stimulants internally, and externally to the affected part. Those most used are mustard, horse-radish, garlic, hartshorn, ether, and oil of turpentine, in their ordinary doses, and to be frequently changed the one for the other. The parts affected with the palsy may be rubbed with the volatile linament, (a mixture of hartshorn and sweet oil,) oils of turpentine, and sassafras, red pepper, and Spanish flies in spirit, powdered mustard, in short, anything may be used that will irritate the skin. In cases attended with loss of appetite and great weakness, give a tea made of wild cherry bark, columbo root, gentian root, or boneset, (cold,) three times a day.

Palsy of the lower extremities frequently arises from a disease in the backbone: and the most successful treatment, is to keep issues or blisters constantly discharging from the surface of the part where the disease commences. Children are most subject to this disease. Sometimes it comes on suddenly, and at others is preceded by a sense of feebleness, languor, and numbness in the extremities—occasional stumbling, and dragging the legs, instead of lifting them properly. If parents would early pay attention to such symptoms in their children, and have blisters applied at once to the back bone, where there seems on feeling to be a little tenderness, they would save many children from deformity for life. However, in many cases, it is caused in children by worms, colds, &c., and in such cases can be removed by a few doses of castor oil, and giving the child some warm boneset tea, to make it sweat freely—taking care that it does not take cold after it.

LEAD PALSY occurs in those who have long been exposed to the influence of the poison, and the majority of those attacked have suffered from lead colic. The attack is preceded by lassitude and a feeling of numbness, and by stiffness of the parts about to be affected, the loss of power gradually coming on. In a few cases, loss of feeling is also observed. Lead palsy is not confined to the hand and arm, but affects other portions of the body, although the former is its most frequent site. The most dangerous form of this disease affects the muscles of respiration, [breathing] which moves the ribs, and proves quickly fatal. It is well here to give a caution to those who are employed amid lead or its preparations, that they should observe the utmost cleanliness, especially at meals, for there is good reason to believe that the poison often finds its way into the system from carelessness on this point.

SHAKING PALSY, in one form, is generally the result of old age; in another it is more traceable to direct disease of the brain and is very apt to occur in those who have drank freely. It comes on very insidiously, and even under the best care is a very hopeless affection as regards cure. This is a different affection from the "mercurial tremour," with which those work in that metal, such as gilders, are liable to be attacked.

SUN STROKE.

This begins by thirst, dizziness, headache, and sometimes there is vomiting or difficult breathing, The symptoms, in fact are pretty much the same as Apoplexy : the patient should at once be taken into a cool shady place, and the first thing have a bucket of cold water poured slowly over his head, and, in all other respects treat the case the same as a case of Apoplexy, *observing* the cautions therein recommended, after the patient recovers, (Shower-bath to the head every day, &c.)

HYDROPHOBIA, OR BITE OF MAD DOG.

Owing to the frequency of this frightful malady, we shall be as explicit in the description, of symptoms, &c., as possible. After a person has been bitten by a rabid dog, the wound heals in the same manner as an ordinary wound from the same cause would. After an uncertain interval of, say between six weeks and eighteen months—the following symptoms begin to be noticeable : The patient experiences pain, or some uneasy or unnatural sensation, in the situation of the bite. If it has healed up, the scar tingles or aches, or feels cold, or stiff, or numb ; sometimes it becomes visibly red, swelled, or purple. The pain or uneasiness extends from the sore or scar toward the central parts of the body. Very soon after this renewal of local irritation—within a very few hours perhaps, but certainly within a very few days, during which the patient feels ill and uncomfortable—the specific constitutional symptoms begin. He is hurried and irritable ; speaks of pain and stiffness perhaps, about his neck and throat ; unexpectedly, he finds himself unable to swallow fluids, and every attempt to do so brings on a paroxysm of choking and sobbing of a very distressful kind to behold ; and this continues for two or three days, till the patient dies exhausted. Hydrophobia has never been cured when once the decided symptoms have shown themselves.

The disease by the inoculation of which hydrophobia may be produced in man, is common in the dog, and it has been communicated by the fox, the wolf, the jackal, and the cat. It has been produced by the saliva (or spittle) of the human being, the horse, the hen, and the duck.

All animals are susceptible of the disease, when bitten by a mad dog, also fowls. The disease cannot be caused by the saliva of a mad dog getting on the skin, unless there be an abrasion, crack, pimples or sore, but when it gets on the mucus membrane, as of the mouth, nose, &c., it almost always produces its peculiar effects. The *scratch of a cat* or dog will not produce the disease, only on account of those animals getting the *saliva on their paws*, which is very often the case.

It is still more interesting to inquire whether the saliva of a human being laboring under hydrophobia be capable of inoculating another human being with the same complaint ? The disease has undoubtedly been so produced. If this be so, the fact will teach us—not to desert or neglect these unhappy patients, but to minister to their wants with certain precautions, so as not to suffer their saliva to come in contact with any sore or abraded surface ; nor with any mucous surface. On the other hand, all carefulness of that kind will be unnecessary, if the disease cannot be propagated by the human saliva.

"*Is a man who has been bitten by a mad dog, and in whose case no precautions have been taken, a doomed man?* Will he be sure to have the disease, and therefore die of it? By no means. But *few upon the whole of those who are so bitten become affected with hydrophobia.*

When a person has been bitten by a dog or cat suspected to be mad the beast ought by no means *to be killed,* but to be *secured* and *kept under surviellance,* and suffered, if it should so happen, to die of the disease. If he do not die, in other words if he be really not mad, that will soon appear, and the mind of patient will then be relieved from a very painful state of suspense and uncertainty, which might otherwise have haunted him for months or years. Should the dog die mad, the injured person will be no worse off than if the animal had been killed in the first instance; nay, in one respect he will be better off, inasmuch as certainty of evil is preferable to perpetual and uneasy doubt.

There are gross errors prevalent with regard to the signs of madness in the dog. *The mad dog never has fits,* in the commencement, he may have *convulsive struggles when dying.* It is a very common belief that a mad dog, like an hydrophobic man, will *shun water;* and if he take to a river, *that* is thought to be conclusive evidence that he is not mad. But the truth is, that the disease in the quadruped cannot be called hydrophobia; there is no dread of water, but an unquenchable thirst; no spasm attending the effort to swallow, but sometimes in dogs an inability to swallow from paralysis of the muscles about the jaws and throat. They will stand lap, lapping without getting any of the liquid down. They fly eagerly to the water; and Mr. Youatt states that all other quadrupeds, with perhaps an occasional exception in the horse, drink with ease and with increased avidity.

There is another superstitious opinion not at all uncommon, viz. that healthy dogs recognize one that is mad, and fear him and run away from his presence. This is quite unfounded. Equally mistaken are the notions that the mad dog exhales a peculiar and offensive smell, and that he may be known by his running with his tail between his legs; except when weary and exhausted he seeks his home.

The earliest symptoms of madness in the dog, are sullenness, fidgetiness, continual shifting of posture, a steadfast gaze, expressive of suspicion, an earnest licking of some part, on which a scar may generally be found. If the ear be the affected part, the dog is incessantly and violently scratching it. If it be the foot, he gnaws it till the integuments are destroyed. Occasionally vomiting and a depraved appetite are very early noticeable. The dog will pick up and swallow bits of thread or silk from the carpet, hair, straw, and frequently he will lap his own urine, and devour his own excrement. Then the animal becomes irascible, and flies fiercely at strangers, is impatient of correction, seizes the whip or stick, quarrels with his own companions, eagerly hunts and worries cats, demolishes his bed, and, if chained up, makes violent efforts to escape, tearing his kennel to pieces with his teeth. If he be at large, he usually attacks only those dogs that come in his way; but if he be naturally ferocious, he will diligently and perseveringly seek his enemy.

Many cautions are annually put forth about the dog days, for muzzling dogs, and so on. Very good and proper advice; but if those who have noticed the statistics of the disease may be depended upon, it would be as appropriate at one period of the year as at another. Some people think this disorder in dogs is produced in warm weather on account of a want of water, the notion is a mistaken one.

When an individual has been bitten by an animal respecting which the slightest suspicion of hydrophobia exists, the one remedy cannot be to quickly resorted to---*complete excision of the bitten part.* Some persons have possessed sufficient nerve to do this for themselves---few perhaps could---but it has been effected by unprofessional persons for others: indeed, there might be more danger in waiting many hours for a surgeon than in submitting to an unprofessional operation. The method of excision most to be trusted, is the insertion of a skewer of wood, made to fit into the wound caused by the tooth, and carrying the incision so far round, that the entire hollow or cone of flesh is cut out along with the piece of wood. This might be done with safety in the thick part of the calves of the legs or the back part of the thighs, &c. Where excision is not resorted to, the free application of lunar caustic or aquafortis, whichever may be most readily procured, would be advisable; or, in lieu of these, a piece of iron, heated to *whiteness*, may be inserted into the wound, so as thoroughly to destroy the surface which may have been poisoned. In the event of none of the above measures being submitted to, or available, the wound may be *thoroughly washed for hours*, by means of a stream of warm water poured upon it from a height; a cupping-glass being applied at intervals, and of course in the meantime procure the services of a physician. When an individual is thought to be attacked with the hydrophobia, if the hope of saving life is small, much may be done to alleviate so terrible an infliction by proper care. The most perfect quietness possible must be observed to prevent as much as may be the recurrence of the paroxysms of suffering. Thirty or forty drops of laudanum given as the occasion may require, are advisable—if the patient can swallow : if not we must depend on chloroform, which can be *inhaled* without difficulty, until rest is obtained—and repeated as may be needed. If ice can be taken, it is said to afford relief put into the mouth in small morsels ; it has also been found of service applied to the back of the neck.

Anyhow the experiment is worth trying. It has been stated that applying ice in bags or sacks to the backbone, has checked the disease ; when it can be had of course give it a trial.

LOCK-JAW, OR TETANUS.

This is an involuntary and almost constant contraction of the several muscles of the body, while the senses remain perfect. It is called *Lock Jaw* on account of the Jaws being locked together, as it were by the contraction of the muscles. The set of muscles most generally affected, after those of the jaws, are those of the back; the patient, by the spasm, is bent like an arch, so that the back of the head and the heels alone touch the bed ; occasionally the body is bent forward. The disease most frequently commences with a sensation of stiffness and soreness of the muscles of the neck and jaws ; the latter become fixed, and the spasm extends more or less over the body. *This* extensive cramp is attended with the most severe pain, which is also, in most cases, experienced severely about the pit of the stomach, being dependent doubtless on the spasm of the diaphragm.

The most usual exciting causes of lock-jaw or tetanus are wounds, especially of a punctured character, but in some persons the very slightest injury is sufficient to develope the disease. It is liable to prevail among the wounded after battles, if exposed to much bad weather; indeed cold will occasionally give rise to lock-jaw independent of injury. When lock-jaw arises from a wound, it shows itself in from four days to three weeks after the injury. It is a very fatal disease, the greater portion of those affected by it dying; some, however, recover.

In the treatment of lock-jaw, begin by giving large doses of Laudanum, say, from thirty to sixty drops if it can possibly be swallowed, and repeated at intervals of from half an hour to an hour, as long as the system remains unaffected by the drug:

if the medicine cannot be given by the mouth, it must be by injection. In addition to the above, the affusion with cold water may relieve. The patient having been taken out of bed; and a quantity of cold water dashed over the body and down the spine, is immediately to be rubbed dry and replaced in bed quiet sleep may possibly follow. While the jaws are firmly closed, nourishment cannot of course be given in the usual way; a medical man will probably administer it by means of a tube passed into the stomach, either by the nose or by mouth, passing it behind the teeth; until his arrival, should that be delayed, the administration of small injections of meat-broth will assist in keeping up the strength.

Mustard plasters applied the entire length of the back-bone or spine, until pain and redness is produced, and after their removal, the application of cloths soaked in a mixture of equal parts of Sweet oil, Laudanum and Chloroform, and over them oil silk or writing paper, is a treatment I would by all means advise. Sometimes large doses of Ipecac or Antimonial wine, given until nausea is produced, will relax the muscles; the bowels should first be moved, however, by injections.

STERILITY, OR BARRENNESS.

This, sometimes, proceeds from defective organization. These cases are, however, very rare, and cannot be cured by art. The next general cause is a torpor, and irregular action in the womb and its appendages, which by proper management ought to be cured.

The most important means of rousing the womb, will be found in exciting the breasts to their natural action. The connection between the womb and the breasts, has often been remarked: it is scarcely possible to excite an action in the one, without affecting the other.

The most natural action for the breasts, is the secretion of milk. They have often been excited to the discharge, without pregnancy. A child losing its mother, and sleeping with a female friend, has been known in the night to get the nipple in the mouth, and to excite the flow of milk by the morning, and the child was abundantly nourished afterwards at the breast of the maid. The idea wished to be conveyed is, that to stimulate the womb of a woman who has been barren or unfruitful, it is only necessary to cause a flow of milk for a short time in her breasts by applying a child, (as in the case just cited) and that the stimulus thus brought to bear on the womb and its appendages, will cause the woman to conceive and bear children. I give this not as my own opinion, but as the opinion of justly celebrated medical men. To those who have been barren and who desire the companionship and blessings of children on whom they may lavish their love, and perchance their property, the experiment is worth trying.

CHILLS, OR FEVER AND AGUE.

This disease prevails mostly in the fall of the year, and near lowlands, marshes, and on the water courses of rivers. The disease occurs at stated and very various intervals—either daily, or every second, third or fourth day.

Ague and fever, is generally divided by writers into three stages :—the cold, the hot, and the sweating, and is thus described.—The cold stage commences with a sense of languor, of weakness, and aversion to motion and to food, with frequent yawning and stretching. The face and extremities become pale; the features shrink, as do all parts of the body ; the skin appears constricted, as if it had been exposed to cold. At length, the patient feels very cold, and universal shaking comes on : breathing is small, frequent and anxious ; the urine is almost colorlesss : sensibility is impaired ; the pulse is small, frequent and sometimes irregular.

These symptoms abating after a short time, the second stage commences with an increase of heat over the whole body, redness of the face, dryness of the skin, thirst, pain in the head, throbbing in the temple, anxiety, and restlessness : the breathing becomes more full and free, but still frequent ; the tongue is furred, and the pulse becomes regular, hard and full ; in cases of great severity, delirium is apt to occur.

These symptoms having continued for some time, a moisture breaks out on the forehead, and by degrees becomes a sweat, which gradually extends over the whole body. As this continues to flow, the heat of the body abates, the thirst ceases, the urine deposits a sediment ; breathing is free and full, and most of the functions are restored to their ordinary state ; the patient, however, being left in a state of weakness proportionate to the violence of the preceding attack.

Although this is the description of a common fit of ague and fever, it is subject to great variations in every stage ; depending on as great variety in causes and peculiarity of constitutions. The treatment, whether the disease recurs every day or otherwise, is the same. Our object is to shorten the duration of the fit when it comes on, and to prevent its recurrence.

The treatment when the fit comes on, is, in the cold stage, to take any weak tea as hot as possible; to apply hot applications to the feet ; and to lessen the shaking, it is of service to grasp the limbs very tight, or to apply tight bandages around them to compress the muscles. When this stage subsides, the drinks should be continued, and twenty drops of spirits of nitre in two tablespoonfuls of water should be given hourly, to favor the sweating. If the symptoms run alarmingly high, as is sometimes the case in *congestive chills*, black pepper tea with a little wine, brandy or whisky in it, should be given occasionally, while at the same time we try to produce vomiting by tickling the throat with the finger or a feather, and use friction with hot cloths, &c. During the sweating stage, the patient should not be kept very warm, but not exposed to such a current of air as might endanger the sudden suppression of the perspiration; when it ceases the patient should be wiped dry with a rough towel, have the clothes changed and partake of some suitable nourishment.

Of course the great object in this disease must be to *break the chill*, or prevent its recurrence. For this purpose nothing is equal to *Quinine*. Commence six hours before the expected attacks of the chill, and take from one to three grains of the quinine every two hours, until the time for having the chill is past. Do the same way the next period, or day when a chill is expected. When the chill has been broken up, then every sixth or seventh day afterwards, take the quinine, until the fourth week or twenty eighth day has passed. The quinine may be taken in powder, mixed in a tablespoonful or two of cold water, or in the form of a pill.

The patient should not eat much the day that the chill is expected and must also avoid exposure to the night air, and wear flannel next the skin. Bathing with a sponge and water, (to which a little salt has been added) night and morning, is advisable. As a strengthening bitter tonic take a wineglassful of tea, (cold) made from wild cherry, or dogwood bark, or columbo root, before each meal.

In the most obdurate cases, I have found the above treatment successful, in my practice in Mississippi and Tennessee, as well as since living in the city of New York. One thing should be observed, however, that before giving the quinine, in the very beginning, it is best to purge the bowels freely. Give five to ten grains of blue mass (blue pill) with half a grain of ipecac, at bedtime, to be followed the following morning by a dose of castor oil or a seidlitz powder, and repeat them, if necessary to work off the blue pill. You are then ready to begin with the quinine treatment as above mentioned.

BILIOUS OR REMITTENT FEVER.

This is nothing more than the Chills, or Ague and Fever, ex cept in this form of the disease there are no *intermissions*, (or entire absence of the symptoms,) though there are *remissions* or partial suspension of the fever, followed by something like a chill or coldness of the nose, &c.

The treatment consists in giving Quinine in the same doses as ordered for Ague and Fever, only it should be given every *three hours*, both *day and night*, until the fever is broken up, *and between* each dose of the Quinine give twenty drops of Spirits of Nitre in half a glass of water.

Cold applications to the head (cloths wrung out of cold water or water and vinegar, are about as good as any,) will be beneficial. Allow cool drinks of lemonade and slippery elm-water, or gum arabic water, &c. Give the blue pill, &c., as before referred to, before beginning the use of the Quinine, and during convalescence use the bitter tonics, &c., the same as recommended in the treatment of Ague and Fever; tea-spoonful doses of prepared chalk and powdered charcoal, given once a day in a little slippery elm-water, will be good to correct acidity of the stomach, and offensive discharges from the bowels.

FAINTING FITS.

These are liable to occur at any time in persons who are subject to them, and always create for the time being, considerable excitement, especially when occurring in a crowded assembly or in the public streets, &c. These sometimes come on suddenly, without any visible warning; and, at others, they are preceded by sickness at the stomach—some oppression in breathing— paleness of face, &c. They are characterised by an entire suspension of all the animal powers; which continuing for a short time, they become gradually restored. In rousing the system to action, we are first freely to admit fresh air, exclude all unnecessary attendants, and see that no tight clothing is interfering with the breathing, &c.—especially removing tight-laced jackets, corsets or cravats. Cold water or vinegar should be sprinkled on the face : strong smelling articles should be applied to the nose—as volatile salts (hartshorn) ether, assafœtida, burnt, feathers, &c. A little wine, or spirit, should-be poured into the mouth, and the extremities rubbed with a coarse brush. The direct cause of fainting is diminished circulation of blood through the brain. It must be obvious, that in the endeavors to restore a person who has fainted, this condition must be altered as quickly as possible ; and for this purpose, *the individual should be laid quite flat down, the head on* a *level with body*, so that the feebly-acting heart may not have to propel the blood upward, but horizontally.

After fainting from excessive evacuations, cordials and stimulating diets should be often given. The patient should be laid down and kept at perfect rest, with hot applications to the breast and extremities.

Remember, though, that when the fainting arises from a great loss of blood, it ought not to be stopped suddenly ; [the fainting fit] ; because during such fainting the blood coagulates and the vessels contract—thereby tending to prevent the continuance of the bleeding.

Persons liable to fainting, or indeed to any kind of fits, cannot be too cautious in avoiding what they have found tending to produce them ; they should never be alone ; because of the danger of falling so that respiration cannot be renewed, and consequently ending in death, when others are not at hand to change the position of the body.

In general, persons subject to fits who are of a weakly, delicate nature, will find relief by leading a more energetic life, occasionally using some of the strengthening medicines mentioned under the head of Tonics in other parts of this work.

ST. VITUS' DANCE.

This is mostly a disease of youth, occurring before puberty, and *usually* disappearing at that period of life, if it has continued so long. It may, however, continue into adult life, but rarely proves fatal. The most prominent symptom of St. Vitus' dance is continued involuntary actions of the muscles, to a greater or less degree—the extent of the muscles affected, and the intensity of their affection, varying with the severity of the disease. The movements, however, generally cease entirely during sleep, and in all cases certainly are diminished. The ordinary voluntary movements are still capable of being performed after a fashion ; that is, in an unsteady, uncertain, and somewhat grotesque manner. It seems as if, after the voluntary impulse had been communicated to them, an additional involuntary one interfered to throw the limb or other part, out of the usual steady movement.

Usually the disease commences with twitching about the face or neck, or in a particular limb, gradually extending to one side of the body, or to the whole body, as the case may be. Pain is seldom complained of, but it does sometimes occur in the head

The appetite may remain quite good, but the bowels are possibly confined, and their secretions unhealthy. To this depraved state of the bowels, or to costiveness, or to the presence of worms, the disease is often traceable. In females it is not unfrequently connected with the menstrual function, (monthly turns) especially if it be delayed, or imperfect. The irritation of the coming of the second teeth, has been assigned as a cause ; and there is no doubt that *imitation*, especially among females, may spread the disease, which is most general, as might be expected, in persons of a nervous tendency. The duration of the attack varies from ten days or a fortnight, to months ; but having once existed, it is, up to the age of twelve or fourteen at least, apt to recur occasionally.

A great many cases of this disease get well without any treatment, but it is not best to trust to nature alone. The bowels should be purged with one blue pill (three grains) at bedtime, and followed next morning by a teaspoonfull of Rochelle salts or Epsom salts in half a gill of water, so as to work off the blue pill. Some mild purgative should also be given once a week afterwards, to keep the bowels open; castor oil will answer. In all cases attention to the general health is required. Good diet, exercise, change of air, and attention to the hours of sleep, putting the child to bed at a stated hour every night, and to free ventilation of sleeping rooms, are all circumstances to be kept in mind in such cases. Bathing by means of a sponge, with salt and water every morning, is of great service. Also give one half to one teaspoonful three times a day of the following mixture ;—tincture of Peruvian bark, tincture of · valerian, spirits of lavender, tincture of henbane, or hyoscyamus, of each one once ; spirits of camphor, half an ounce, all mixed together and kept in a closely-corked bottle.

EPILEPTIC FITS.

This disease consists in a sudden deprivation of the senses : accompanied with a violent convulsive motion of the whole body. It attacks by fits, and after a certain time goes off ; leaving the person in his usual state, excepting a sensation of languor and exhaustion.

The fits, or convulsive seizures of epilepsy, are most varied as to the time of their occurrence. Frequently the interval is one of months, but again, daily fits, or even two or three times a day, are the rule, in the worst cases. The attack of epilepsy is for the most part sudden : the individual, in the midst of some accustomed occupation, or while holding active communion with persons around, sud-

denly utters a loud—a fearful—cry, and, if unsupported,
falls to the ground; the eyes are staring or rolling; the head,
or rather chin, is drawn toward one shoulder, the countenance
becomes dark or purple, the veins of the face and temples filled
with blood, and the features are thrown into convulsive movement; there is frothing at the mouth, while a kind of choking
noise is often made in the throat; the limbs are also more or
less convulsed, and the excretions are often expelled involuntarily.
The tongue very often suffers from being bitten, and the teeth
have even been fractured during the fit. Gradually, these convulsive movements diminish, and the person awakes to consciousness, with a heavy stupid look, or falls into a deep leth-
argic sleep, which continues for some hours; but even when
roused from this, there often remains slight temporary suspen-
son of the activity of the brain. Such are the symptoms of a
severe epileptic paroxysm; the disease, however, occurs in much
milder forms, even in those who at other times suffer from it
in greater intensely. A slight temporary unconsciousness
may be the only symptom, with or without the slightest approach to convulsive movement, as evidenced by the twiching of a
finger, the roll of an eye, or slight spasmodic action of the muscles
of the face; the patient may fall gently as in a faint, or remain
standing as it were asleep for a few moments. As there is every
variety in the nature of the attacks, so is there likewise in their
duration; from a few moments to the average period of from
five to eight minutes, but sometimes much longer.

The attack in many cases appears to bystanders to come on
suddenly and without warning, but most epileptic patients are
sensible for some time previously of the approach of the paroxysm, and even for twenty-four hours are aware that a fit is at
least probable, although its direct accession may not be certainly
known until just previous to its occurrence. It may, however,
happen that these symptoms will pass off without a fit, either
independently of any effort of the patient to ward off the attack,
or in consequence of some of those measures found to be effca-
cious, and adopted by epileptic patients for the purpose.

Low spirits, or unusual irritability, sometimes an increased
energy, dizziness, noises in the ears, floating specks before the
eyes, and many other signs connected with disorder of the
nervous system, are the precursors of the epileptic paroxysm. But
the most generally marked and remarked precedent is the epileptic "aura," a sort of creeping sensation, which is described by
the patient as arising at some particular part of the body, such
as the extremity of a limb, and gradually ascending upward to

the trunk or head, till the individual loses his consciousness in the convulsion.

Epileptic seizures are very frequent in the night-time, just as the person is falling asleep ; but they may occur at any period of the twenty-four hours, and may be induced by causes affecting the nervous system ; the excitement of joy or passion, or depression of grief, intoxcation and sexual excesses, are most frequently not only actual exciters, but also predisposers, to the attack of epilepsy.

Epilepsy may be a congenital disease, that is, the child is born with the tendency, and becomes subject to the fits, either with or without any apparent cause, early in life. Intoxication is a cause of epilepsy, and delirium tremens may be complicated with it. Strong and prolonged mental exertion may induce epilepsy. Fright is another and very frequent exciting cause. Worms and irritations in the bowels, indeed whatever can irritate the nervous system, may induce the disease in question. Imitation, or at least witnessing an individual in the epileptic paroxysms, has been known to give rise to the fits in others ; but they were most likely predisposed, or at all events of nervous and susceptible temperament; for this reason, such persons, young females and children especially, should never, if possible, be permitted to witness an epileptic fit. The premonitory cry is so terrifying that it has been known to affect even the lower animals.

When means will allow of it, the epileptic ought to have an attendant constantly with them. When an individual is seized with a fit of epilepsy, but little can be done for its immediate relief ; the chief thing was to prevent the inflicting injury upon himself, by striking against surrounding objects, and also to protect the tongue. Those who are much in attendance upon the epileptic ought always to have on hand a piece of India-rubber, or a thick India-rubber ring—such as is used for children teething—to insert between the teeth. All fastening about the body, such as the neckcloth, &c., ought to be loosened, and air freely admitted ; the head should be raised, and cold wet cloths *may* be applied to it if there is much heat. It has been advised to cram the mouth full of salt as soon as the fit comes on. Dr. Watson, who had the plan tried in hospital, thought it seemed to curtail the duration of the convulsion.

The most important treament is *during* the *intervals* ; mix together half a teacupful of ground mustard, two table spoonfuls of laudanum, three table spoonfuls of tincture of cayenne pepper, half pint of vinegar, and same amount of cold water and alcohol, put into a well-stopped bottle—shake well and apply this as a *liniment* along the backbone (or spine) with a woolen cloth, night and morning, wearing a flannel shirt next the skin all the time. Also, take *inwardly* the following ; tincture or extract of skullcap, tincture of valerian, tincture of hyosciamus, (henbane) spirits lavender, of each one ounce, dose, one teaspoonful three times a day.

Keep the bowels open, take plenty of exercise in the air and avoid all kinds of excitement. A sponge bath every morning, is a good remedy, also pouring water [or a shower-bath] on the head at the same time, is recommended by physicians.

DROWNING.

This is of such frequent occurrence that every man, woman and *child*, should know what to do in cases of apparent death from drowning—in fact it should be a part of the education of our chidren, from the fact that it often happens that children are the only persons present when one of their little playmates falls into the water, and while going for help the sufferer dies for want of the proper attention.

The first thing to be done when a person apparently drowned is rescued from the water, is to wipe and cleanse thoroughly the mouth and nostrils—the next to apply warmth to the body. This last cannot possibly be done as long as it is covered with *wet clothing ;* and if this is the case, it should be removed, *cut off,* if necessary for haste—as quickly as possible. If there is a house or shelter of any kind very near the spot where the body is got out, it may be taken to it at once, and before the clothes are removed ; but if such is not the case, *provided dry coverings are at hand, the wet clothes should be stripped off on the spot.* Wrapping the body in *blankets* is always to be preferred. In removing the body it is best done by laying it on the back or side, on some flat board, such as a door or shutter, the head and shoulders being well raised ; but if there is nothing at hand on which the body can be laid, care should be taken in carrying it that *the head is well supported* neither allowed to *fall back,* nor *forward upon the chest.* As soon as may be, warmth is to be applied to the entire external surface ; if a warm bath is available, it should be used, if not, the body is to be covered up with warm things ; bags of hot bran, hot salt, or sand, or any other convenient vehicles for heat, are to be placed wherever they can be without interfering with the necessary manipulations : to the pit of the stomach and to the feet especially, their application is to be used. Frictions with stimulants of some kind, such as camphorated oil, brandy or any other spirit mixed with oil, of–turpentine, should, any of them, be used warm, and be rubbed in with a flannel; a warm stimulant injection, consisting of gruel, containing a tablespoonful of turpentine, or double the quantity of brandy, may be given, and strong-smelling salts held to the nostrils *at intervals.* Artificial breathing, recommended by some, is condemned by others. Certainly the old method of using bellows and other means to inflate the lungs was much more likely to do harm than good, particularly in the hands of the unprofessional, who would be much more likely to inflate the stomach, and thus impede the breathing. Attempts to imitate the natural process of respiration may, however, be made, by pressing inward the ribs and pit of the stomach, and allowing them to rise again by their own elasticity, repeating this process twenty times in the minute.

External warmth and continued friction, with care taken that the shoulders and head are raised, the mouth and nostrils free, and carefully keeping the patient wrapped up in blankets, are of the greatest importance. For the more easy application of remedies, the body should be laid on a table of convenient height.

It is necessary to notice also *what ought not to be done ;* for many old and most injurious modes of treatment are still apt to be resorted to by the ignorant and prejudiced. Most of these have originated in the idea that water swallowed was, or had something to do with the cause of death ; hence patients have been hung up by the heels, rolled on barrels, choked with emetics, under the idea of making them disgorge the water.

If there is much water swallowed—as sometimes happens— it would be better to remove it ; but any means which unprofessional persons can use for its removal would only be a worse evil. If a medical man is present, and thinks well to use the stomach-pump quickly, remove the water and replace it with a small quantity of hot brandy and water, it may be of service, but no attempts should be made to give any thing by the mouth as long as unconsciousness continues.

Never despair of these cases, for when all hope seems to have vanished and no sign of life been given for one, two, or four, six, or even eight hours, the perseverance of those around has been at last rewarded, and life preserved.

After a person has been restored to consciousness, there may be considerable congestion of blood about the head, which may require leeches or scarifying the temples. In all cases of recovary, the greatest care must be taken to preserve the re-excited actions ; if stimulants are thought requisite they must be given cautiously ; guard against any excitement from friends or relatives ; and support the strength by tea, coffee, soups, &c.,

and let the patient be warmly wrapped in blankets in bed until the perspiration is started, after which they may be considered safe.

HANGING.

This may cause death in three ways : by simply compressing the wind-pipe, by which death is caused by suffocation ; by apoplexy, from compression of the veins of the neck, or by dislocating the neck, which is not often the case.

Recovery from hanging must, in some degree, depend upon the completeness or not, of the interruption to the passage of air through the wind-pipe for any time ; it is not likely that resuscitation will be effected if this thing has continued *four minutes.* The first thing to be done when a person is found hanging is, of course, to cut them down at once, to loosen the material around the neck, to dash cold water over them, and to bleed from a vein in the arm or foot, or temple. In such an emergency, a person would be justified in cutting across the temple, where the artery runs, (or beats), with a sharp knife of any kind, allowing the blood to flow freely. The bleeding could be checked when desired by pressing with the thumb over the part or putting burnt alum into the wound. With the excepttion of applying heat, and removing the clothing, the treatment of hanging is very much the same as in cases of apparent death from drowning. I should advise, however, that as soon as life is thoroughly restored, and after the patient has had time to think over his folly a little, in cases of attempted *suicide*, by whatever means may have been selected, that the person so offending should be thoroughly flogged, and then made to do the state some service for a month or two. The prevalence of this sin, and the consequent notoriety that is given to deaths from this cause in the newspaper press of the day, should be looked upon by every body with disfavor. No man or woman who thus tries to rush unbidden into the presence of God, deserves a falling tear, or a newspaper notice, unless insanity has been the cause.

CHOKE-DAMP,

Or Apparent Death from Inhaling Carbonic Acid Gas.

How many sad instances of deaths from this cause every year, and yet a little knowledge ,such as we have tried plainly to impart in this book, would have prevented an occurrence of this kind. Old wells, brewers' vats, the holds of ships, &c., are all liable to become the receptacles for carbonic acid gas, which, formed form some decomposing vegetable matter, lies like a

stratum of water at the bottom. Should any one incautiously
descend, so as to become enveloped in the carbonic acid atmosphere, breathing is either instantly stopped by spasmodic closure
of the chink at the upper portion of the windpipe, and complete suffocation is the consequence; or the gas, if sufficiently
diluted with air to be drawn into the lungs, speedily manifests
its narcotic effects upon the system, and the person quickly falls
in a complete state of stupor. The breathing becomes difficult,
and after a time ceases; the countenance is purple or pale, and
there may be convulsion and frothing at the mouth. In such a
case, the body of the individual must be removed, if possible,
and as soon as possible, from the poisonous atmosphere, or the
latter must be destroyed or dispersed.

The many fatal accidents which have occurred from persons
venturing rashly into wells, and such like places, might be a
warning for the future, and prompt the invariable employment of the simple test of lowering a lighted candle into
the suspected place. If the flame be extinguished, the atmosphere is *destructive to life;* if it burn even with a *feeble and
diminished intensity, there is danger.* Of the various modes for
destroying a carbonic acid atmosphere, none is more speedily
effective than the introduction into it of *newly slaked lime,*
either spread upon a board, or mixed with water, and dashed
into the place; fresh lime, having a powerful affinity for carbonic acid, quickly absorbs it. In the absence of lime, a
quantity of *fresh water dashed freely down,* so as at the same
time to absorb the gas and promote circulation of air, will be
serviceable; or large bundles of combustible material, which
will cause currents of air, may, when blazing freely, be thrown
in. Caution in the first instance is the best preservative; but
in the event of an individual dropping in an atmosphere of
choke-damp, it is perfectly useless for others to rush in to bring
him out; they can no more exist in it than he could, and in
stooping to lift a fallen body, they become all the more
thoroughly immersed in the poisonous gas. Instead of rashly
sacrificing life in the ill-directed endeavor to rescue another, let
those who are present dash bucket after bucket of water or
weak lime and water into the place, and on the fallen person,
until the unextinguished flame tokens that the fatal atmosphere
is weakened at least; and when they do venture in, *tie over the
mouth a cloth soaked in lime-water,* or of simple water, if the
other cannot be obtained.

When from any cause, a person gets into choke-damp as
above described, cold water should be dashed freely over the individual as soon as removed into the open air, and this measure,

succeeded by heat applied to the surface, stimulant embroca-
tions to the chest, spine, &c., stimulant injections, and ammonia
held *at intervals* to the nostrils, while artificial breathing (as
described under the head of Drowning) is at the same time
brought into action, and steadily persevered in for some hours.

Carbonic acid is produced during fermentation, or by slow decomposition of vegetable matter, such as *damp straw*, *sawdust*,
wood-chips, &c. It is the gas disengaged in effervescing liquors
generally ; it is also produced, along with other vapors of
which carbon forms a constituent, in the burning of charcoal.

Poisoning by charcoal fumes, either by design or accident, is
not an unfrequent occurrence. In the latter case it usually occurs from persons ignorantly retiring to sleep in a closed-up room,
in which burning charcoal is used as a means of warmth. The
carbonic acid and other fumes disengaged, act slowly and insidiously, and exert so powerful a narcotizing or stupifying effect, that
those exposed to the influence are quickly rendered unable to remedy the circumstances, and perfect insensibility ensues. Too often it happens that the discovery of the accident does not take
place until morning, long after it is too late to remedy the fatal
effects ; the sufferers being usually found dead. If living, they
will probably be perfectly insensible ; the countenance pale and
livid. Immediate removal to the open air, and free exposure to
its influence by removal of the greater part of the clothing, is
the first proceeding, when the treatment recommended in cases
of poisoning from choke-damp should be followed. Carbonic
acid is largely given off in the process of lime burning, and persons who have incautiously slept in the immediate neighborhood
of a lime-kiln, have been destroyed by it.

STROKE OF LIGHTNING.

Every summer tells the story of death from lightning. There
can be no doubt that in many instances life could be preserved
by the application of the proper treatment in time. Persons
who are stunned, but not killed, by lightning, generally remain
in a state of insensibility for some time, the breathing being
slow and deep, the muscular system relaxed. In such cases it
will be proper to use means for preserving the animal warmth,
which has a tendency to become depressed, to keep up artificial
breathing as recommended under head of *Drowning*, to use mustard-plasters to the spine and pit of the stomach, to administer,
from time to time, a little hartshorn in water, if the patient can
swallow—if not, to give a warm injection, containing half an ounce
of turpentine—or to use such other means as are recommended
under *Drowning* and *Choke-damp*, which may seem adapted

to the case. It is a common idea, that persons who have been killed by lightning do not stiffen, and that the blood remains fluid, but this is erroneous. It would, considering how often the fact is reiterated, seem almost superfluous to point out the ordinary precautions which those who chance to be exposed to a storm of thunder and lightning ought to adopt ; but not a summer passes without lives being lost from sheer ignorance. Harvest laborers and others *will* persist in sheltering under trees ; people *will* continue to put up even iron umbrellas in the midst of a thunder-storm, and mowers walk unconcernedly home with their scythes over their shoulders. If an individual is overtaken by a thunder-storm in a place where trees abound, he should avoid them as much as possible. A thorough soaking will be rather a protection than otherwise. If, on the contrary, the position is on a wide plain, where the body is the highest object, *lying down* is the safest thing that can be done. In any case, metallic objects, such as sickles, scythes, &c., being laid aside at considerable distance. Under shelter, the most hazardous position appears to be in a draught or current of air, such as between a door and window, or, as is often the case, females sewing near a window or door. The fine pointed needle having, of course, a powerful attraction for the electric fluid. Every house should be protected by a lightning rod, it costs but a trifle and may save many valuable lives.

CRAMPS.

This is a painful contraction of various muscles, mostly of the muscles of the legs and arms. They may be confined to one or two muscles, such as those of the legs, or may be more general, as happens in cholera. The affected fibres are drawn in hard, knotty contractions, and maintain this condition for a longer or shorter time. The most frequent causes are the presence of indigestible food in the stomach, or of acid in the bowels, or the pressure exerted on the nerves by overloaded bowels. The weight and pressure of the child, acts in a similar manner in pregnancy and labor, and occasions painful and troublesome cramp. The disorder is also often associated with the presence of worms. When cramp affects the arms and fingers, it may be connected with disease of the heart and great blood-vessels of the chest. The power of the application of sudden and prolonged cold in producing cramp, is often sadly exemplified in the case of bathers. The best immediate remedy for cramp is friction with the hand, or, better still, with a mixture of half a pint of vinegar, two tablespoonfuls of ground mustard, and one of cayenne pepper, to be rubbed on with a woolen cloth.

Shake well before using—one tablepoonsful of laudanum added to the above, will be beneficial if it is to be had. When the legs are affected, it is always expedient to take medicine, say ten grains of rhubarb and ten of magnesia, with fifteen grains of carbonate of soda and a little ginger ; and afterwards, to clear out the bowels with some active aperient, such as castor-oil, especially if there is any existence of costiveness, or a possibility of their being loaded. Some persons find relief from the immediate attack of cramp, by tying a band of some kind tightly round the limbs, between the affected part and the body, while others are in the habit of standing upon some cold substance. The first process is perfectly safe, and may be tried ; the second certainly is often effectual, but it is not devoid of danger. Active friction with the dry hand, warmed, is the best temporary remedy. Cramp affecting the arms is always to be regarded with suspicion, if it occurs often. There is reason to fear some disease of the heart, lungs, or liver, the best plan would be to get a physician's advice, at once.

EXTERNAL VIOLENCE.

This is likely to occur from accident or design, at any time, and we should always be ready to meet these emergencies. It is my desire that every purchaser of this book will *carefully read it*, especially those parts of it which treats of things requiring *immediate attention*. A stroke or injury on the head may cause merely bruising of the scalp ; if more severe, concussion or injury to the brain, or fracture of the skull. The latter accident is most likely to happen at the side of the temple, where the bone is thin, ; but severe injury to the brain frequently occurs from blows at the under and back parts of the head. A severe blow on the back may cause paralysis (loss of feeling or motion, or both) of the lower limbs, with or without fracturing the bones. When a blow, even comparatively slight, is inflicted upon a spot immediately over a collection of nerves, most distressing effects, and sometimes immediate death, may result. Such is the case from blows on the neck, on the pit of the stomach, or over the region of the heart. The deadly faintness which ensues, should instantly be combated by stimulants—ammonia, ether, or spirit of any kind—which can be procured. Cold water should be suddenly dashed over the surface or down the back. If this is unsuccessful, the patient is to be put into a warm bed, and artificial breathing, as mentioned under the head of *Drowning*, employed along with external heat, mustard-plasters to the back and pit of the stomach, and stimulating injections of two or three tablespoonsful of whiskey, or brandy or spirits turpentine, with a like amount of sweet oil and a gill of warm water.

CRAMP IN THE STOMACH FROM DRINKING COLD WATER.

There are a great many diseases, properly speaking, brought on by drinking *ice-water* and eating *water-ices* and *ice cream*, when the body is *over heated*, such as diarrhœa, dysentery, inflammation of the stomach, &c., but these will be treated of in their appropriate places. Our purpose here is to treat of "cramps" in the stomach only. There are three circumstances which concur to produce disease or death from drinking cold water : the patient is *extremely warm*, the water is *extremely cold* ; and a large quantity of it is *suddenly* taken into the body. The danger from drinking is in proportion to the degrees which occur in the three circumstances mentioned.

Soon after the patient has swallowed the water, he is affected by dimness of sight ; he staggers in attempting to walk, and, unless supported, falls to the ground ; he breathes with difficulty ; a rattling is heard in his throat ; his nostrils and cheeks expand and contract in every act of breathing ; his face appears suffused with blood, and of a purple color ; his extremities become cold, and his pulse imperceptible ; and unless relief be speedily obtained, the disease terminates in death in a few minutes. This description, of course, refers to the worst cases. More frequently the patient is seized with acute spasms in the breast and stomach. These spasms are sometimes so painful as to produce fainting. In the intervals of the spasms, the patient appears to be perfectly well. The intervals between each spasm become longer or shorter, according as the disease tends to life or death.

Punch, beer, toddy and various other fancy drinks fixed up in drinking saloons, when drank under the same circumstances as cold water, have all been known to produce the same dangerous and fatal effects. The means to be tried for giving relief is strong stimulation, by large doses of *laudanum*, ether, spirits, &c. ; and, above all, it is necessary that the patient should not be permitted to remain for an instant in a recumbent posture ; *but should be kept in constant motion until relieved.* A mixture of whiskey, brandy or gin, two to four tablespoonsful, cayenne pepper, half teaspoonful, and of laudanum thirty drops, to be taken in a little water as hot as it can be drank, will usually give speedy relief. The dose may need repeating in half an hour. For children reduce this dose according to age. At the same time hot applications to the pit of the stomach, such as hot salt, sand, meal, ashes, &c., or a mixture of ground mustard and pepper, with warm vinegar, applied in the same way.

When heated, persons should abstain from drinking *very cold* water. Gargling the throat, or washing out the mouth in cold water, will allay thirst until the person has time to cool off.

Where the powers of life appear to be suddenly suspended, the same remedies should be used which have been so successfully employed in recovering persons supposed to be dead from drowning. Care should be taken in this, as in all cases of apparent death, to prevent the patient's suffering from being surrounded or attended by *too many people.* The act is kindly meant but it is very dangerous for the patient.

BITE OF VENOMOUS SNAKES.

This may very properly be called a poisoned wound, for the the poison is inserted at the same time the bite or wound is inflicted. The wound in itself is generally trifling, perhaps not more than a scratch, but speedy death may follow.

Immediately after being bitten by a poisonous snake, the parts begin to swell, and there is terrible and speedy depression of the vital powers of the system generally. When an individual suffers from a wound known or *believed* to be poisonous, immediate steps should be taken to prevent if possible, the poison being *aborbed into the system.* The steps to be taken are sufficiently detailed in the article on hydrophobia.

In addition to the local treatment of the wound (mentioned under head of *Hydrophobia*) continued friction with some oily material appears to be most generally useful, while, at the same time, stimulants are freely given internally to counteract the depression. Of course, any stimulant first attainable should be used; but hartshorn, is most highly recommended, taken in teaspoonful doses every ten minutes, in a half a gill of water, until reaction has been established, and the patient is better ; being free from poison and the swelling checked. However, hartshorn may not be so handy to get at. Then take half a glassful of brandy, whiskey or spirits of any kind, every fifteen or twenty minutes, until the patient is fully under its influence, which will be when he is "dead drunk." It should not be forgotten that the part bitten should *in a moment be cut out*, and then freely washed with water until hartshorn or caustic can be applied, which can not always be obtained on the spot, but as soon as they can be, either of them, apply to the bitten part.

NEURALGIA.

This is a most prevalent disease of this fast age in which we live. Of its nature but little can be told, only that it is seated in the nerves, though having its *origin*, often, no doubt, in the excessive use of tea, coffee, rum and tobacco. However, some of the worst cases have been connected with *diseased growth of bone* in different parts of the head or face, especially about the canals through which the nerves pass : other severe cases have been found to depend upon irritation excited by foreign bodies acting upon some of the nerve branches ; *decayed teeth* are not unfrequently connected with the disease. The most general seat of neuralgic pain is in the head or face ; but the fingers, the chest, the abdomen, &c., may any of them become affected.

Persons afflicted with this disease have described it as a "plunging," darting pain of the most intense and agonizing kind ; but, except in long-continued cases, there is no external mark—no redness, swelling, or heat, to indicate the disorder to others, and many a sufferer from this disease has been taunted with playing " Old Soldier,"—a most cruel and unkind procedure. After a severe attack of neuralgia, the skin is often left tender, and when the pain has recurred frequently, exquisitely tender swelling of the part has been known to come on. The access of the pain is usually sudden, its remission equally so, and it is generally periodical in its attacks : it is suspended during sleep.

Among the exciting causes of neuralgia are damp and cold, or damp alone, if combined with malaria, such as cause ague ; exposure to currents of cold air, more particularly if the individual is heated. Debility of constitution renders the individual much more susceptible to those and other exciting causes; it has often, too, been traced to anxiety of mind.

It has been observed sometimes, that sudden attacks of neuralgic pain in various parts of the body, have been traced to temporary stomach disorders, such as superabundant acid, &c.

If the patient is resident in a climate or situation likely to excite it, some change should, if possible, be made : this will probably be most beneficial if the removal be to a dry, warm air ; but should disease have commenced in a cold, dry district, change to a moist, but warm one, will probably offer most advantage. If disorder of the stomach exists, it must, of course, be rectified, by giving purgatives, such as any ordinary good purgative pill, or a dose of Rochelle salts, Rhubarb, Seidlitz powder, or Castor oil, two or three times a week. After that, if the disease still continues, quinine, given in one or two grain doses, every six or eight hours, will most probably be of service. Carbonate of iron, in from half drachm to drachm doses, is a most useful remedy, especially in weak constitutions. Blisters behind the ears, or at the back of the neck, are often valuable aids in the treatment of neuralgia of the face. To relieve the paroxism of pain, a sponge, or piece of flannel, dipped in boiling water, or vinegar and water, and applied as hot as it can borne over the site of the pain, will often allay its severity, or remove it altogether. Opium internally may be given in very severe cases, or rubbed on the part. Chloroform applied to the affected part by means of a piece of lint soaked in it and covered with oil Silk, is a very successful application, and should be tried if it can be procured. A liniment made of Tincture of Arnica and Laudanum, in equal parts, applied often to the part affected will afford speedy relief. Five grains of morphine, thoroughly rubbed up into a tablespoonful of lard, to make an ointment, to which add ten drops of oil of lemon, while mixing, makes a good application to the painful parts, put on with the ball of the fingers every two or three hours. Friction to the affected parts with the dry hand, three times a day, during the intervals, should be strictly attended to, with a view to break up the disease. Electricity is often beneficial, when all other means fail.

SEA-SICKNESS.

If any of my readers have ever known what it is to be " Sea-sick," they can, no doubt, say, with the author, who dreads the Sea on that account more than any thing else, that a sail on the ocean wave is more sick than romantic.

This sickness is considered to be dependent on some peculiar affection of the brain, produced by the rocking motion of the vessel. The affection is more readily caused by long heaving waves, than by a short rough sea. The best preventives of sea-sickness seem to be the horizontal posture, as near the centre of the vessel, and therefore of the centre of motion, as possible—that is, where the motion is least. Exposure to the open air renders the liability less. Stimulants, combined with sedatives, certainly appear to have considerable effect in preventing or alleviating the affection. A pill, composed of four grains of cayenne peper, with two or three of extract of henbane, taken at intervals, may be found useful. Creasote is also an excellent antidote.—one or two drops made into a pill with bread crumb, to be repeated, if necessary, once or twice during the day. Some persons find themselves less liable to sea-sickness if they take food freely—with others the reverse is the case; the effect probably depends upon the state of the digestive powers of the stomach, temporary or permanent. If these are vigorous, the excitement of digesting food acts probably as a counter-agent to the cause of the nausea. Sea-sickness, of itself, is rarely injurious, but it should be a subject of consideration with persons who are liable (or likely to be) to head-affection, who are the subjects of rupture, prolapsus, &c., how far they should incur the risk of these being aggravated by the mechanical action of vomiting. Some who do not suffer from sickness while on the water, experience nausea and other uncomfortable sensations after landing—an effect, doubtless, due to a partial disturbance of the digestive organs, and probably to biliary disorder.— One or two doses, of Rochelle salts, or Castor oil, will, generally remove the inconvenience.

Some persons have been greatly, and often permanently relieved of sea-sickness, by taking from three to five drops of chloroform, put on a lump of sugar and swallowed *immediately*, then going to bed, and, if possible, going to sleep. With children, rubbing the pit of the stomach with a sponge dipped in a few drops of ether, chloroform, or laudanum, or the three mixed together, will afford relief ; or a cloth with some of this mixture dropped upon it and laid on the pit of the stomach, will answer.

Varicocele.—Almost the first symptom that is observed in the genital organs, produced either by masturbation or excessive indulgence with females or other causes, is a flabby, relaxed condition of the privates; the testicles hang lower than usual, the spermatic cord is relaxed, and there is a dull, aching, heavy, dragging feeling in the parts, with sometimes, in more advanced stages, pains shooting up occasionally into the groin and lower part of the abdomen, and also a heavy aching feeling in the small of the back. There is enlargement of one side, usually the left, and the scrotum feels like a bag of worms. There is pain, and at times a coldness and numbness in the privates. The disease is quite prevalent : the author has had no less than three hundred cases within the last few years, and has treated them successfully.

All that can be done in domestic practice, is to bathe the privates two or three times a day in cold water, and procure a proper suspensory bandage, which should be worn during the day time, and left off during sleep.

HYPOCHONDRIA, or Lowness of Spirits.

MILD cases of this disease are called very often the " *Blues*," sometimes *Hypo*, etc. It is a condition of the mind, produced by real or imaginary causes, which should be remedied as speedily as possible, before it gets too firm a hold on the system. The common symptoms are, loss of, or a variable appetite, a troublesome flatulency in the stomach or bowels, sour belch ings, costiveness, a copious discharge of pale urine, spasmodic pains in various parts of the body, giddiness, dimness of sight, palpitation of the heart, general sleeplessness, and often an utter inability of fixing the attention upon any subject of im- portance, or engaging in any thing that demands vigor or courage. The mental feelings, and peculiar train of ideas that haunt the imagination and overwhelm the judgment, exhibit an infinite diversity : sometimes the hypochondriac is tormented with a visionary or exaggerated sense of pain, or of some concealed disease ; a whimsical dislike of particular persons, places, or things ; groundless apprehensions of per- sonal danger, or poverty ; a general listlessness and disgust, or an irksomeness and weariness of life. In other instances, the disease is strikingly accompanied with peevishness and general malevolence ; the patients are soon tired with all things ; discontented, disquieted upon every light occasion, or no occasion ; often tempted to make way with themselves ; they cannot die, they will not live ; they complain, weep, lament, and think they lead a most miserable life : never was any one so bad.

The whims that are sometimes seriously entertained under this complaint are of the most ludicrous description. Men have imagined that they were a lump of butter, and were afraid to go into the sunshine or near the fire for fear of melting ; others, that they were continually in some place of dan- ger, and likely to be killed at any moment ; some have thought they had toads, snakes, and no telling how many imaginary things in their stomach that would sooner or later destroy them. As to the causes of this disease, there may be a strong constitutional predisposition, or the disease may be the conse- quence of a sedentary life of any kind, especially severe study protracted to a late hour in the night, and rarely relieved by social intercourse or exercise ; debauched, dissolute habits; great excesses in eating and drinking ; the immoderate use of mercury, violent purgatives, the suppression of some habi- tual discharge or long-continued eruption. Some peculiar affec- tion, such as congestion or fullness of one or more of the im portant organs within the abdomen, is a frequent cause.

The principal objects of treatment in this disease are, to remove the indigestion, to strengthen the body, and to enliven the spirits; and one of the best plans with which we are acquainted, for the fulfillment of these intentions, is, *constant exercise and change of place*, with a sponge bath about three times a week, early hours, regular meals, and pleasant conversation, the bowels being at the same time carefully regulated by the occasional use of gentle purgatives, and the stomach strengthened by some appropriate tonic medicine. A tea made from gentian root, dog wood, or wild cherry bark, half a glassful before each meal, taken cold, is recommended as about the best. Exercise in the open air, whether walking or on horseback, is the best; this should be combined, if possible, with agreeable company and constant change of air and scene. Traveling is a powerful remedy in this disease, since it is often one of the most effectual means in removing indigestion, of strengthening the body, and exhilarating the spirits; and where the patient's circumstances will permit, it ought invariably to be one of the first measures resorted to, as it will undoubtedly be found one of the best. The patient should accustom himself to early rising, and regular meals of nourishing and easily-digested food. The bowels are almost always torpid in this disease, and will, therefore, require constant attention in selecting articles of diet which are of an opening quality, such as stewed fruits at meals, ripe fruit uncooked, between meals, roast apples after tea, etc., with an occasional employment of medicine if needed.

Regular daily friction over the limbs and bowels, with the flesh-brush, is advisable.

The *moral management* is of very great importance in this disease, for assiduous kindness and consoling conversation produce a deep effect. The patient should rarely be opposed in the expression of his sentiments, and never with ridicule. A very important object is to gain the patient's confidence, and in order to effect this, we must humor his foibles, and seem to fall in with his views. When he is dwelling upon some imaginary disease, it must be prescribed for, and should his anxiety pass in succession from one complaint to another, they ought all to be prescribed for in their turn. That is better than trying to *argue* him out of his " notion."

HICCUP, or HICCOUGH.

This is an affection too well known to require any description. It is a sudden jerking spasm of the diaphragm (or midriff), expelling the air from the lungs with a peculiar sound

Acidity of the stomach or eating some article of food which disagrees with the stomach is usually the cause of the disease. Generally it is a trivial and transient inconvenience, but its occurrence in the last stages of acute disease is a grave (often fatal) symptom, indicative of approaching death. There are cases however, differing from either of the above; it sometimes occurs in the persons, more especially of young females, of an hysterical tendency, and may continue for weeks without cessation, except during the hours of sleep, in spite of all kinds of treatment. The causes of ordinary hiccup are generally fasting, or some sudden stimulant taken into the stomach, such as highly seasoned soup; and the affection generally subsides of its own accord. When inconvenient, nothing is so likely to remove it as some active emotion of the mind suddenly excited. Startle the patient by slapping on the back, clapping the hands unexpectedly behind the head, chase around the room as if in pursuit of a rat or mouse, etc. Any thing to engage the patient's mind quickly. The continued sipping and swallowing of cold water is a frequent domestic remedy, a few drops of peppermint in a glass of water or a teaspoonful of soda in a glass of water will often remove it. If further treatment is necessary, *press firmly on the two collar bones* for a few moments: this will often work like a charm. A few drops of laudanum, hartshorn, or paregoric, in water are also good remedies.

GIDDINESS or DIZZINESS OF THE HEAD.

There are few diseases of a simple character, that are productive of more unpleasantness to an individual than a "swimming" in the head, as it is often called. Some disorder of the circulation of the blood in the head is a probable direct cause of giddiness, and this is most palpable after persons have been confined to bed or to the horizontal position for a short time: on first assuming the erect posture, giddiness is generally experienced. Intoxication is an example to a certain extent of the same thing. A mere passing giddiness is probably owing to some cause which a little attention to the state of the stomach will correct; but repeated attacks, especially if accompanied with palpitation of the heart, or pain and heat about the head, indicate danger, and no time should be lost in consulting a physician. The treatment of course must be according to the *cause* of the disease. If produced by too long application of the mind to any subject or study, *leave it off*, and take recreation. If caused by dyspepsia, reduce the diet, and leave off the use of indigestible articles of food. If

costive, take purgatives two or three times a week ; a dose of
Rochelle salts or a Seidlitz powder on an empty stomach, or
five grains of blue pill, with five of Dover's powder made into
a pill at bed time, followed the next morning by the Rochelle
salts or Seidlitz powder, will have a most happy effect, and
eat such articles of food as will keep the bowels open, (see
Hypochondria.) Abstain from the use of coffee, tea, and intoxicating liquors, take plenty of active out-door exercise,
avoid meats for supper, bathe the head often in cold water,
and keep the feet warm and dry.

H E A D A C H E .

It is scarcely necessary for me to say that it is unwise ever
to *neglect headaches.* They are sources of great suffering, and
often lead to serious derangements of the general health.
Headaches are more common among civilized than uncivilized
nations, more frequent among females than males, among
those of sensitive feeling than otherwise, and among the sedentary than those who are more in the open air, or who take
active bodily exercise.

HEADACHES are of various forms, dependent on their various
causes, such as *Plethoric* headaches in those who are stout,
robust, or full-blooded ; sometimes it lasts but a few hours,
or it may continue for several weeks.

SICK HEADACHE—so called from the nausea or sickness at the
stomach, which attends the pain in the head.

BILIOUS HEADACHE—occurring most in summer or fall ; it
affects persons mostly who have dark complexion, dark hair
and melancholy disposition. *Nervous headaches* are more
common among females than males. The pain is usually acute
and darting, and is made worse by light, and with a feeling as
if the temples were being pressed together, and a " swimming"
in the head.

There is sometimes a sense of sickness, with a dread of falling and great despondency, or restlessness. The bowels are
generally costive and the sight dim. The pain comes on mostly of a morning and lasts through the day, going off in the
evening.

Besides the above-named varieties, there are also *Hysteric*
headache, headache from *Exhaustion*, *Brow Ague*, *Rheumatic* headache, &c.

The great point about these different varieties of headache
is, that there is a *cause* for them, if you will but find it out,
and most of them therefore can be cured. Not merely *relieving the headache when you have it*, but preventing the recur

rence of the disease. How many persons with premature grey hair, a care-worn and wrinkled face, an oldish look and loss of memory, &c., from some form of headache which has lasted for years.

The great point in treating cases of headache, is to find out the *cause* if possible. However, in all cases except in head-ache from *exhaustion* or *weakness*, or during pregnancy, you can safely apply the following remedies:—As soon as it is ob-served that a headache is going to "spoil a day's work or stop a day's pleasure," take half a teacup full of water as warm as it can be comfortably swallowed every fifteen minutes, until vomiting occurs, or the headache is abated. Then take a tea-spoonfull of Rochelle or Epsom salts in a glass of water, or a Seidletz Powder, every three hours until the bowels operate. When the bowels are opened, the headache will be, as a gene-ral thing, cured. In most of cases, it will be necessary to live on a lower diet for a while, especially in ruddy, stout, robust persons who are full-blooded. A shower-bath over the head every morning, or a vessel of water poured from a distance, of two or three feet above, on the head, in most cases of head-ache, is a good *preventive* remedy. *Costiveness* is a very pre-valent cause of the various cases of headache, which must be avoided by purgatives and diet.

Sometimes swallowing the *juice of a lemon*, in which is mix-ed a little sugar, will relieve an ordinary headache in a few minutes—or when a lemon is not handy, heating a little vine-gar in a sauce-pan, and inhaling the vapor as it rises, up the nose, will answer;—or a teaspoonfull of vinegar in a little water taken every hour will answer in many cases. For pro-curing sleep and rest in cases of headache use the following mixture:—Tincture of valerian, one ounce, tincture of hyos-cyamus, (Henbane) one ounce, paregoric, two teaspoonsful, spirits lavender, one ounce. Mix and take a teaspoonful every three hours till rest is procured.

Persons who are subject to headache in any form, should avoid *eating between meals*, and by all means avoid *hearty suppers;* eating meat for supper, or any indigestible article, or a plate of ice cream, *just before going to bed*, being very apt to be followed next day by an attack of headache. Meat ice cream, &c., taken at *proper times*, are all good in their place, but remember that there is " a *time* for all things."

HERNIA, or RUPTURE.

It is highly important that this dangerous affection sh'uld be understood, from the fact of its frequency, and that it is so often badly treated, or not treated at all. By the term Rup ture, we understand a protrusion of any internal organ from its cavity or where it belongs; but the term is generally re- stricted so as to mean no more than a protrusion of the bowels through the walls of the belly. If the abdominal walls are weak, from any cause, no matter what—lifting, straining, or making violent muscular exertion of any kind, will then cause the bowel to force itself through at the weakest spot, and push- ing the lining of the belly, (the PERITONEUM,) along before it, a bag or sack is formed in which the projecting bowel is en- closed, forming an external tumor, lump, or swelling.

Hernia occurs in both males and females, and children are often born with it; or it may occur at any period of life, from infancy to old age.

Whenever a lump or tumor appears in any part of the bel- ly, in the groin, or upper part of the thigh, or vicinity, it should at once be attended to. When there is an enlargement or fullness in the scrotum or bag, VARICOCELE or HERNIA should at once be suspected. A swelling coming suddenly in the groin, or at the navel after considerable exertion, may be taken for Rupture without much fear of mistake.

Hernia, or Rupture, is never free from danger until the pa- tient obtains a *properly constructed Truss*—the low-priced Truss, which is too often sold in drug stores, &c., being worse than none at all.

Rupture is a very common affliction; it has been estimated that every fourth person is more or less affected. Females, from motives of delicacy, are apt to conceal the misfortune, and not seek advice—this is wrong; it exposes them to danger.

The complaint being discovered, the bowel should be put back in its place, and a Truss at once obtained. In the case of young persons a Truss will often effect a cure, but that it may do this it should be worn night and day, except just time enough for cleaning it, &c., and then the patient *should be in bed.* Pieces of old knit stockings are the best to put between the pads and the skin, renewed every few weeks. The parts should be bathed or washed in cold water night and morning, the Truss being removed at the time, and the patient lying down until the Truss is put on again. To tell a Rupture from Varicocele, let the patient lie down on the back for a few minutes, then press with the two middle fingers pretty firmly over the region each side of the genital organs, (pubes,) and

Double Inhaler, for the Cure of Catarrh in the Head.

Many years ago, I became convinced that the ordinary treatment for CATARRH IN THE HEAD, prescribed by some of the very best men in the medical faculty, was of but little, if any, practical utility. I therefore made it a matter of study and various experiments, and soon satisfied myself that "Medicated Inhalation" was the only rational and successful remedy. To apply this properly, I invented the "DOUBLE INHALER," which is represented by the above engraving. Directions for use will be found under the head of "Catarrh in the Head."

·

rising to the feet keep up the pressure. If it is Rupture, the tumor or enlargement will not make its appearance in the scrotum or bag—if it is Varicocele, it will immediately become full again, and feel like a bag of cords or worms. Persons who have Rupture must not allow the bowels to become costive, as straining at stool is highly injurious.

CATARRH IN THE HEAD.

This disease is usually the result of a cold in the head, which through mismanagement, inattention, constitutional predisposition, an enfeebled state of the system, a scrofulous taint of the blood, or other unfavorable circumstances, runs into a chronic state. From this apparently trifling complaint, with which every one is forced to make acquaintance, at some season of the year or other, is developed a disease as frequent and offensive as it is dangerous, and which in its progress is apt to affect seriously some of the most important organs of the human economy, and to be instrumental in causing the impairment or loss of smell, taste, or hearing, and even loss of life itself. This disease is popularly recognized by the name of Catarrh in the head.

The first sensation is usually a feeling of dryness and heat in the nose, and a frequent inclination to sneezing. There is an inability to breathe freely, as the nose becomes stopped up, sometimes on one side, sometimes on the other.

Soon a clear, watery, irritating discharge makes its appearance, excoriating the nostrils and edges of the lips, which become red and somewhat swollen. After a few days the discharge becomes thick, yellowish, extremely frequent, and continues to be a marked feature of the disease, and a source of much danger and the greatest annoyance. After more or less time, it becomes thick, very disagreeable, and assumes an extremely offensive odor. It is usually so profuse as to require, when confined to the nose, the frequent application of the handkerchief, or if it drops into the throat, which is more particularly the case, while laying down, a constant coughing, and sometimes both.

Sleep is frequently disturbed by a sensation of choking, caused by the presence of the discharge in the throat. Owing to the heat in the head, the watery portion of the discharge often evaporates, and, assuming a condition of solidity, is deposited upon the membrane of the nose and upper part of the throat, in the shape of crusts or hardened lumps. The accumulation of these incrustations produces a feeling of discomfort, and narrows the passages so as to embarrass respiration.

Therefore, frequent efforts have to be made to remove them, either by forcibly blowing the nose or by persistent hawking. During sleep these incrustations accumulate more rapidly, and the feeling is therefore most uncomfortable in the morning. Sometimes all efforts to clear the throat are futile until after breakfast, or after something warm is swallowed. The discharge, which is at first without smell, assumes in the progress of the complaint an excessively offensive odor, the breath participates in this, and becomes occasionally so revoltingly offensive as to render the patient an object of disgust to himself as well as to others. Ulceration of the mucous membrane of the nose takes place occasionally. The accumulation of the discharge, together with the thickened condition of the mucous membrane, renders breathing through the nose very difficult, and oftentimes impossible, necessitating respiration principally through the mouth—a method very deleterious to the general health, but more particularly so to the lungs. Sometimes the voice loses its musical quality, and assumes a discordant, harsh and nasal character; the sense of smell becomes much impaired or entirely lost, and the same effect, though less frequent, is produced on the sense of taste. Occasionally, while blowing the nose, a crackling or bubbling sound will be heard in the ear, and hearing will be found quite thick and stopped up, but returns suddenly with something like a snapping sound. This is sometimes repeated, until, at one time, hearing does not return, and remains permanently injured. Noises in the head of every conceivable description will make their appearance and add to the distress of the sufferer, and hearing may be lost so gradually that a considerable degree of deafness may exist before the person is really aware of the fact. The eyes are apt to become weak, irritable, and disposed to water on exposure to cold and wind, or after the slightest exertion. A pain, more or less acute, or a distressing feeling of *pressure* is experienced over the eyes, and sometimes on the top or back of the head, and also pain in the face, closely resembling neuralgia, for which it is very often mistaken. The distress in the head weakens the memory and produces irritability and moroseness of disposition. The stomach generally suffers more or less, is weak and irritable; the appetite is variable, and is nearly always bad in the morning. In severe cases, the system becomes feeble and prostrated, and there is an aversion or inability to either physical or mental exertion. Not unfrequently catarrh proves fatal, either by debilitating the system and wearing out the patient, or by traveling downward and producing throat affections, bronchitis, and finally

CONSUMPTION. It may be safely asserted that after hereditary predisposition, catarrh is the most frequent and important cause of this fatal complaint.

The symptoms of catarrh vary considerably in different individuals, and the degree of their severity depends upon constitutional peculiarities and various external influences. With some the complaint continues for a number of years in a mild form without causing any of the injurious results above described, while with others all the worst effects are produced in a very short space of time; and cases, apparently most harmless, may, through imprudent exposure, additional cold, or unfavorable changes of the weather, suddenly exhibit all the violence and malignity which characterizes the severest ones.

In the treatment of this disease but little can be done without medical treatment by a physician who can take into consideration the age, occupation, how long the disease has existed, the condition of the general system, &c. The author of this work has had extensive experience in the treatment of this disease during the last few years, having treated patients from most of the States of the Union; and I was led to examine very closely into its nature, and the best treatment adapted thereto; and having seen tried the various popular "Catarrh Remedies" sold by those who knew nothing of medicine, as well as the usual treatment of the medical profession, without satisfactory results, I came to the conclusion that the inhalation of medicated vapors was the best treatment. I invented an Inhaler, which is now extensively used by the profession, and is sold in many of the wholesale drug stores. It has two elastic tubes, with glass nose-pieces, to fit one in each nostril; and also a centre glass tube going nearly to the bottom of the inhaler.

Some hot water is poured upon a handful of bitter herbs, finely cut, such as boneset, sage, horehound, green or black tea. or such other article as may be thought advisable. After standing a few minutes to cool, the inhaler is half filled with this mixture, and the vapor is inhaled through the tubes up the nostrils, and then passed out at the mouth, continuing the inhaling from ten to thirty minutes. It should be used three times a day. When one of these inhalers cannot be obtained, inhaling the vapor from the spout of a tea-pot (not too hot) is better than none.

At the same time the bowels must be kept open, cold water often gargled in the throat, and the head bathed also in cold water every morning. Give at same time some vegetable tonic to strengthen the system.

SPITTING OF BLOOD.

This is a more serious discharge of blood than is generally supposed, and many cases of this kind suffered to linger along unattended to, terminate in consumption. The discharge is of a bright red color, brought up by hawking and spitting, frequently preceded by a saltish taste in the mouth, a sense of heaviness about the heart, difficult and painful breathing, and dry, tickling cough. It differs from blood brought from the stomach; that from the latter is of a more dark and clotted appearance. It most commonly occurs at ages from fifteen to thirty, and may be occasioned by any violent action of body or mind: by the suppression of accustomed evacuations: by a rarified air; and most frequently takes place in persons of long necks and narrow chests: often in families subject to similar complaints.

Bleeding from the lungs may occur in every degree, from a mere tinge of the expectoration, to the copious coughing up of fluid blood. The blood is *coughed* up, whereas, when it comes from the stomach, it is *vomited*, a distinction which *appears* evident enough, but which is not always readily made in practice. The management of hemorrhage from the lungs must be that recommended for hemorrhage generally. Until medical assistance can be procured, perfect quiet is to be observed, cool air, especially on the chest, freely admitted, and cold, or iced and acidulated drinks given plentifully. Alum will also be found useful; either letting it dissolve in the mouth and swallowing it, or dissolving a teaspoonful in a little water, and taking it as may be needed. Should the attack continue and medical assistance still be absent, cupping on the chest, or between the shoulders, might be had recourse to; or in an extreme case, when medical aid is far distant, one grain and a half of sugar of lead may be given, made into pill with crumb of bread, every two, or three, or four hours, being washed down by a draught of *vinegar and water*. The expressed juice of the common nettle is sometimes popularly used, and, it is said, efficaciously, to check bleeding from the lungs; the dose is one teaspoonful three times a day. The inhalation of the smoke from the burning leaves of the belladonna is said to check the immediate flow of blood from the lungs. For this purpose one teaspoonful of the cut and dried leaves is to be thrown upon live coals, and the fumes inhaled. Another valuable, because such a ready means of checking bleeding from the lungs, is to eat freely of salt, or drink salt and water every ten or fifteen minutes. When there is troublesome cough, it should be relieved by thirty or forty drops of paregoric, given occasionally, as required. Persons of scrofulous constitution, or who have any malformation (imperfectly formed) of the chest, are most liable to suffer from it. It rarely occurs in children. The *exciting* causes of this form of hemorrhage are such as call the lungs into active, strong, or continued exertion, such as violent bodily movements, much loud exercise of the voice, playing on wind instruments, &c. ; these things must, therefore, be sedulously avoided by those who have any tendency to the disorder. Temperance and moderation, strict attention to the condition of the bowels, and to all things necessary to preserve health, as detailed in the first part of this work, will be necessary. While treating of this subject, it should be mentioned that persons are often needlessly much alarmed, from thinking they are expectorating blood, while the fluid simply comes from the *throat or gums*, or, it may be, is the consequence of blood from the nose trickling down the back of the throat. To test the matter, a slight attempt at coughing should be made; if the bleeding is *not* from the lungs, there will be no fresh blood coughed up.

TYPHUS, OR TYPHOID FEVER.

This has been called by different names, such as malignant fever, continued fever, slow fever, &c.; but for all practical purposes, in a work like this, we prefer the name most familiar with the people, so they will better understand our meaning. This is an affection of the whole system, and by medical men is subdivided into several types or forms; but it would serve no good purpose to enter into these here. The management of a disease so gravely important as fever can never be legitimately undertaken by unprofessional persons, if medical assistance is procurable; but as a provision for circumstances when this is absent, the less complicated the account both of the disease and its treatment, the more likely it is to be managed with advantage. The first symptoms of incipient fever are usually displayed through the nervous system. The individual feels an unaccountable languor, and complains of headache and shivering, cannot exert his powers either in the duties or pleasures of life, is easily tired, sleep is disturbed, the appetite is impaired, the skin looks dusky and the eyes heavy, the pulse quickens, and at length the feeling of general illness drives the patient to bed. The attack, however, may commence much more suddenly—a shivering, or, as the people in many places call it, an "ague fit," may be the first symptom; or severe headache, or vomiting, or fainting, or even convulsion may be the first symptom of the impending malady

When fever is fairly established, the pulse ranges above 100 the tongue is furred, probably brownish and dry, sleep is disturbed or supplanted by delirium, the muscular power is diminished, and the mind indifferent to passing circumstances: dark incrustations collect about the teeth; the patient sinks down in bed, and perhaps passes the natural evacuations unconsciously, thus displaying the most evident signs of debility. This condition may increase till it terminates in death, or tends toward recovery, either by some marked crisis, such as profuse perspiration, or by an almost imperceptible amendment. Tranquil sleep, improved aspect of the countenance, the skin cooler and with more tendency to moisture, the tongue cleaning at the edges, and a natural desire for food, all give signs that the disease is passing away; on the other hand, if a fatal issue is approaching, the general weakness increases, the patient slips down in the bed in consequence, and lies in a state of dreamy muttering; there is convulsive starting of the fingers or other parts of the body, picking at the bedclothes, the insensibility to external impressions increases, and probably stupor closes life.

The above are the general features of fever, whether simply continued, or when it runs out to the more serious forms of typhoid, low, nervous, or typhus fever. There are many other indications which occur, but which it would serve no good purpose to detail here; all that is required is that the disease should be recognizable, so that its general management may be properly and intelligently conducted when it falls to the lot of an unprofessional person to have the direction.

In the first place it must be remembered, that for continued fever we have no cure, that is, we have no medicine which we can give with the tolerable certainty of removing the disease, as quinine removes ague: it must be vanquished by the powers of the constitution, by the tendency to health, and our endeavor must be to place these powers in the most favorable condition possible for the struggle, and, where they appear to be insufficient, assist. Sometimes the constitutional power will throw off fever at the very onset.

Probably few medical men have not experienced in their own persons, when attending fever patients, that they had contracted the disease, and that after all its symptoms had been in course of development for four-and-twenty, or even eight-and-forty hours, it has been cast off, probably, by perspiration or diarrhœa, and health restored. From this almost ephemeral attack, to the week after week of continued fever, the disease may be thrown off at any period of its course. A person attacked with fever ought to be placed in as roomy and well-aired a situation as possible—better even in a barn than in a close or crowded room; the greatest cleanliness as regards everything around must be observed, and perfect quietude; if thirst is present, it should be liberally indulged with simple acidulated drinks, such as lemonade and toast-water, mixed; flax-seed tea, cold, &c.; if nourishment is taken, it should be given in moderate quantities, and consist principally of milk and farinaceous preparations, corn starch, farina, or grapes, oranges, and ripe fruits, if they do not create flatulence or diarrhœa, are allowable. If the skin is hot or dry, it should be sponged with water; this practice is beneficial, more or less, in most cases. By these simple means of management, almost without medicine, beyond some gentle purgative, as castor oil, &c., at intervals, to keep the bowels perfectly free of their necessarily depraved contents, many a case of fever may be well conducted to a favorable issue, with much more certainty than under a more *meddlesome treatment*—care being taken when signs of amendment show themselves, that there is not *too great hurry in giving or permitting strong nourishment.*

In more serious forms of fever, the same principle of treatment must be kept in view, but more urgent symptoms may call for more active interference; violent delirium may require the treatment pointed out under the article devoted to the subject; difficulty of breathing and cough may render a blister on the chest desirable, or tenderness of the bowels on pressure, particularly in the right lower side of the abdomen, near the hip bone, may call for the application of half a dozen leeches. Diarrhœa may require to be checked, by small doses of paregoric, or drinking occasionally a wine glassful of allspice tea, with a little prepared chalk in it, or constipation removed by gentle aperients; castor oil or rhubarb or senna will generally be found safest and best, or injections of weak soap-suds. Sleeplessness at night, with convulsive starting of the fingers, may require laudanum, ten to forty drops; or the general sinking of the powers, the pulse becoming feeble and easily extinguished, may call for the administration of wine or brandy in teaspoonful doses, every hour or two, according to symptoms, with strong meat-broth, or gravy, in frequently repeated small quantities. At this time care must be taken to observe whether urine is passed; if there seems to be difficulty, a bag of hot bran on the lower part of the body will possibly make it easier; if it dribbles away, means should be taken to protect the back and hips of the patient from being wetted with it. This may be done in various ways, either by waterproof material, or by constant renewal of dry cloths; it is much better effected, however, by bags of bran, or a sponge, so placed as to absorb the urine as it comes away. If with every quart of bran, four ounces of the *diluted* sulphuric acid be mixed, it will neutralize the ammoniacal emanations which so quickly arise wherever urine collects.

It is often the case, that patients who lie long in fever become liable to bed-sores, or ulcerations on the prominent parts of the body which are subject to pressure as they lie; these are, especially the back and hips, points of the shoulder-blades, back of the head, tips of the ears, &c. When these ulcerations form, they not only add materially to the sufferings of the patient, but may become the cause of a fatal termination to a case that might otherwise have recovered. They should, if possible, be prevented. The parts named above should be frequently examined, and on the slightest appearance of redness, the skin at the spot should be rubbed with whisky, brandy, bay rum, or spirits of camphor. When the skin has actually broken, it may be dressed with simple spermaceti ointment spread on linen; lead plaster spread on soft leather is often useful, (see directions for making under head of " Medicines,") or the white of egg beat up with alum. Both as a preventive and as a remedy after the sores have formed, the parts should be relieved from pressure as much as possible by various arrangements of cushions, &c., the elastic ones made for the purpose being the most suitable.

The foregoing *are the general principles on which a case of fever is to be managed ; by attention to them an unprofessional person will be much more likely to do good than by meddlesome interference.* Attention to the ventilation of the room, to the perfect cleanliness of the patient and of every thing around— free supply of diluent drinks, as previously stated, and care that the bowels are duly, but not *forcibly* relieved of their always depraved contents, ought to constitute the chief resources of the domestic management of fever.

Again it is repeated, fever is not a disease to be *cured,* but to be *guided to a safe termination.*

In regard to the causes of fever, they may be enumerated as follows : Predisposing—whatever lowers, either temporarily or permanently, the standard of the general health ; and Direct—contagion and the products of animal or vegetable decomposition.

Attention is called before closing to the employment of *fresh* yeast in cases of fever, particularly of a low, malignant or putrid tendency, in which it is most useful. It is given in tablespoonful doses, repeated every three or four hours.

It should also be borne in mind in the treatment of these fevers, that the patient should be allowed to partake freely of gum arabic dissolved in cold water ; or slippery elm water, adding a few drops of spirits nitre occasionally. Also let him have a piece of some pleasant bark, root, or herb in his mouth constantly (if agreeable) to chew on ; it promotes the flow of saliva or spittle, and saves much suffering from dryness of the mouth. A piece of orange or lemon peel, sassafras root, an gelica root, dogwood bark, or things of that kind, are admissible. When in the course of the disease there is a tendency to *sinking,* do not be afraid to give one or two grains of quinine every three or four hours. to be given in smaller doses or left off entirely if much headache should follow its use.

In the *early period* or *forming stage* of this disease, an emetic will often be found advantageous. A dose of ipecac may be administered, and its operation promoted by giving the patient large quantities of weak camomile tea or warm water. After the patient has rested from the effects of the emetic, a dose of castor oil should be administered, and followed by purgative injections if necessary. The purgative should be again repeated, after a short interval, provided the first dose has not the effect of producing full and free evacuations from the bowels. When these are obtained, they relieve the oppression of the stomach ; render the tongue clean and soft, mitigate the thirst and restlessness, and the morbid heat of the surface, and prevent that formidable oppression of the brain and nervous system upon which the symptoms of colapse, which attend the second stage of the disease, depend.

Whenever the symptoms of excitement run high, the patient feeling hot and restless, and the skin being universally and steadily above the natural temperature, and at the same time perfectly dry, no remedy has been found to act with so much promptitude, and to be productive of greater benefit, than the application of cold water to the surface of the body. It may be used in the form of ablution, or sponging ; the patient is then to be wiped perfectly dry, and covered up in bed ; after which a bowl of warm tea or thin gruel is to be given him. It can only be employed, however, during the height of the fever, when the heat of the skin is steadily above the natural temperature, and no tendency to perspiration is present. When the head is much affected, cold should be applied to the scalp in the same manner as was directed in bilious fever

YELLOW FEVER.

This is a disease peculiar to warm climates, and is more prevalent, and in the worst form in proportion to the greater heat of the season; however, heat *alone* is not sufficient to produce yellow fever. It is attended with great diversity of symptoms in different cases. In some patients it commences with symptoms which would appear to indicate a perfectly mild disease. In general, it attacks suddenly, with a chill, pain in the head, back and limbs, and occasionally with a nausea or vomiting. The eyes appear red and inflamed, and feel hot and painful, the pupils are sometimes dilated, but more generally contracted. The chill is commonly of very short duration; as it goes off, the pulse becomes, in general, full and quick, the skin very hot and dry, the face flushed, the eyes red and watery. The face has an expression similar to that of a person intoxicated; there is great oppression and tightness at the pit of the stomach, with constant restlessness and frequent sighing; the bowels are costive, the tongue white and coated, or of a bright red color. The heat of the skin and pains in the head and limbs augment rapidly during the first thirty-six hours, and then gradually decrease, so that at the end of the third day there is either a very great remission of symptoms, or even complete intermission of the disease, terminating in the recovery of the patient. When an imperfect intermission only takes place, it is in a few hours succeeded by pain and a sense of burning in the stomach, constant nausea, with efforts to vomit, discharging at first only a little thick green mucus. The pulse is now small, quick, and irregular; the stomach sore to the touch, the bowels costive or griped, and the tongue brown and dry. The symptoms rapidly increase, until at length the sense of heat and pain of the stomach cease, and vomiting of a black, flaky matter, resembling coffee-grounds, takes place, called "black vomit." The patient now often feels quite easy, thinks he is getting better, rises out of bed and walks about, but soon becomes exhausted, and falls into convulsions, or into a state of lethargy, terminating in death. Some patients become drowsy, and die without a struggle; in others, death is preceded by the discharge from the mouth, nose and ears, of a dark-colored blood. The skin of most becomes of a bright yellow before the black vomit occurs. In some instances which terminate fatally, there is no vomiting of black fluid at *any period of the disease.*

The predisposing and exciting causes of yellow fever are the same as in other fevers : namely, exposure to heat, fatigue, cold, intemperance, fear, anxiety, etc.

In regard to the treatment of yellow fever, this must vary according to the stage of the attack, and the violence of the symptoms by which it is attended. In the early or forming period of the disease, marked by a slight disorder of the stomach, headache, pains in the limbs, lassitude, and a sense of chilliness, much advantage may be derived from the use of the warm bath, brisk frictions of the surface, and the administration of calomel, followed in the course of a few hours by castor oil, ten to fifteen grains ; injections of soap suds also will be beneficial in procuring a speedy evacuation of the contents of the bowels. At a later period of the disease, when all the prominent symptoms are fully developed, when there is heat of the skin, burning pain and sense of distress at the stomach, with soreness upon pressure, a mustard plaster applied for a few minutes at a time, every hour, will often afford relief : or leeches or cups may be applied over the stomach.

Cold water applied to the surface during the height of the fever, by sponging, should never be neglected. It is among the most powerful means we possess of reducing the violent excitement by which this fever is attended. When the head is much affected, after the application of leeches or cups to the temples or nape of the neck, keeping the head constantly wet with cold vinegar, water, or the application to the shaved scalp of a bladder partly filled with powdered ice, will be productive of the most decided benefit.

The thirst of the patient is to be quenched by small and repeated draughts of some cold beverage : iced water or lemonade, toast or barley water, with the addition of ice, will be found among the best. In cases attended with great irritability of the stomach, we know of no remedy from which more advantage will be obtained than from a teaspoonful of powdered ice occasionally repeated.

It is to be recollected, that the foregoing remedies are only adapted to the first stage of the disease, and if judiciously and vigorously employed, the danger of the second stage will be avoided, and the most unpleasant symptoms diminished within the first twenty-four hours.

In the second stage of yellow fever, our hopes of arresting the fatal progress of the case are but slender. The same treatment as to the gum arabic and slippery elm water, for cooling drinks, etc., also, the quinine, when symptoms of sinking appear, which was recommended in typhoid fever, is as applicable in yellow fever. In fact, I would recommend in the very beginning, as soon as the bowels have been opened, to give a mixture composed of two grains of quinine, five grains of powdered charcoal, and one grain of powdered sassafras bark, to be repeated every two hours. This will not interfere with the other treatment, and will by its power of *keeping up the strength* on the one hand, and its antiseptic properties on the other, be well worthy a trial.

VOMITING OF BLOOD.

This is usually preceded by a feeling of heaviness and pain about the stomach, but without cough. Generally it is preceded by nausea, loss of appetite, etc. There is often pain or uneasiness in the left side, with anxiety and a sense of tightness in the chest. The blood discharged is generally dark colored, clotted, and often mixed with some of the contents of the stomach. In some cases there is also a discharge of dark colored blood by stool. The amount of blood vomited is various. Sometimes a large quantity is thrown up at once, when the disease ceases, and does not again occur; but in general the vomiting is repeated at short intervals, until the patient is completely exhausted. Vomiting of blood may occur in persons of a full habit and robust constitution, but is most common in those who are weakly, or who have labored for a considerable time under a disease of the digestive organs.

The most frequent causes appear to be grief, or other depressing or violent passions; costiveness, especially if occurring in a constitution in which the stomach is particularly irritable; blows on the region of the organ affected; fullness of habit, combined with an intemperate mode of life; the use of intoxicating drinks; the suppression of the menstrual flux, or of the discharge from bleeding piles; acrid or irritating substances taken into the stomach, and the abuse of emetics and active purgatives.

It is, in general, easily distinguished from spitting of blood, by the blood being here brought up by vomiting, and by its being of a darker color. It is also generally mixed with some of the contents of the stomach. In spitting of blood, on the contrary, the fluid discharged from the lungs is brought up by *hawking or coughing,* and is generally of a *bright red color.*

When a person is attacked with vomiting of blood, he should be kept perfectly quiet, in a room, the air of which is rather cool than warm, and his dress should be loosed so as to prevent any pressure upon the stomach. If the complain' occur in a person of a full habit, and possessing considerable general strength, cups or leeches should be freely applied over the stomach. In cases occurring in debilitated habits, or where the discharge of blood has already lowered considerably the strength, cups alone over the stomach will be proper, and these should be applied *without scarifying.*

If the bowels are costive, they should be opened by an injection of warm water mixed with thin gruel and a little soap suds. The thirst in this complaint is always considerable; it may be allayed by the patient taking small quantities of gum arabic, slippery elm, or toast water *perfectly cold.* In many cases where vomiting is incessant, advantage will be derived from the administration of an occasional spoonful of iced water, or even of powdered ice.

When the vomiting of blood has been produced by suppressed menses or a cessation of the discharge from bleeding piles, leeches should be applied about the upper part of the thighs, or to the anus (outlet of the bowels)—the patient's feet should be immersed in hot water, to which salt or mustard has been added, and afterwards mustard poultices applied to the ankles. If it be necessary, from the great discharge of blood, promptly to put a stop to the vomiting, twenty-five grains of ipecac should be taken, in a little warm water, and if the first dose fail to stop it, a second may be administered after an interval of two or three hours. Ipecac is frequently of very great service in this complaint. It is applicable to the case of strong as well as of weakly persons.

Sugar of lead is, in many cases, a medicine of great value here, as well as in all other profuse bleedings. Combining it with the ipecac is often advantageous: we may give two grains of the sugar of lead with two of the ipecac. Calomel in grain doses, every two hours, will in some instances speedily arrest the vomiting; it should be given combined with five grains of gum arabic, and mixed with a little water.

After the blood has ceased to flow, and the patient begins to recover himself, the further treatment of the case will depend in a great measure upon the nature of the symptoms which remain. The occasional application of cups or mustard plasters to the stomach will still be demanded, in most cases.

The diet recommended under *Dyspepsia* will, in genera', be the most proper.

DROPSY.

THE disease to which the term *dropsy* is most usually appli
ed, is that general swelling over the whole body, of a soft and
doughy feel, accompanied with great weakness, and other
symptoms to be hereafter described ; or it is that swelling of
the belly, from the accumulation of a watery fluid within, by
which it is often distended to a prodigious size.

GENERAL DROPSY is a swelling of the external parts of the
body, from a collection of fluid chiefly under the skin ; the
swelling first appearing on particular parts only, but at length
gradually extending to the whole surface. The swelling in
dropsy is always soft and uniform over any member ; and
when pressure is made upon the skin with the finger, a pit or
hollow is formed by the water being pressed out of some of
the cells beneath, into the neighboring ones. Soon after the
pressure is removed, the swelling returns to its former fullness. Generally, the swelling appears first on the lower extremities, and that only in the evening ; it is seldom, at a i
early period of the disease, very perceptible in the morning.
The more a person has been in the erect posture through the
day, the greater is the swelling towards evening. It is easy to
be seen, that this is owing to the water making its way downwards by its own weight ; while the posture during sleep
allows it either to diffuse itself equally over the whole body,
or if the quantity be great, and the disease far advanced, to
accumulate in the upper parts of the body, and to occasion the
swelled face, and closed-up eyes, which some dropsical patients exhibit in the morning. Sometimes the fluid which is
accumulated immediately under the skin, oozes out through
the pores of the skin ; sometimes being too thick to do so, it
raises the outer skin in blisters. Sometimes again, the skin
not allowing the water to pass through it, is hardened by distention, and gives the swelling an unusual degree of firmness.
General dropsy is almost always attended with scantiness of
urine, which is generally high-colored, and after cooling, lets
fall a copious reddish sediment or settlings. There is also an
unusual degree of thirst ; the appetite is generally bad ; and
there is a feeling of debility, with sluggishness, drowsiness,
and disinclination to motion.

Exhausting diseases of various kinds, copious and long-continued discharges of blood, or any other weakening cause, may
produce a debility of the system, which leads to dropsy. Intemperance in the use of strong liquors, especially dram-drink-
ing, is one of the most common, intractable, and fatal causes

of dropsy. Profuse bleeding has been already mentioned as causing general debility, and as producing dropsy.

From the account given of the causes of dropsy, it is evident that, while they continue to act, it will be useless to attempt carrying off the accumulated fluid, and therefore one of our first objects must be to put a stop to those causes. In a dram-drinker, or an indolent, debilitated person, it will be in vain to give drugs, or to direct any particular diet, however salutary, till these bad habits are given up. We are next to attempt to get rid of the water already accumulated. Sometimes very strong purgatives, particularly those which produce large watery stools, procure a very rapid discharge of the accumulated fluid; of this kind are gamboge, scammony, and the like; or jalap in combination with aloes, scammony, gamboge, or cream of tartar. A powder for this purpose may consist of eight grains of aloes, ten of jalap, and six of gamboge or scammony, to be taken in a bolus, or suspended in syrup or mucilage, to be repeated, if necessary, every two or three days. Or ten grains of jalap, with a teaspoonful or two of cream tarter, may be used, given in a wine glass of cold water, on an empty stomach. Another class of remedies much used in dropsy, are medicines that act on the kidneys, and could we insure the success of their operation, we should be better pleased to carry off the dropsical waters in this way, than by any other method whatever. Cream of tartar is, perhaps, one of the best diuretics in general dropsy. Dissolve a large tablespoonful of cream tartar in half pint of water, and take a tablespoonful of the mixture every hour or two. Or mix one tablespoonful of spirits nitre with half pint of water, and take one tablespoonful every two hours.

It is an important improvement in the cure of dropsies, that the patient is not restricted in the quantity of fluid which he chooses to drink, but that a plentiful allowance of watery liquors is considered as rather conducive to a cure, by conveying to the kidneys any diuretic we mean to employ, and even as of itself greatly promoting their action. Friction is another means of promoting the action of the absorbents; and exercise, if the patient can take it, may have the same effect; and when the swelling is abated in the morning, skilful and equable bandaging (commencing at the toes and going upwards to the knees, or further,) will prevent the swelling of the legs towards night. When by these or other means, we have managed to get rid of the water already effused, our next object is to prevent its re-accumulation; and by strengthening the system, to complete the cure of the disease. Exercise, and the proper regulation of the diet, are important items in this plan; and are to be accompanied, in the debilitated, by tonic medicines, as Peruvian bark tea, (one tablespoonful of the bark to half pint of hot water,) of which take, when cold, one-third of this mixture before each meal. Or the tea of wild cherry bark may be given in the same way. Great attention is to be paid to the state of the bowels, to keep them open and free; and we must not neglect to keep up a proper action of the skin and of the kidneys, by wearing flannel next the skin, and giving some water-melon seed tea, occasionally through the day or a few drops of spirits of nitre in a little water, three or four times a day. Also bear in mind that the sponge-bath of salt and water, over the entire surface of the body and limbs, once a day, to be followed by a rough towel and flesh brush, must not be neglected, and should be continued for two or three months.

ASTHMA, OR PHTHISIC.

This disease is an affection of the lungs, which comes on by spells, most generally at night, and is attended by a frequent, difficult, and short respiration, together with a wheezing noise tightness across the chest, and a cough; all of which symptoms are much increased when the patient is lying down.

Changes in the weather, peculiarity of situation, errors in diet, anxiety, fatigue, mental excitement, may any of them produce a paroxysm of asthma in the predisposed. The attack itself is indicated by feelings of indigestion and flatulence, headache, chilliness, languor, and drowsiness. After having experienced these sensations during the day, the asthmatic individual is probably awakened from his first sleep by a distressing sensation of constriction of the chest; he is forced to sit up in bed, laboring for breath, or, it may be, to seek an open window. The distressed state of the breathing, if not relieved by remedies, continues for some hours, and at last gradually subsides; the characteristic wheezing becomes less; the cough, almost impossible before, is now brought out, and sleep, never more welcome, comes on. The latter stage of the paroxysm of asthma is generally accompanied with expectoration of mucus—sometimes it is not; and upon this a distinction into dry and humid (or moist) asthma is founded. Confirmed asthmatics have a distressed cast of countenance, and acquire a peculiar rounding or elevation of the shoulders, perfectly characteristic. Asthma may occur at any period of life, but is more general about the middle; and men are more commonly the subjects of it than women. The remedies are various; what gives immediate and full relief to one person tota y fails in another. The practice of smoking the leaves and stems of stramonium, or thorn-apple, is now extensively and popularly resorted to; with some it succeeds admirably; to others it seems to be hurtful; it may be tried. Æther and laudanum is a favorite combination; half a teaspoonful of the former along with twenty drops of the latter being given in a wine-glassful of water. A teaspoonful of hartshorn may be substituted for the æther, but is not so efficacious. Twenty grains of powdered ipecac, given in half a wineglassful of water, to act as an emetic, may be of service, particularly if the attack has come on after a full meal, or if there is any suspicion of the stomach being loaded. Some experience much benefit from strong coffee, drank without milk or sugar. The inhalation of chloroform, a few drops sprinkled on a pocket-handkerchief, has lately been found to afford relief; the fumes of burning nitre (saltpetre) diffused through the air of the apartment, by means of pieces of blotting-paper dipped in a thorough solution of the salt, and dried, is also useful. One of these, about the size of the hand, ignited and placed upon a plate in the room, quickly diffuses its fumes throughout the apartment. When there is great dryness and deficient expectoration, steam from bitter herbs, (as in catarrh in the head,) inhaled either simply or with a few drops of sulphuric æther, is worth a trial. If there is much acidity of stomach, ten or fifteen grains of carbonate of soda with a teaspoonful of hartshorn in a little water should be given. Indeed, when an attack of asthma is threatened or has come on, care must be always taken, as far as possible, to remove any existing disordered action. A cup of warm tea and retirement to a warm bed will remove the chilly sensation. Costiveness of the bowels ought to be relieved by a gentle dose of castor oil, or of rhubarb and magnesia; flatulence or acidity is corrected by a mixture of water, in which some essence of peppermint and prepared chalk have been put, given every hour. Flatulence particularly must oc obviated, and all sources of it avoided Effervescing draughts, soda-water, and such-like, are almost always hurtful. The effects of situation and of atmospheric peculiarity upon asthmatics are most varied; some can breathe freely in clear dry air, which drives others into a damp cellar for temporary relief: a close, warm room which suits one will be unbearable to another. Individuals who are never free from asthma in some situations, lose their attack as soon as they are removed.

These are peculiarities of which all should be aware. Certain odors produce asthmatic breathing in the predisposed;—

the powder of ipecac is notorious for this effect ; and the smel. of new-made hay, so pleasant to most persons, produces hay asthma in a few unfortunate individuals.

Sponging the chest and shoulders every morning with cold salt water, friction being afterwards made with a rough towel or flesh brush, is a practice to be recommended, provided no other predisposition forbids.

A few drops of æther, put into a gill of cold tar-water, and inhaled by means of one of the Inhalers recommended in catarrh in the head, will be of service in some cases.

SCALDS AND BURNS.

The frequency of accidents of this kind makes it obligatory for every individual to be posted as to the best treatment in such cases, as it is necessary to apply our remedies *at once*, to be the most successful, for in all accidents from scalds and burns, it seems to be of the utmost importance to apply a remedy at the instant; for by this means the violent anguish is allayed, and blistering, which in scalds, at least, is usually so considerable as to lay the foundation for a tedious curative process, is in a great degree prevented.

In the case of slight burns and of scalds, generally the best application is the cotton wadding in sheets ; it should be used to envelop the injured parts, double if possible, and bound or bandaged on with moderate firmness. If this mode of treatment be resorted to within the first twenty minutes after the injury, nothing more need be done ; the cotton may be allowed to remain on from twenty-four hours to three or four days, according to the severity of the accident. Under its use blistering rarely occurs, and if it has commenced before the application, it subsides quickly and painlessly. For the first ten minutes after the cotton-dressing is put on, the pain of the injured parts seems increased, but ere long it diminishes, and the inflamed skin appears to relieve itself by gentle perspiration. In the cases such as above named, when cotton is to be procured—and no house in the country ought to be without one or two sheets of it—it is perfectly unnecessary to use any other measures.

A lotion made with six tablespoonfuls of vinegar to a pint of water may with advantage be kept constantly applied to a burn, *if it be not extensive*—a saturated solution of carbonate of soda has likewise been recommended. Cold water is perhaps the most directly grateful application to a burnt or scalded surface, and if *continued sufficiently long*, will undoubtedly restore the usual condition of the part, but it must be persevered with for many hours; and when a burn or scald is extensive, this is a serious objection, in consequence of the extreme *constitutional depression* which so often follows the accident, especially in the young. We would warn parents of the necessity of watching closely the effects of even slight injuries of this kind upon children, particularly when the chest or abdomen are the seat of the accident : extreme depression—requiring the use of stimulants—may unexpectedly come on, and death, from an apparently very slight cause, be the result. *When cotton is not readily procurable,* flour sprinkled over the surface is an admirable substitute, *even in slight burns,* but is more useful still in those severe effects of heat in which the parts are deeply destroyed by the action of fire ; in these cases, flour applied at once, and repeated again and again for days together, whatever slight moisture seems oozing through the caked covering it forms—is the most generally applicable, pleasant, and safest remedy ; a little fresh sweet-oil applied to the surface in the first instance will make it adhere. Applying a thick coat of white lead paint has been found an excellent remedy. It should not be washed off, but suffered to remain on until cast off by nature's own process.

Whatever application is used in the treatment of a burn, should be calculated to *exclude the action of the external air ;* it ought to be one, also, which does not require *frequent changing ;* the more extensive the surface involved in the accident, the greater care should be taken not to expose it to air, which increases pain, and adds to the general depression of the system. This depression must always be carefully watched, and combated by the use of ammonia, wine, or spirit, sufficient to support without stimulating. When pain is excessive, and is irritating the nervous system, a gentle opiate is required ; but in some of the severest burns, the sensation, not only in the injured part, but generally, is either wholly or partially abolished, in consequence of the shock to the nervous system at large. This is always a bad symptom. In the less severe forms of injury from heat, if the cotton, the flour, or cold water, have been properly used, little after-treatment is necessary ; but when a burn has been neglected or badly treated, the blisters broken, and when the true skin beneath is inflamed and secreting matter, a simple tepid bread and water poultice should, in the first place, be applied for six or eight hours, and after it an ointment composed of one tablespoonful of white lead paint, rubbed up with same amount of perfectly fresh lard. This ointment spread on linen quickly relieves the very painful condition of the injured surface. One teaspoon ful of powdered sugar of lead will do when the paint can not be had, rubbed up with one tablespoonful of lard.

In cases of deep burn, with destruction of the parts, after the flour has been applied some days, it begins to be pushed off by the matter formed underneath : at this time poultices are to be continued until the caked flour is separated, and the surface below exposed, after which the simple dressing with tepid water will generally be the best and safest application ; or, in a later stage, if healing is slow, use the lead oint ment recommended above.

During the cure of burns involving contiguous parts, such as the fingers, care must always be taken to keep the surfaces asunder by the interposed dressings : otherwise they may grow together. After extensive burns or scalds, the constitution requires attention—the stimulating treatment of the first few hours or days must be dropped when feverish symptoms come on, and mild and cooling diet, gentle purgatives, and cooling drinks, as lemonade, apple tea, toast-water, &c., administered : opium or laudanum, in ordinary doses, being given if requisite, to allay pain or nervous irritation. This system will again require to be changed for one of stronger nourishment—meat soups, meat and wine, or other stimuli, if there is continued discharge. The use of stimulating diet, however, requires caution, on account of the tendency to inflammation of the lining membrane of the stomach and bowels, which exists during convalescence from injury to the skin by heat.

The sores left by burns have some peculiarities. They shoot out *fungous*, or *proud flesh*, as it is vulgarly called ; they are difficult to heal : and when they do heal, contract so much as often to produce great deformity.

To arrest the growth of this *over-growing flesh*, sprinkle the part with burnt alum, once or twice a day, or apply lunar caustic as often as may be required.

I will state, in concluding this subject, that during my residence in Mississippi, during the years 1849–50, I had occasion in my practice very often to try the application of lunar caustic to recent burns and scalds, with the happiest effects—the pain ceasing almost immediately, and the parts healing in a short time. It was applied with a camel's hair brush or linen mop at once, and again in a few days, if healing did not seem to be going on satisfactorily. Twenty grains of caustic dissolved in an ounce of water, were the proportions used.

RHEUMATISM.

T. is disease is divided into the *chronic* and the *acute*; being known by the former appellation when there is no great degree of inflammation or fever, but merely pains; and by the latter when both fever and inflammation exist.

It may arise at all times of the year; but the Spring and Autumn are the seasons in which it is most prevalent. It attacks persons of all ages, but very young people are more exempt from it than adults. Those whose employments subject them to exposures to great heat and cold, are particularly liable to Rheumatism.

Obstructed perspiration, occasioned by wearing wet clothes, lying in damp linen, sleeping on the ground, or in damp cellars or basements, or by being exposed to cool air when the body has been heated by exercise, or by coming from a crowded room into the cool air, the causes which usually produce Rheumatism. Those who are much afflicted with this complaint, are very apt to be sensible of the approach of wet weather, by wandering pains about them at that period.

Chronic Rheumatism is attended with pains in the head, shoulders and other large joints, which at times are confined to one particular part, and at others shift from one joint to another without occasioning any inflammation or fever; and in this manner the complaint continues often for a considerable time, and at length goes off, leaving the parts which have been affected in a state of debility, and liable to fresh impressions on the approach of bad weather.

Acute Rheumatism usually commences with languor, chilliness, succeeded by heat, thirst, restlessness, and a quick pulse; there is also a sense of weight, coolness of the limbs, and confined bowels. In the course of a day or two, inflammation, with acute pain and swelling, makes its appearance in one or more of the larger joints of the body. The pain is frequently transitory, and apt to shift from joint to joint, leaving the part previously occupied, swollen, red and extremely tender to the touch. The pulse now becomes full and hard; the tongue preserves a steady whiteness; the bowels are commonly very costive; the urine high-colored; and often there is a profuse sweating, unattended by relief.

Sometimes, however, the pain is the first symptom, and the fever follows. When the pain is not very severe, and confined to a few parts, the fever is slight; when it is severe and felt in many parts, the fever is more considerable, and it is most so when the pains extend over the whole body. Both the pain and fever generally suffer an increase in the evening, and a remission towards morning. The pains are much increased on the slightest motion requiring the action of the muscles affect ed, and are most severe, as well as most apt to shift their place, in the night time. The fever abates sooner than the local symptoms, and is rarely protracted beyond a fortnight or three weeks. The pains, for the most part, are the last symptom which leaves the patient.

When an individual who has either suffered from an attack of acute rheumatism, or is hereditarily predisposed to it, or indeed when any one, after exposure likely to produce an attack, suspects the disease to be impending, the first effort should be to excite the free action of the skin. If a warm or vapor bath can be procured, it is highly desirable; if it can- not, the best substitute will be a well-warmed bed with hot bran bags, or hot bottles, and the free use of warm drinks. A mixture composed of half an ounce of spirit of nitre, one drachm of paregoric, and fifteen drops of wine of ipecac, in a wineglassful of water, may be given every four or five hours. To the above draught, ten grains of the bicarbonate of potash may be added with advantage. Under the above circum- stances, any stimulant sweating may be given with benefit, even a little gin, or other spirit, or wine, *well diluted* with hot water; these stimuli being used, of course, *only at first*, and while *fever is not yet present*.

When an attack of acute rheumatism is established, if medi- cal assistance is not immediately procurable, the patient must be kept in bed, *moderately warm*, the thirst quenched by the free use of simple warm drinks, and the diet reduced to a very low scale, any thing like alcoholic stimuli, or animal prepara- tions, being strictly forbidden, except in the case of very debilitated persons, when animal broths, such as beef-tea, may be permitted in moderation. If fever runs high, tartar emetic, in from an eighth to a fourth of a grain dose, may be given every four, five, or six hours, and with this, from six to ten drops of laudanum may be combined, to alleviate the pain. Dover's powder, in doses from ten to twenty grains, may be given at bed time, and followed by a purgative of castor oil next morning. The above measures might with safety be adopted, under peculiar circumstances, in the absence of medi- cal assistance. There are, however, many other modes of treatment. That by large doses of nitrate of potash, or salt petre, has had its advocates, and may be tried: the mode of administration, as laid down by Dr. Basham, is to dissolve two ounces of the saltpetre in two quarts of water, and to give this quantity in the course of twenty-four hours. This treatment, which is said to be very successful at times, might be available in the absence at least of other remedies; of course, if symptoms indicative of irritation of the stomach or bowels came on, it would require to be abandoned. More recently the treatment of acute rheumatism by lemon-juice has come into practice, and seems in many cases to answer extremely well. This treatment has the advantage of being perfectly safe, and therefore, where the lemon-juice can be procured, may, without danger, be pursued in the absence of a medical man. One tablespoonful, or half an ounce of lemon-juice, is to be given every four hours. The "alkaline treatment" of acute rheumatism is followed by some fifteen to thirty grains of bicarbonate of potash being given, well diluted in water, every four hours.

As regards the local treatment of the inflamed joints, little is to be done in a disease which shifts its site as rapidly as acute rheumatism; for even if it can be driven from one joint, it must, as long as the poison is in the constitution, show itself elsewhere, it may be in the heart.

When the joints are much swollen and painful, much ease may be given by enveloping them in a large quantity of the soft carded cotton—"cotton wool"—over which there is wrapped *completely* a piece of oiled silk, or oiled paper. By this air-tight covering, the joints are kept in a perfect vapor bath, and when it is removed after twelve or twenty-four hours, the wool will be found saturated with moisture which is strongly acid.

The causes of acute rheumatism already pointed out will suggest to most persons the precautions to be adopted, especially when liability to the disease exists. Cold and wet are particularly to be guarded against, and, after exposure, the preventive measures already laid down adopted. Flannel or woolen, worn next the skin, must always be regarded as one of the chief preventives; it should of course be proportioned in thickness to the season and temperature.

Persons of full habit, liable to rheumatic attacks, should avoid malt liquor generally, should take animal food sparingly, and avoid violent exertions which heat the body. Persons of spare or feeble habit, may live better, and indeed require to keep up the condition of the body to as good a pitch as possible.

In chronic rheumatism, instead of heat, there is often a sensation of cold around the affected parts. The chronic nature of this disease must generally place it under proper medical control; the chief efforts of the unprofessional must be to correct any slight deviations from the general health, to pro tect the affected parts especially from cold by means of warm clothing, and to use friction, either with simple oil, or by means of liniment of camphor, sweet oil, and laudanum, in equal parts. Much comfort is not only derived from friction, but, if combined with proper exercise of the joint or joints, it may do much to prevent permanent deformity. In chronic rheumatism, warmth of climate is of much importance, and as much should be done toward the attainment of this as circumstances will permit.

Chronic rheumatism, properly so called, is such as above described; but the term rheumatism, or rheumatic pain, is also used to a great variety of anomalous pains, and from this has arisen considerable confusion. The best marked of these is "muscular rheumatism," which affects chiefly the thick muscles, such as those of the shoulders, arms, neck, loins. &c This form of rheumatism often comes on suddenly, after exposure to a current of cold air—sometimes after cold bathing; its chief characteristic is severe pain, when the affected muscles are thrown into action. This muscular rheumatism seems to be a purely local affection, and is generally removable by purely local remedies. Of these, the best is a large hot bran or other poultice, or some other means of applying heat and moisture, applied over the affected part, for a few hours; this often at once cures—care of course must be taken to protect the part to which the heat has been applied, by a covering of flannel. After the hot application is removed, if the heat does not entirely cure, and even instead of it at times, a liniment of camphor, sweet oil and laudanum, in equal parts, or volatile liniment, mixed with laudanum, either alone or combined, with one-third of turpentine, may be used to advantage; two or three teaspoonfuls being well rubbed into the part every few hours.

Any notice of rheumatism at the present day must be imperfect, without some allusion to electric and galvanic agencies, galvanic rings, electric chains, &c. That these appliances are at times of apparent service in cases of chronic rheumatism is undoubted, and if such is the case, we are not justified in rejecting their aid because we cannot exactly explain the why and wherefore of their action.

The author has found, as a general thing, that the wine of colchicum, given in from two to five drop doses, three times a day, gradually increasing the dose until sickness at the stomach has been produced, or looseness of the bowels, or both; and then going back again to the small doses, (after a day or two,) and gradually increasing again if necessary, has been the most reliable remedy in rheumatism. A liniment of whisky, (or alcohol) keroane oil, laudanum, sweet oil, and turpentine, in equal parts, applied as needed, is about the best local application.

G O U T.

This disease is marked by violent pains, principally in the ball or joint of the great toe, with redness, swelling and general feverishness, returning at intervals. It is preceded by symptoms of derangement of the digestive organs. When the attack passes off, the patient is left in apparent good health. Gout sometimes comes on very suddenly, particularly in its first attacks. In general, however, the inflammation of the joint is preceded by various symptoms, indicating a want of vigor in different parts of the body. The patient is incapable of his usual exertions, either of mind or body ; becomes languid, listless, and subject to slight feverish attacks, especially in the evening. He complains of pains in the head, coldness of the feet and hands, impaired appetite, flatulency, heartburn, spasms of the stomach, and the usual symptoms of indigestion. He is oppressed with heaviness after meals, and a disturbed, unrefreshing sleep ensues. The bowels are seldom regular, being either costive or too much relaxed ; the mind at this period, being generally irritable, anxious, and alarmed at the least appearance of danger. A *deficiency of perspiration* in the feet also, with a *distended* state of their *veins,* cramps, numbness of the feet and legs, and other strange sensations, often presage the approaching fit. The duration of these symptoms, previously to the fit occurring, is various; sometimes only a day or two, at other times, many weeks.

The fit sometimes makes its attack in the evening ; more commonly, about two or three o'clock in the morning. The patient goes to bed free from any suffering more than the symptoms above alluded to, but is awakened about this time by a very acute pain, generally in the first joint of the great toe, the pain often resembling that of a dislocated bone, with a sensation as if hot water were poured on the part. It sometimes extends itself over all the bones of the toe and fore part of the foot, resembling the pain occasioned by the stretching or tearing of a membrane. Cold shivering is felt at the commencement of the pain, which is succeeded by heat and other symptoms of fever. The pain and fever increase, with much restlessness, till about the middle of the succeeding night; after which they gradually abate, and in the most favorable cases, there is little either of pain or fever for twenty-four hours subsequently, at least in the early period of the disease. The patient, as soon as he obtains some relief from his pain, generally falls asleep, a gentle sweat comes on, and the part which the pain occupied, becomes red and swelled. In most cases, however, the pain and fever return on the succeeding night with less violence, and continue to do so for several nights, becoming less severe till they cease.

The symptoms here described constitutes an ordinary fit, simple *acute* gout. But it often happens, that after the pain has abated in one foot, it attacks the other, where it runs the same course; and in those who have labored under repeated attacks of the disease, the foot first attacked is often seized a second time, as the pain in the other subsides, which is again attacked in its turn, and they are thus alternately affected for a considerable length of time. In other cases, it seizes on both feet at the same time. After frequent returns, it begins to seize upon the joints of the hand, and at length the larger joints. When the gouty tendency is very great, almost every joint of the body suffers; the pain, when it leaves one, immediately fixing in another.

The whole fit is generally finished in about fourteen days in persons of good constitution. In the aged and those who have been long subject to the gout, it generally lasts about two months; and in those who are much debilitated, either by age, or the long continuance of the disease, till the summer heats set in. In the first attacks, the joints soon recover their strength and suppleness; but after the disease has recurred frequently, and the fits are long protracted, they remain weak and stiff, and at length lose all motion.

Chronic gout, (which is by some physicians called irregular gout,) is the disease of a worn-out or debilitated constitution. Here the inflammation and pain are more slight, irregular, and wandering, than in the acute; there is only faint redness of the affected joint, or no change at all from the natural appearance of its surface; there is much permanent distention of parts, or continued swelling, with impaired moving power; no critical indications of the disease terminating, present themselves.

The subjects of chronic gout are generally such as have, for a considerable time, labored under regular attacks of the acute form of the disease; this, however, is not universally the case.

Retrocedent gout is that form of the disease, in which the morbid action is suddenly transferred from the joint, or other external part affected, to some internal organ, as the stomach, intestines, head, &c.

Whatever tends to produce an unhealthy fullness of the blood-vessels, disorder the digestive organs, and impair the vigor of the system, may be ranked among the causes of the gout. Perhaps the principal causes are an indolent and luxurious life, or a sedentary and studious one; hereditary pre-disposition; anxiety or vexation of mind, excessive avocations of any kind; cold. improper diet, or immoderate indulgence in fermented or acid liquors; the suppression of any accustomed discharge; sudden exposure to cold when the body is heated; wet feet, &c.

Gout is distinguished from rheumatism by the previous and accompanying symptoms of *indigestion* above noticed, which do not necessarily occur in rheumatism; by the pains attacking particularly the *smaller joints*, while rheumatism occupies the *larger;* by the deeper redness and greater swelling of the parts affected in the gout than those which are the seat of rheumatism; and by the age of the patient, his habit of body, and mode of living. When a patient is warned of the probable approach of a gouty paroxysm, by the occurrence of drowsiness, heartburn, flatulence, costiveness, and other premonitory symptoms, which the subjects of the disease are well acquainted with, it will be advisable to attend to these signs, and resort to suitable remedies,—such as low diet, purgatives, a dose of paregoric or laudanum, at bed time, &c.,—thereby the threatening attack may frequently be averted; and even if this object can not be fully accomplished, the paroxysm, when it does occur, will be thereby rendered milder, and probably shorter. In all cases, a complete abstinence and the use of some mild diluent, as toast or barley water, are all-important measures. In the young, the robust and plethoric, or whenever there is considerable hardness or fullness of pulse, purgatives must be given. A tablespoonful of Rochelle salts, in half-pint of water, and repeated in two hours if it has not operated, will be a very good remedy for the purpose; or Epsom salts will answer.

After the employment of purgatives, if the local affection still continue with any violence, leeches or cupping-glasses may be applied in the vicinity of the latter, and followed by a blister. The affected part should be freely exposed to the cool air, and while we avoid every excitement of body or mind, the patient is to be strictly confined, so long as the fit continues, to low diet and mild teas, for drinks; *acidulated* fluids should not, however, be allowed him.

Under symptoms of very severe suffering, it may frequently be advisable to apply some anodyne directly to the part affected, in addition to the internal use of opium; and the extract of belladonna appears to be a very appropriate and efficacious application for this purpose. A drachm of this extract may be mixed with an ounce of lard, or simple cerate.

and a sufficient quantity of this mixture, to cover the affected part, spread on lint, and applied over the seat of pain. In urgent cases, it may be repeated twice or three times in the twenty-four hours, if necessary; and sometimes its tranquillizing effects will be augmented by covering it with bread poultice, made with whisky and spirits camphor.

A fluid diet was long ago pronounced the proper one during a fit of the gout; and such a diet is still recommended by the best informed of the profession. Under very acute symptoms, the nourishment must be wholly *fluid*, unstimulating, and rather small in quantity, until the severity of the inflammation, &c. has been subdued, and the patient is beginning to recover. The best food for the patient in this stage, is bread and milk, light bread puddings, gruel, barley-water, vegetable soups, and rennet whey. Roasted apples, grapes, and oranges, are likewise generally admissible; and when the patient begins to recover, an egg may be added to the above, and sometimes a little bit of chicken or roast mutton for dinner.

In a severe fit of the gout, and during the height of it, the patient is of necessity confined to his bed in a helpless state, and then the affected limb must be carefully placed on small pillows, in the most easy position; but except under such extreme circumstances, the patient ought not to indulge in bed beyond what is unavoidable. When able, he should every morning leave the bed for the couch or the chair, having his legs raised and supported in the most easy position; and, in proportion as pain and inflammation abate, should gradually employ such further exertion as relieves rather than produces irritation. Subsequent stiffness and debility of the limbs are invariably to be counteracted, in a great measure, by a gradual and systematic use of the limb, out-door exercise, &c. To prevent a return, live on a plain, substantial diet, avoid intoxicating drinks, take active exercise in the open air, and use friction with the hand to the lower extremities twice a day.

SCROFULA, OR KING'S EVIL.

The children most commonly attacked with scrofulous diseases, are those of a soft fine skin, fair hair, and delicate complexion; but it is sometimes seen in those of a more robust make and darker complexion. Children having a tendency to rickets, as indicated by a large belly, large joints, and prominent forehead, very generally possess the scrofulous habit. Those who reside in damp, uncomfortable dwellings, exposed to many privations, who are badly clothed, who live on scanty and unwholesome food, deprived of exercise in the open air, and who are inattentive to cleanliness, are those who are most subject to the disease. The countries where scrofula is most prevalent, are those of a cool, moist atmosphere, where the seasons are variable, and the weather unsteady. Seasons and weather, which remain cold and wet for a considerable time, often prove the occasion for an attack of scrofula.

This disease is hereditary from parent to child; and families, the members of which have a tendency to scrofula, ought to be particularly careful as to the manner in which they bring up their children. Since the malady is not always in active operation, it becomes a matter of great importance to know whether it can be kept from appearing and committing its destructive ravages; whether any management in early life, or in more advanced years, will protect the lively child or the beautiful youth from the dangerous enemy which has attacked his fellows, and whether an early and assiduous care may not counteract the hereditary constitution of scrofulous families.

Children who show any predisposition to scrofula, should be brought up on *plain*, but *nourishing* and easily digestible food; such as good broth, with a moderate allowance of solid meat; but pastry, heavy puddings, and the like, should be avoided. Their clothing should be warm, and they should use much active exercise in the open air when the weather is temperate and dry; they should use the sponge salt-bath, and, in the proper season, sea-bathing.

The disease most commonly first shows itself between the third and seventh year of a child's age; but it may arise at any period before the age of puberty, (about twelve years of age,) after which it rarely makes its first appearance, at least *externally.*

The attacks of scrofula usually begin some time in winter or spring, and get better or disappear in summer or autumn. The first appearance of the disease is the occurrence of small round tumors under the skin of the neck, about the ear, or below the chin, *without any pain or discoloring.* In some cases, the joints of the elbow or ankle are the parts first affected. In this case, the swelling surrounds the whole joint, and impedes its motion. After some time, the tumors acquire a larger size, the skin which covers them becomes more purple and they inflame, come to a head, and break into little holes, from which a mixed pus-like fluid, intermixed with curdy-looking matter, at first proceeds, which soon changes into a thin watery discharge. These ulcers spread unequally in various directions; some of them heal, but other tumors form, followed by other ulcers. In this way, the disease continues

a number of years, and at last the ulcers heal up, leaving
behind them very disagreeable scars, which are often the cause
of great deformity.   In some scrofulous habits, the eyes and
eye-lids are the principal seat of the disease, shown by the incessant inflammation of the ball, and the raw and painful state
of the lids of the eyes.   Diseased spine is also much connected
with a scrofulous constitution.   Many internal parts are subject to disease in scrofulous habits.   The mesenteric glands,
(situated along the inside of the back-bone) through which the
fluid destined for the nourishment of the body has to pass,
become obstructed, inflamed, or break; the consequence of
which is, a swelling of the belly, while the rest of the body is
wasting; hectic fever, disordered bowels, and gradual decay.
The lungs of scrofulous persons have, almost universally, tubercles or little whitish knots in them, which inflame and break,
and are the commencement of fatal consumption.   Water in
the head, which carries off so many children of the same family, is believed to be connected, very generally, with scrofulous taint.

It is certainly proper, as far as we can, to prevent the tumors
from coming to a suppuration or head; and for this purpose,
we are to endeavor to promote their *dispersion* by the prudent
use of gentle friction with any mild liniment, or the dry hand,
aided by proper diet, with occasional purgatives; taking care
to avoid all exposure to cold and moisture, and to keep the
swelled parts covered with flannel.   Flannel should be worn
for under-clothing at all seasons of the year.   When we find our
attempts to promote a resolution (or scattering) of the tumors to
be unavailing, we must apply to them flaxseed poultices; and
at the same time give nourishing diet to invigorate the system,
and bring it to a head.   It becomes a matter of importance
how to treat the abscesses when matter is formed; whether to
let them break, or to open them with the lancet.   Whichever
way they are opened, there is a probability of a long-continued discharge; but by allowing the matter to be discharged
by a lancet, a small and effectual opening can be made;
whereas the matter, if the swelling *be left to itself*, will perhaps break, and discharge at several different places; and
nothing will be gained, with respect to the continuance of the
after-discharge, or the prevention of unseemly scars.   When the
ulcers remain open and spread, a variety of applications will
be necessary.   Sometimes a stimulant dressing is required, as
the ointment of verdigris, (ten grains to the tablespoonful of
lard,) or basilicon ointment; at other times, simple dressing,
as lard, or cerate, is all that can be borne.   Sometimes a de-

gree of inflammation will suggest the propriety of a poultice; but this must not be *continued long*, lest we induce a relaxation of the parts around. We must vary our treatment also, by the application of different washes, astringent or cooling, as sulphate of zinc, (one teaspoonful to a pint of water), or sugar of lead, in the same proportions. Under every treatment, scrofulous ulcers are apt to disappoint our hopes, and continue open for a tedious time ; and at length, in many cases, without any perceptible cause, they suddenly put on a healthy action, and heal up, not to break out any more.

The constitutional treatment, during this period, should be as invigorating and as little stimulating as possible; a good, wholesome, but light diet, pure air, and active exercise are necessary ; a residence in the country, and sea-bathing are useful auxiliaries.

Iodine in its various preparations, (see list of " Medicines, their doses and uses,") especially that of the *iodide of iron*, also iron itself, with tonics generally, and, above all, the use of codliver oil, are the principal remedies. From five to ten drops of tincture of iodine, mixed with a tablespoonful or two of codliver oil, three times a day, is about the best preparation, given in syrup. In some countries there is considerable importance attached to the contagiousness of scrofula. It cannot be considered contagious in the ordinary acceptation of the word, but it must be always advisable, especially for those predisposed to the disease, to avoid close contact with the affected.

A B S C E S S.

THIS may be properly called the collection of pus, or matter in a cavity formed in the substance of any soft parts of the body. The contained matter, or pus, may be either of a healthy or of an unhealthy character ; if the former, it is of a yellowish-white color, cream-like in consistence, and possesses a faint sickly odor ; in the latter case, it may resemble whey, with bits of curdy substances floating in it, or it may be bloody and offensive.

Common abscess. An abscess formed on some external part of the body, which has been previously the seat of inflammation. Whenever inflammation occurs, the most proper course is to endeavor to remove it by an early resort to appropriate remedies, before it causes suppuration, (or the formation of an abscess ;) but if this latter cannot be prevented, we must endeavor to accelerate its progress by warm fomentations of bitter herbs, such as hops and vinegar, boneset, sage, or horehound, and hot water, &c , and by poultices. These are to be made of bread and **milk**; oat-meal or linseed-meal and **water** or slippery elm; and to promote the ripening of an abscess in the inside of the mouth, a roasted fig or apple may be used. When the tonsils (or almonds of the ear) are tending to suppuration, (or coming to a head), the patient should draw into the mouth the steam of boiling water, alone, or with a little vinegar and hops added to it. In general, those poultices are best which retain their heat the longest, and they should be frequently changed, to prevent their becoming cold, and thus having a contrary effect to what we wish them to have. The tendency to suppuration may be known by the inflammation continuing long; by the stretching pain becoming less; by a throbbing sensation, and the patient's being affected with cold shiverings. When an abscess forms in a place under our inspection, it is accompanied with swelling, whiteness, or yellowness of the skin, and a soft feel, as if there were a thickish fluid in a bag. When matter is formed, it must be discharged, and nature endeavors to accomplish this by causing the matter to have a tendency to the nearest outlet: thus an abscess formed in a fleshy part of the body will point to the skin; one in the lungs will burst in the air-cells; and one in the liver, either into the belly, or externally through the side. When the abscess is quite ripe, which is known by the pain being lessened, and the matter pointing or coming to a particular spot, it is, in general, best to give it vent by opening it with a lancet, or other clean cutting instrument; and this, in the position or situation which is *lowest*, on purpose to let the abscess *empty itself* by the weight of the fluid. It is better to have a free vent *of our own making*, than to allow the matter to find its way under the skin, to distant and inconvenient parts, or to allow the matter to discharge itself by a *ragged and irregular opening*. It is almost always proper to make the cut *large*, as a small one is nearly as painful, and as it is liable to close too soon, and thus occasion the necessity of repeating the operation. When the abscess is large and deep, a small piece of lint should be put between the lips of the wound, that it may close from the *bottom*, and this is to be renewed at each dressing. The poultices are still to be continued till the thick yellow appearance of the matter changes into a thin watery discharge; after this it is proper to discontinue them, and to dress with simple cerate, made of wax, lard, and sweet oil melted together, in equal parts by weight, or any good healing ointment.

When it is wished, as much as possible, to exclude the air from the inside of an abscess, an opening is made by passing a broad cutting needle, for the purpose of bringing through the sides of the abscess a skein of silk or cotton. This is desirable, particularly in lumbar abscesses, or those collections of matter which come from the inside of the belly, and point at the upper part of the thigh; as air admitted into them is often found to be the cause of hectic fever, and consequent generally ill health.

The degree of danger which attends an abscess, depends on its situation and its consequences. If it is situated in the lungs, it may burst into the air cells, and prove fatal by suffocation; or if in the liver, &c., it may be effused within the cavity of the peritoneum (or lining membrane of the bowels) and excite inflammation there. Large abscesses are dangerous by the wasting discharge with which they are accompanied; and by the slow fever and general symptoms which, in certain constitutions, follow them.

BOIL.

A Boil is a small tumor common to every part of the surface of the body; hard, tender to the touch, and coming to a head, with a hard *core* in the centre. It is chiefly found in persons of a full habit, and great vigor; but is sometimes met with also in debilitated patients, who are evidently suffering from ill health.

When it occurs in strong and vigorous patients, they should be put on a low diet, and some cooling opening medicine should be given, as a dose of salts, or cream of tartar, or senna-tea with the addition of salts. If the boil be large, and attended with considerable swelling, pain and fever, small doses of ipecac, half grain to one grain, should be given every three hours. A common poultice should be applied to the boil till it suppurates and breaks, when it may be dressed twice a day with any good healing ointment, (see Medicines, their doses, uses, and manner of preparation,) till it heals. If the ulcer gets into an indolent state, and wants stimulating, in order to its healing, we may apply the basilicon ointment. Boils not unfrequently arise in weakly habits, and where the constitution is evidently in an unhealthy state; in such cases, the patient should be put on a nourishing diet; he should take daily exercise, if possible, in the open air, and use the sponge-bath and frictions to the surface every other day. The compound decoction of the of sarsaparilla is an excellent medicine, and may be taken three times a day, in one gill doses.

GUM-BOIL.

GUM-BOILS are sometimes limited to the substance of the
gums, and sometimes connected with the decay of a tooth, or
its socket.   In the first variety, it is a disease of only a few
days' duration, and ceases almost as soon as the boil bursts, or
is opened; in the second, it will often continue troublesome
till the decayed tooth is extracted.

Gum-boils, when connected with decayed teeth, rarely dis-
perse without coming to a head, and it is, therefore, generally
better to encourage this process, by the use of warm fomentations, (previously alluded to under head of Abscess,) or poultices, than to repel it.   An *early opening* of the tumor is of
importance, as, from the structure of the parts concerned, the
walls of the abscess are mostly tough and thick, and the confined matter seldom obtains a natural exit with sufficient freedom.   A little mild opening medicine, as salts, cream of tartar,
or senna-tea, every other day, will be found useful ; and after
the abscess has burst, or being opened, washing the mouth
once or twice a day with an astringent lotion, will tend materially to make the cure permanent.   Twenty grains of sulphate
of zinc, dissolved in half a pint of rosewater, will be a suitable
lotion for this purpose, or cold green tea will answer.

MAMMARY ABSCESS or GATHERED BREAST.

THIS is an abscess seated in the female breast, affecting
chiefly women after confinement, or during the period of suckling.   Previous to the birth of the child, a great quantity of
blood is sent to the womb to supply materials for the growth
and nourishment of the child ; but when the child is born, and
requires food of another sort, the blood then flows in great
quantity to the breasts, and occasions, in some constitutions,
feverishness, known by the name of *milk-fever ;* and, in others,
severe local pain of the breasts, followed by suppuration, or
coming to a head.   Independent of the milk-fever, inflammation and abscess of the breast may arise from checking the
flow of milk at too early a period ; from exposure to cold,
fright, mental anxiety ; too great motion of the arm, when the
breast is large and distended ; blows, and pressure from tight
clothes.   But the abscess of the breast often occurs, when no
obvious cause can be assigned.   The pain arising from the in
flammation of so large and tender a structure is very great
and occasions very severe distress.   The breast sometimes puts
on the appearance of several distinct swellings, has a knotted
feel, and the pain often extends to the armpit.   At first, we
must endeavor, if possible, to put a stop to the inflammation

and to prevent its coming to a head. This is to be done by giving frequent doses of cooling laxatives, as of salts; by applying cold or tepid fomentations to the breast, and by having the milk regularly drawn off. We are also to apply leeches in great numbers, and to rub the breast gently with a little warm sweet oil and turpentine. The diet is to be very spare and cooling.

If we fail in relieving the inflammatory state, our next endeavor is, to bring it to a head, by poultices, and to discharge the matter, when ripe, by a large opening. When a suppurating breast is left to itself to break, it too frequently allows the matter to work itself into various winding holes, and to make its way out by different openings, occasioning a long and wasting discharge; to prevent this, there is no method so sure as making a large and free incision, and thus lay open, through all their depth, the hollows from which the matter flows. When a hardness remains in the breast, after inflammation and abscess, it is to be dispersed by frictions with camphorated oil and spirits of turpentine; attention being paid to the avoiding of external injury from tight clothing, &c., and the general health and state of the bowels are to be attended to.

LUMBAR ABSCESS.

This is a very troublesome complaint. It is a collection of matter forming at the loins internally, and making its appearance at the upper part of the thigh. At the commencement of the disease, there is some difficulty in walking, and an uneasiness is felt about the loins; but, in general, there are large collections of matter formed without much previous pain, and without any indication of disease, till it begins to show itself by an external swelling.

It is sometimes connected with disease of the bones of the spine; but in many cases there is no such combination. It very often occurs in scrofulous constitutions, without any obvious cause, and it may proceed from blows on the back and loins, and from exposure to cold and damp, as by lying on wet ground.

Should there be any symptoms to lead us to suspect the complaint coming on, we are to endeavor to prevent it, by keeping the patient at rest, and by the application of leeches or cups to the loins, by blisters, and purgative medicines. The great difficulty in the treatment of this disease is, to determine on the mode of opening the abscess, when we have decided that such a measure is proper. It is found by very general experience, that when these large collections of matter are freely opened,

and admission is given to the external air, very terrible consequences ensue; and that the patient suffers from slow fever,
wasting discharges of matter, and, at length, death is caused.
A plan which has been adopted with success for opening these
abscesses, is, to make an oblique opening, large enough to
discharge the flakes of matter and clots of blood from the
cavity, then to cover the wound carefully, and get it to heal
as quickly as possible. When the matter collects again, a fresh
opening is made, and the same method pursued as before.
When the abscess has been opened, or when it has burst,
which we must always endeavor to anticipate and prevent, the
strength of the patient is to be supported by nourishing diet, by
a liberal allowance of a tea made from Peruvian bark or one
grain of Quinine, before each meal; and, in some cases, the
use of wine; at the same time, moderating the slow fever, by
sponging the body with vinegar and water, and by paying a
proper attention to the action of the stomach and bowels.
When possible, a physician should be had to attend to such a
formidable disease as this.

CARBUNCLE.

In this distressing disease, for it is one attended with much
suffering, the first symptoms are great heat and violent pain in
some part of the body, on which arises a kind of pimple, attended with great itching; below which a round, but very
deep-seated, and extremely hard tumor may be felt with the
fingers. This tumor soon assumes a dark red, or purple color
about the centre, but is considerably paler about the edges.
A little blister frequently appears on the top, which, as it occasions an intolerable itching, is often scratched by the patient.
The blister being thus broken, a brown watery fluid is discharged, and a scab makes its appearance. Many such pimples
are sometimes produced upon one tumor, in consequence of the
patient's scratching the part. Considerable local pain always
attends the disease. As the complaint advances, several openings generally form in the tumor. Through these, there is
discharged a greenish, bloody, offensive, irritating matter.
The internal sloughing is often very extensive, even when no
sign of it can be outwardly discovered.

With regard to the local treatment of a carbuncle, the grand
thing is to make an *early* and *free* incision into the tumor, so
as to allow the sloughs and matter to escape readily. As much
of the contents as possible is to be at once pressed out, and then
the part is to be covered with a poultice. Fomentations will
also be found to afford considerable relief, both before and

after an opening has been made. As the discharge is exceedingly offensive and irritating, it will be necessary to put on a fresh poultice two or three times a day. The use of the poultice is to be continued, till all the sloughs have separated, and the surface of the cavity appears red, and in a healing condition; when soft lint and a pledget of some unirritating ointment should be applied, together with a tow compress and a bandage. Basilicon ointment, mixed with a little white vitriol, or alum finely powdered, will answer. The dreadful manner in which the disease is protracted by not making a proper opening in due time, can not be too strongly impressed upon the mind of every patient, and it may justly be regarded as a frequent reason of the fatal termination of numerous cases.

With respect to the **treatment for invigorating** the general health, we should remember that the disease usually occurs in two very opposite states of system—in those of full habit, and in those of broken down constitution. In the former, eight or ten leeches may with advantage be applied round the base of a large incipient carbuncle, and free purging, cooling saline medicines, such as Rochelle, or Epsom salts; and a low diet be resorted to. In those of broken down constitution, the opposite treatment will be requisite—all unnecessary loss of blood must be avoided, and the system must be soothed by opiates, and supported by quinine, one grain three times a day, or tea of Peruvian bark, a gill before each meal, along with strong meat-broths, wine or porter. In such constitutions, a carbuncle of any size is a serious, and not unfrequently a fatal affection.

It often happens that a *large* carbuncle has been preceded by two or three *smaller* ones, or boils, in succession. The occurrence of these ought always to be taken as a warning; the man of full habit should reduce his diet, meat and stimuli in particular; take exercise freely, and five or six grains of blue pill every second night, for a week, followed by a Seidlitz powder next morning. A tendency to carbuncle in the delicate or aged should always be seriously regarded, not only as indicative of disorder in the system generally, but from the direct danger arising from the disease itself. In such cases consult a physician at the earliest moment.

SICKNESS FROM UNHEALTHY MEATS.

THAT sickness, in various forms, such as boils, abscess, carbuncle, typhoid fever, ulcers, sores, sorethroat, diarrhœa, inflamation of the bowels, and other affections, are produced by the use of *diseased* or *unhealthy meats*, there can be no reasonable doubt. I mention it in this place to put persons on their guard against eating meats of this kind. Always be careful to select meats that are free from the least unpleasant odor, or when cooked looks unnatural, or that has any peculiar unpleasant taste. A whole family has been prostrated with sickness by partaking of a joint, or a piece of meat, or a fish, a little tainted or having some unpleasant taste. As it was cooked and on the table, rather than throw it away, a dinner has been made off it, costing much suffering, and perhaps, death. The *slightest taint* of this kind in meat shows that it would be unsafe to eat it.

PILES.

THERE are two kinds of tumors observed in piles, the first being an enlargement of the veins, caused by the obstruction of the blood, which are in the immediate vicinity of the lower bowel. They are of a dark bluish color, soft and elastic to the touch, and considerably lessened by pressure; they occur in regular clusters, and extend high up in the bowels; these sometimes inflame, burst, and discharge blood profusely.

The other variety of tumors consist of a thickened, condensed, hard lump, of a pale red or brownish color, situated a small distance above the margin of the rectum, or outlet of the bowel; the latter are firm, and more fleshy than the former; these may inflame, ulcerate, or remain entire, and give rise to very painful and distressing sensations.

When Piles are accompanied by a discharge of blood, they are called *Bleeding*, if not, *Blind Piles*.

A common consequence of this affection is a kind of bearing down sensation; there is also heat, and throbbing in the part, varying from a moderate degree of these sensations to the most excruciating suffering; these are caused by the great flow of blood to the parts. Sometime the inner coat of the bowels protrude at every evacuation, forming what is called Prolapsus, or falling of the bowels; this is the effect of long continued irritation and weakness of that organ. In some instances the patient experiences nervous pains, which are indescribable, and known only to the sufferer, which commence immediately after an evacuation, and continue from thirty minutes to several hours; these sensations are very annoying and sometimes very distressing. This disease, when of long continuance, is attended by pain and weakness in the back, irritation of the kidneys and bladder, and other organs in the vicinity, pain and numbness in the legs and feet, a sense of straitness about the chest, and unnatural fulness of the abdomen, accompanied with palpitation of the heart, and oppression Individuals sometimes experience previous to an attack of Piles, symptoms denoting great derangement in the circulation; there is a sense of weight and pressure in the abdomen, with a peculiar feeling of uneasiness in the bowels, costiveness, and a sensation of bearing down in the parts, attended with pain in the back and loins, nausea, and slight pains in the stomach, scanty and high-colored urine, pale countenance, confused sensations in the head, weariness, and irritable and discontented state of mind, and a sense of fulness and oppression in the region of the stomach.

This disease is caused by drastic purgatives, torpidity of the liver, costiveness, sitting on cold stones, on the wet ground, &c. Sedentary occupations, and high living, of course render the above causes more liable to produce the disease; some individuals afflicted with the Piles, are in the habit of remaining at stool, and after the bowels have been evacuated of their contents, they frequently experience sensations similar to those preceding the calls of nature; they feel as though there was something more that should be removed, and with this impression, great efforts are made to expel the delayed contents of the bowels, but all in vain; the efforts are ineffectual, and are followed with pain and anguish.

Let the patient remember that these sensations are produced by the descent of internal Pile Tumors, while the bowels are being evacuated, and that very violent expulsive efforts protrude them still farther. Use as little effort as possible, and you will facilitate the cure, and save yourself much unnecessary suffering

After the bowels have been moved, remain quiet for a short time, to allow the parts to resume their natural position. Much walking or long standing is very injurious, and should be avoided as much as possible.

In cases of Piles, accompanied by falling down of the lower part of the bowel, so as to protrude externally, the following treatment should be resorted to;—Take of white oak bark one ounce, boiling water a pint; boil the bark in the water in a tin vessel, and let stand one hour, then strain through muslin. When cold, inject about a gill up the bowel morning and night, or it may be used as a wash to the protruding part of the bowels. Or the following may be used. Take one ounce of lard, nutgall finely powdered two drachms, fifteen grains of sugar of lead, ten grains of powdered opium, (or four grains of morphine which will answer as well.) Mix all thoroughly together; of this put about a teaspoonful or two up the bowel with the finger two or three times a day. When the bowel is down, first put it up by pushing it back with the index finger well oiled, before using the ointment. This ointment, or the wash can be used in all cases that are very painful, or even ice applied to Piles when they come down and are painful, will be found very soothing; to be applied constantly while the pain lasts. Where ice cannot be had the coldest water will be found a good substitute, by applying linen cloths dipped in the water, to be removed as often as they become warm. The oak-bark wash will be the best to use in cases of excessive bleeding from the bowels in the Bleeding Piles, to be injected into the bowel every three hours.

A handful of hops, boiled in a quart of water for half an hour, then allowed to cool, and applied as a wash or an injection is also a good remedy.

The most thorough cure of this disease, however, is between the attacks, to inject half to one pint of cold water up the bowels night and morning, at same time washing the outer parts in cold water. Besides this, *keep the bowels open* by eating freely of stewed fruit, such as dried apples, peaches, figs, raisins, &c., and if needed, a Seidlitz powder, or dose of Rochelle salts, or Castor oil. In persons who are of a bilious temperament a Blue pill at bed time occasionally is of great service, always giving the next morning some mild purgative, being careful always to *avoid Aloes*, except a sudden or imprudent suppression of the hemorrhoidal flux is followed by violent headache, pain of the chest or abdomen, the premonitory systoms of apoplexy, or a discharge of blood from the lungs or stomach, then the remedies are bleeding from the arm in *bad cases;* active purgatives, as aloes, soap and gamboge combined; purgative injections into the rectum; warm fomentations to the parts, either by poultices, or the patient sitting over the steam of hot water, also putting their feet into a mustard bath, &c.

STINGS OF POISONOUS INSECTS.

Under this head we will include the stings of the bee, wasp, hornet, musquito, &c. The poison from all insects of this kind may be looked upon as of an acid nature, which coming in contact with the veins, or blood vessels, causes very painful symptoms, even dangerous, sometimes. In some it is trifling, in others it is very great, and in a few individuals it extends over the entire body, while at the same time there is much sick faintness, &c., requiring the administration of hartshorn and other stimulants. If the sting has been inflicted about the

FRACTURE OF THE ARM.

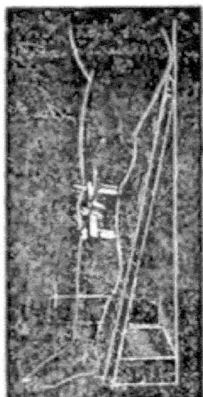

FRACTURE OF THE KNEE-PAN.

FRACTURE OF THE LEG.

FRACTURE OF THE THIGH

throat, the swelling has been known to prove fatal. The domestic local applications to stings are very numerous. Oil is frequently applied, and gives relief; but alkaline preparations certainly appear to be most serviceable. Soda sometimes is employed; but ammonia or hartshorn (the weaker solution) is the best form of alkaline application, and it may be used alone or mingled with oil. However, before any remedy is used, care should be taken to ascertain that the *sting does not remain in the wound*. If it does, it must be extracted by tweezers or by squeezing. If the pain and swelling remain severe, a solution of one teaspoonful of sugar of lead in a pint of water and kept constantly applied to the parts affected, by means of cloths dipped therein, will be found beneficial. Sometimes poultices have been used with advantage in these cases.

FRACTURES.

THE bones most frequently broken, are those of the extremeties. It is called a simple fracture, when there is no opening from the fracture externally; but a *compound fracture*, when there is an external communication. In old age, as well as in particular diseases of the constitution, bones are more liable to be broken: and also, in winter. This formerly was supposed to arise from the influence of cold; but it is now believed to be the result of extraordinary *muscular action*, excited by the exertions to avoid falling on the frozen and slippery places or pavements. For this reason, persons whose muscles are relaxed, as in a state of intoxication, much less frequently have their bones broken from a fall, than those who are sober and very solicitous to guard against tumbling.

The symptoms of fractures are, severe sudden pain, *alteration in the form of the part*, sometimes a *shortening of the limb*, an *inability to move* the limb without severe pain at the injured part, an *inequality of the skin* covering the bone: a grating, (called *crepitation*,) of the edges of the bone against each other; a motion and noise not to be mistaken for any other. By taking hold of the limb above and below the fracture, and moving the fractured extremities of the bone, the noise is produced, and the existence of the fracture rendered unquestionable. It is well here to remark, that the fewer of these attempts made, the better; as it is injurious that the edges of the bone should be much rubbed over each other. When the parts are much swelled before examination, the difficulty of ascertaining the fracture is increased.

The union of fractured bones is effected nearly in the same manner as that of the soft parts. The inflamed vessels pour out the matter necessary for the union, and the absorbing vessels take up the unnecessary parts. The matter poured out for the union, is called *callus;* it is at first soft, but gradually becomes firm and hard.

In the treatment of fractures place the parts of the broken bone as near as possible in their original position, and keep them so until union is effected. The first is done by moderately extending the parts, so that the edges may be made to come in contact; the other is done by the application of splints and bandages.

The treatment of the patient, as it relates to the constitution, is to be regulated by circumstances. A certain degree of inflammation is essentially necessary for the process of restoration. If it be too violent, instead of the formation of callus for the reunion, common matter will be formed; it will come out, and, thereby making an opening, will convert a simple into a dangerous compound fracture. The inflammation must be regulated, not so much by *purges*, as generally it is very inconvenient; but chiefly by small doses of ipecac when necessary, and low diet; but when the inflammatory stage is passed, the individual should return as much as possible to his ordinary food. Care must be taken, if the person has been accustomed to much alcoholic stimulant, that it is not *unduly* abstracted, otherwise the powers of the constitution will be so reduced that the reparative process cannot take place, and the fracture will remain ununited.

When the soft parts are much injured, greater attention is necessary to keep down the inflammation, than when only the bone is broken. In cases where there is much swelling, or much effusion of blood, applications of sugar of lead and water, one teaspoonful to the pint, and local bleeding by leeches, can alone prevent the formation of matter. In every case where it is proper for the patient to remain in bed, it is necessary to have a bed pan for the evacuation of his bowels: if a good one cannot be had, a better mode will be to have the patient laying on a mattress, with a hole in it of a proper size; the mattress should be on a plank bottom, and a door, in which another hole is to be, correspondent to that in the mattress; in these holes are to be suitable stoppers. And when the patient wishes to evacuate, the door and all are to be elevated, the stoppers removed, and a pot placed underneath to receive the discharge, which being finished, the whole is to be replaced. In some places, bedsteads with screws and pullies, are made to effect these objects: but as they cannot be had in the country, a contrivance can be made to answer in the way I have suggest ed, requiring no skill in construction, and the labor of only one or two to place and replace it. Under any circumstances, it is improper to place the patient on a *feather bed*, on account of the irregularities necessarily ensuing. If a hair or straw mattress cannot be procured, it is better to substitute a few folded blankets or quilts on even boards.

After a fracture has once been set, it should never, if possible, be disturbed again. This does not mean that the appliances are not to be removed, and the progress of the case inspected; for if this be not done, and if by any chance the proper position should have been disturbed, the bones may become solidly fixed in an improper manner, and deformity result, or the skin may become ulcerated. But the appliances should not be removed, if possible, before the end of the first week, and if all seems going on well, not moved again for ten days at least, unless for some special purpose. If a fracture is often disturbed or pulled about during the process of consolidation, it may chance that this will only be effected imperfectly, and what is called a false joint formed; that is, the broken part, instead of being firm, moves like a joint, and the limb is useless. It had better be crooked or shortened.

Much care is always required that a limb which has been fractured is not used too soon after the accident; otherwise it may be either snapped again, or it may be bent.

FRACTURE OF THE LOWER JAW.

This bone is liable to fracture in all its parts. The symptoms are, severe pain at the time of the accident; an inequality is perceived in passing the fingers along the bottom of the jaw; the teeth, on examination, are found unequal; and on taking the two sides in the hands, it is easy to reduce the teeth to their proper level, and in doing so, the grating motion is perceived.

To reduce the fracture, nothing more is necessary than to shut the mouth, and forcibly push upwards the lower fragment, until the teeth contained in it come in contact with those of the upper jaw, when it is to be supposed the parts are in proper place. The simplest and best plan to keep the parts in place, is to avail yourself of the support given by the teeth in the upper jaw, by binding the fragments firmly against them and this can be very conveniently done by means of a simple roller of common cotton muslin passed repeatedly round the top of the head and under the chin. It may be further secured by passing a few turns of it round the back of the neck and in front of the chin.

The patient should be nourished fifteen or twenty days on spoon victuals, sucked between the teeth ; and the only additional remark we have to make, is that when the teeth at the fractured bone are loose, they are not to be touched ; much less removed, as that would convert the simple into a compound fracture, or, in other words, admit the air to the broken parts of the bone.

The following treatment is recommended by some surgeons, as being the most simple :—Two narrow wedges of cork, about an inch and a half long, a quarter thick at the base, and sloping to a point, are placed between the teeth, one on each side ; a piece of pasteboard softened in warm water, or of gutta-percha, is then to be moulded round the jaw, and fixed, either by a bandage or handkerchief going over the crown of the head. By this method, space is left between the front teeth for the administration of liquid nourishment. The sufferer should rinse the mouth frequently with tincture of myrrh and water.

FRACTURE OF THE RIBS.

These are generally broken near the middle. The fracture is ascertained by a severe pain felt at the injured spot in every motion of the body, even in breathing : by careful examination with the hand ; and by feeling the grating of the bones, particularly when the patient coughs. This grating sensation may be felt by another person laying the hand on the injured parts. The only treatment necessary, is to pass a roller about six inches wide repeatedly around the chest, and as tight as the patient can suffer it to be drawn. It is to be prevented from falling down by shoulder straps.

When the edges of the bone have wounded the lungs, sometimes there is spitting of blood ; and violent, if not fatal, inflammation may follow. On this account, an individual who has suffered from fractured ribs should be especially careful, and for some little time after the accident should reduce his usual diet considerably. The application of the hot bag of bran for some days after the accident will afford much relief, and it may be used over the usual bandage. When ribs on both sides of the chest are injured, this, with leeches if requisite, should be the sole application ; the patient being confined to bed for at least a fortnight or three weeks, in the posture found to be the easiest, which will probably be a half-sitting one, supported by pillows, or some other means. When the ribs on one side only are injured, less confinement is required, but the chest should be encircled, as firmly as can be borne comfortably, with a band of stout calico, from eight to ten

Inches wide. and double; this should go once and a half round, and be sewed. A month will probably be required for the cure

FRACTURE OF THE ARM.

This most generally occurs about half way between the shoulder and elbow. When it is broken directly across, and near the middle of the bone, no great derangement takes place; the limb preserves its length, and its form too, unless it be moved. The mode of treating this fracture, is to set the patient on a chair; one person is to hold the body, with his arm around the chest; another the forearm (between the wrist and elbow) bent at right angles over the breast, and to raise it a little from the side. and extend it, when the operator is to place the two ends of the bone in contact; to pass around the arm, from the elbow to the shoulder, a roller, moderately to compress the part without impeding the circulation. He then applies a splint of wood. firm paste-board, or of raw hide, on the top of the arm, from the elbow to the shoulder; then another on the outside, of similar length; and in the inner side of the arm another, from the armpit to the lower part of the elbow. These are to be secured by another roller or bandage. Folds of flannel are to be placed in the armpit, to give some support, and the forearm is to be suspended in a sling. At the expiration of a week, the parts are to be examined, and, if found out of place, to be rectified; the joint at the *elbow* should be *gently* and *carefully moved*, in order to *prevent stiffness;* and this ought to be repeated, after the first week, once every other day. At the end of three weeks, it is recommended to alter the dressing and substitute splints, which instead of keeping the arm bent at *right angles*, will keep it *nearly extended.* This is done to prevent its partial deformity.

FRACTURES OF THE FOREARM.

The forearm (extending from the wrist to the elbow), is composed of two bones; and sometimes only one of them is broken, at others both. The symptoms are, great pain at the time of the accident. increased by motion of the hand; an inability to turn the hand either up or down, and the grating common to all moveable fractures.

To reduce the fracture, the forearm is bent to a right angle with the arm; an assistant takes a firm hold of the arm just above the elbow; a roller is to be applied, extending from the hand to a little above the elbow; two splints, broader than the arm, made of either of the materials before mentioned, are to be applied, one on the inner and the other on the outer side of the hand, extending from the fingers up to the elbow, leaving the thumb upwards, projecting between them. The hollow places should be filled up with tow or cotton, and a bandage applied around, to preserve the whole in place.

In about ten days the parts are to be examined, to see that all is going on right. In thirty or forty days, the union is generally completed. The elbow and wrist should be moved every other day, as in fractures of the arm, *after the first week.* When only one of the bones is broken, the treatment should be the same.

When the bone at the elbow is fractured, it is drawn up by the muscles attached to it, leaving a considerable space between the broken parts. The proper treatment of this accident is to extend or straighten the arm, and apply a broad bandage around the arm, one beginning at the fingers and extending *up* to the elbow joint, and the other beginning at the shoulder and extending *down* to the elbow joint, to compress the muscles, and thereby prevent their action in separating the broken part. The separated bone is to be pushed down to its natural place, and a long compress placed on it, over which is to be applied a roller extending the greater way over the arm and forearm, (as explained before); a large long splint is to be put on, extending from the inner part of the arm to the hand, and a roller is to be applied from beginning to end, commencing at the hand. It should be so passed around the elbow as to form a kind of figure of 8, in order that the upper part of the bone may be kept down by its oblique compression. In about thirty days, the joint may be very gently moved by the hand of an assistant—but it is not to be attempted by the patient for nearly double that time, as the parts otherwise might be ruptured again.

In fractures of the bones of the hand and fingers, all that is requisite is to restore the parts as accurately as practicable to their natural position, and preserve them in as quiet a state as possible; taking care, as in all other cases, to prevent the inflammation faom extending too far.

FRACTURES OF THE THIGH.

There can be no doubt, but that in all such cases when surgical aid can, it will be obtained. But in case of this being imposs.ble, it may be of service to add, that the patient is to be laid on a mattrass on boards, with a hole in it for the evacuation of his bowels.—That instead of the splints, usually recommended on such occasions, it may suffice to have a box made without a top, just wide enough to receive the limb, and

of length to extend six or nine inches *beyond the foot*, up to
or near the crotch, with the outer plank, or side of the box,
to extend by itself up to or near the *armpit.* In this part
there is to be a couple of holes, for tieing a bandage securely
The limb is to be placed in the box, with a pocket handkerchief so equally applied to the foot, that it shall not bind too
much on any one part. Another pocket handkerchief is to
be applied between the thighs; one end to go under the
crotch, up to the hole in the outside plank of the box; the
other end is to go up in front to the other hole, and the two
ends are to be tied together. The next operation, is to pul.
the limb downwards, and put the bones in place: and the
handkerchief around the foot is to be extended and secured
to the end of the box by any contrivance that will hold it.
Thus the handkerchief between the thighs will preserve the
extension above, and that at the foot will extend the lower
limb, and it may occasionally be drawn tighter.

This rude sketch, will enable you to do some good on such
occasions. The box ought not only to be made of firm materials, but should be well lined or stuffed in every part where
it touches the flesh of the patient, with finely carded and
smoothly placed cotton or tow, to prevent excoriations.

FRACTURES OF THE KNEE PAN.

When this accident occurs, the patient generally falls
though sometimes he gets up and by dragging the limb side-
ways, may be able to walk, taking great care not to bend the
knee. A depression or hollow at the place or fracture is perceived, and commonly the upper is found considerably drawn
up from the inferior part, which is fastened to the lower
bone.

The great object to be attended to in the treatment of these
fractures, is to preserve the fragments as near as possible together, so that the substance connecting them may be as short
as possible, and the motions of the joint be perfectly preserved.
—In proportion to the violence producing the fracture, should
be the attention to keep down inflammation.

The local treatment consists in keeping the limb at rest, in
an extended posture : and by a splint and bandages preserving the contact of the fragments. This is done by taking a
piece of plank about half an inch thick and three inches wide,
and extending from the upper end of the thigh to the heel.
Upon this splint, covered with folds of cotton or flannel so as
to fill up the inequalities of the limb, the patient's leg and thigh
are to be placed. A common roller is to be carefully applied

from the foot to the knee, and one from the *top* of the thigh down to the knee, so as to equally compress all parts, leaving none exposed; but you are to observe, in passing the roller over the knee, as in the case of the elbow, it is to be so done as to press down the upper to the lower part, making, as in the other case, the figure 8; so that the roller as frequently passes one part below, the other above the fragment several times, when it is to be continued to the thigh. A compress of folds of flannel, should previously be put over the knee.

The limb is then to be equally bandaged to the splint underneath. Some surgeons recommend too slips or bands of doubled muslin, each a yard long, to be nailed underneath the splint, at a distance of six inches from each other, and about the middle of the splint, or just so as to be underneath the knee. These bands are to be passed—the lower one above the upper fragment, and the upper one below,—so as still to make the figure 8, and press down the upper to the lower part. The joint should gently be moved by an assistant about the thirtieth day, to be continued moderately every other day, to prevent stiffness.

FRACTURES OF THE LEG.

In fractures where the bone is broke directly across, it is merely necessary to reduce the fracture with the hand, when the limb is extended, and apply a roller from the foot to the knee; then two splints, one to each side, of pasteboard or thin plank, extending from the knee to the sole of the foot; and over these splints another roller is to be applied, to preserve them in their places. The bandages should never be drawn *too tight*, as the limb will swell, and they may do injury by compression. The limb is now to be placed on a pillow, and put in a box, or, what will answer, between two long slips of plank, which are to be tied around, so as to support the whole.

When the fracture, instead of being directly across, is in an oblique direction, it is necessary to dress the limb in the same way as in fractures of the thigh bone; excepting that, when the fracture is not near the knee, the upper part of the box may be fastened a little below or around the knee; and therefore its external side need extend no farther. There should be a hole in the upper part of both the inner and outer side of the box, and the handkerchief or band should be so made as to pass from the knee through each of these holes, where it is to be fastened. Another handkerchief is then to be applied around the foot, which is to be extended; then the ends м the handkerchief fastened to the end of the box, so as to preserve the extension of the limb. Of course, a roller is first applied around the leg from the foot to the knee; and the box is to be well supplied with cotton or tow, to fill up the inequalities of the limb.

FRACTURE OF THE COLLAR BONE.

When it is broke, the part nearest the shoulder is drawn downwards by the weight of the arm; the arm of the affected side falls over upon the breast, and the patient is unable to raise his hand upon his head. He leans to the fractured side; the grating of the bones may be perceived; and the finger passing over the bone, will readily detect the place of fracture.

This fracture has frequently been successfully treated, by simply keeping the patient laying down, with his arm so placed that the broken edges of the bone may be in contact. But most commonly it is dressed in the following manner:—A bolster or pad is to be made of quilted cloths, in the shape of a wedge, about as long as the arm, four inches wide, and at least three inches thick. This is to be put under the arm, the base close to the armpit, the point down the side; and it is here to be well secured by a roller passing around the body, and so turned over the shoulder that it cannot be displaced: a contrivance which any one can make, who will exercise common sense. The patient being seated on a stool, and held by an assistant, the operator is to bend the elbow at right angles, and the forearm is to be supported by a sling around the neck, the arm and elbow are to be pressed to the side; the wedge above acts as a point for the extension of the broken bone, and it is to be bound down in that state by a wide roller passing around the body and over the elbow. It is impossible to give an accurate description of the particular manner of applying the bandages. By the exertion of sound sense, it can be done to effect the main objects, which are, first, to preserve in place the wedge underneath the arm; second, support the arm bent on the breast; and third, to press down and keep the elbow on the side, so that it shall cause the extension of the upper part of the arm, and consequently the broken bone. A slip of adhesive plaster will be sufficient to cover the broken bone.

DISLOCATIONS, or OUT OF JOINT.

THE necessity for the speedy reduction of a dislocation is great, from the fact that every day increases the difficulty of its performance; and when a certain time has elapsed, no force which can be exerted—consistent with safety to life and limb—will be adequate to return the displaced bone, partly owing to the resistance of the muscles, but also to obliteration or doing away with the cavity which formed the one portion of the joint. When dislocation occurs, two different actions take place; one, that by which the bone is driven from its usual position; the other, the action of the muscles, which tend still further to draw it from its proper site as soon as the balance of resistance of bone against bone is removed. It is, too, in most cases, the action of the muscles which tends to keep the bone displaced, and to resist the efforts made to replace it. This is evident from the fact, that if a person be seen immediately after a dislocation, and while suffering from the faintness which almost invariably accompanies the accident, and while the muscles are necessarily in a state of weakness and relaxation; the dislocation may often be reduced with the greatest possible ease, even by the unskilled; and further, when the surgeon has to deal with a case of dislocation in a strong and muscular subject, he endeavors to produce this faintness—if that following the accident has passed away—by bleeding, nauseating medicines, warm baths, &c.; chloroform and aether inhaled are used by surgeons and physicians, but are not safe in inexperienced hands.

When, therefore, a dislocation occurs, the bone is not simply pushed out of its place, but is drawn for the most part upward, or toward the body; the dislocated bone of the finger is drawn upward over its fellow; the arm-bone, in dislocation of the shoulder, may be drawn upward, or into the arm-pit—in this case downward, it is true, as regards the joint, but still toward the body; and the same will be found to be the case in most forms of dislocation. The first object, therefore, in treating a dislocation, must be to draw it down *from* or out of the situation to and in which it has been drawn and is retained by the muscles of the limb, and to get it as near the corresponding part of the joint, or, in other words, as near the part from which it has been dislocated, as possible. If the dislocated bone is thus drawn down to, or near to the level of the other portion of the joint from which it has been removed, the muscles will of themselves tend to draw it into its old position. A good deal is often said about the adjustment, &c., &c., of the bone in reducing dislocations; and though, perhaps, useful in some cases, in many nothing of the kind is required, at least unprofessional persons should not attempt it all that is to be done is, give the muscles the chance of drawing the bone into its old place, by bringing it to a position in which this can be effected. This is often exemplified in cases in which much force is used in the reduction of a dislocation ; if the force be kept up strongly, the bone cannot be drawn into its socket, because the force is stronger than the muscles of the patient ; but relax the external force for a moment, and without any fitting or adjustment, the bone is instantly drawn into its proper position by the power of its own muscles. There is yet another important principle involved in the reduction of dislocations. It has been pointed out how the bone farthest from the body—which is usually drawn up—is to be drawn down ; but, that this may be done properly, the bone above it must be *fixed*, otherwise it will be drawn down too. This is easily effected in such cases as the ancle or the wrist, by any one grasping and holding firmly either the leg or the forearm ; but in the case of the hip or the shoulder, more management is requisite.

Again, in " making the extension," that is, using the forcible effort to return the dislocated bone to its place, the extending force will best be made in the direction in which the limb is fixed, and in the manner most likely to bring the joint portion, or " articulation " of the displaced bone, as near to the old position as possible, and it must be applied directly to the bone which is displaced. Thus, in dislocation of the shoulder, the reducing force is applied to the arm-bone; in dislocation of the hip to that of the thigh. This extending power may simply be by the hand, but a cloth, or band of some kind, put round the member to be replaced, is often more advantageous. The particular dislocations most likely to be recognized and to be remedied by unprofessional persons, are those of the small joints, such as fingers and toes ; of the wrist and ancle ; of the elbow, shoulder, and lower jaw.

Dislocations of the fingers or toes may generally be made out by any person, and should, if possible, be reduced at once ; the dislocated bone being grasped, and forcibly pulled into place ; or a noose, made with a piece of tape, may be used. Dislocation of the thumb, it should be known, is extremely difficult of reduction, and should this not be effected *at once*, the attempt ought to be given up until the surgeon's arrival ; it is, moreover, one of the dislocations which may be left unreduced with less subsequent inconvenience than many others. Dislocation of the ancle is very generally accompanied with fracture, but the distortion is often so great and evident, and the suffering so severe, that when the accident does occur far from proper aid, some attempt ought to be made to put the displaced parts into better position. For this purpose, while one individual grasps the leg firmly, another, putting one hand on the heel and the other on the instep, should endeavor, while steadily pulling downward, to bring the joint into its natural position.

DISLOCATION OF THE WRIST is produced by the forearm being tightly grasped by one individual, the surgeon laying hold of the patient's hand in his, and endeavoring by steady pulling downward, and *slight* up and down movement, to bring the joint into its proper condition.

DISLOCATION OF THE ELBOW, if attended to quickly after the accident, may often be easily reduced by seating the person in a chair, carrying the arm well behind the back, and pulling, not very forcibly, upon the forearm.

Both these dislocations—of the wrist and elbow—may be suspected, when, after violence—particularly such as is *calculated* to push either the hand or lower arm upward—inability to use the limb below the seat of the injury, and distortion and impaired motion of the joint, are unaccompanied with any grating sensation, such as occurs when a bone is fractured.

DISLOCATION OF THE SHOULDER is generally occasioned by violence applied to the elbow, or by falls, while the arm is not close down to the side of the body. Sometimes the exact discrimination of an injury to the shoulder joint is a matter of much difficulty, for fracture alone or fractures with dislocation may occur. At other times, particularly in thin persons, it is tolerably easy made out—more so if the examination is made before swelling comes on. In addition to the general symptoms of dislocation already enumerated, the injured shoulder will be perceptibly altered in shape; it will appear more depressed and flatter than the sound one, and if the hand is placed upon the spot which ought to be occupied by the round head of the arm bone—and this may be discovered by examination of the uninjured shoulder—it will be found hollow ; and further, if the arm be now gently moved about, and its bone traced up toward the shoulder, it will be found moving in some unusual position, most probably in the arm-pit. Supposing, therefore, that the case is sufficiently clear, the means for the reduction ought to be set about as speedily as possible—if it can be, while faintness from the injuries continues. One method of reducing dislocation of the arm-bone into the arm-pit frequently employed is for both patient and surgeon to lie down upon the ground side by side, but with their heads different ways, and so that the surgeon having previously taken off his boot, can place his heel in the arm-pit of the patient, while he grasps the hand, or a towel fixed to the arm of the effected side; in this way, while the heel is used to push against the displaced bone in the arm-pit it, combined with the pulling exerted by the surgeon upon the limb of the patient, tends to give a leverage by which the bone is so placed that it can be drawn into the socket by the muscles. This method may be a convenient one, when only one person is in company with the individual to whom the accident has happened. The following is the most useful and most generally resorted to method of reducing a dislocation of the shoulder. The patient being seated on a chair, a large towel or a table cloth, folded broad, is to be passed round the chest, close under the arm-pit of the affected side, crossed over the opposite shoulder, and held either by a strong assistant or fastened to some fixed point. By this application, the shoulder blade is fixed; the arm itself is then to be pulled, chiefly in the direction in which it has been fixed, *firmly, steadily, and slowly;* this being done, either directly by the hands of assistants, or by a towel fastened round the arm by the hitchnoose. If when this steady pull has been persevered in for some time, the displaced bone does not get into place, the effect of suddenly taking off the attention of the patient may be tried, either by some sudden exclamation, or by dashing a little cold water in the face. By such a proceeding, the muscles which resist the reducing or pulling force applied to the arm, are for a moment, so to speak, thrown off their guard, and that moment may suffice to permit the bone to pass into its socket.

DISLOCATION OF THE LOWER JAW.

This accident, in most cases, is produced by yawning, or opening the mouth excessively wide. It is sometimes produced by a blow upon the chin while the mouth is opened. The symptoms of its occurrence are, an inability to close the mouth; immediately before the ears an empty hollow space is perceived; the cheeks and temples are flattened; the spittle flows from the mouth; speech and swallowing are difficult, and the chin projects forward. When the jaw continues dislocated several days, these symptoms are not so strongly marked, though they are still in greater or lesser degree.

To effect the reduction, the patient is to be seated on a low chair, his head supported against the head of an assistant; the operator is to defend his thumbs with a piece of leather or linen he is then to place them as far back on the jaw teeth as he can the fingers are then placed under the chin; and while he presses down the back teeth with his thumbs, he at the same time raises up the chin with his fingers; and then the chin is pushed backwards, when the parts become replaced very suddenly. As this is done, the operator is as quickly to move his fingers from under the teeth to the cheek. After the operation, the patient should for some days live on soup, &c., in order that the jaw may be at rest to recover its strength.

DISLOCATION OF THE COLLAR BONE.

THIS accident is generally occasioned by falling on the shoulder, and is ascertained by examining with the fingers; the end of the bone being found under the skin covering the elevated point of the shoulder bone, there causing considerable projection. The patient inclines his head to the affected side, and moves the arm and shoulder as little as possible. The treatment of this accident is precisely the same with that of fracture of the collar bone.

CONTUSIONS or BRUISES.

THE bad consequences of bruises are not invariably proportioned to the force which has operated; much depends on the nature and situation of the part. When a contusion takes place over a bone which is thinly covered with soft parts, the latter always suffer very severely, in consequence of being pressed at the time of the accident, between two hard bodies. Bruises of the shin thus frequently cause death of the soft parts, and troublesome sores. Contusions affecting the large joints are always serious cases; the inflammation occasioned is generally obstinate, and abscesses and other diseases which may follow, are consequences sufficient to excite serious alarm.

In the treatment of bruises, the practitioner has three indications, which ought successively to claim his attention in the progress of such cases.

The first is to prevent and diminish the inflammation, which, from the violence done, must be expected to arise. To effect this, the bruised parts should be kept perfectly at rest, and covered with linen, constantly wet with cold water, or sugar of lead and water, one teaspoonful to the pint. When the muscles are bruised, these are to be kept in a relaxed position, and at rest, until the effects of the bruise are entirely removed.

If the bruise has been very violent, it will be proper to apply leeches, and this repeatedly. In every instance, the bowels should be kept well open with saline purgatives, seid lit powders, Rochelle and **Epsom** salts, etc., and the patient put upon a low diet.

A second object in the cure of bruises, **is** to promote the absorption of the extravasated or bruised blood by liniments, etc. These may at once be employed in all ordinary contusions, not attended with too much violence; for then nothing **is so** beneficial as maintaining a continual evaporation from the bruised part, by means of cold applications, and at the **same** time, repeatedly applying leeches. In common bruises, however, a solution of salammoniac in water, or vinegar and water, is an excellent application; but most surgeons are in the habit of ordering slightly stimulating liniments for all ordinary contusions, and certainly they do much good in accelerating the absorption of the bruised blood. The soap and camphorated liniments are as good as any that can be employed.

In many cases, unattended with any threatening appearances of inflammation, but in which there is a good deal of bruised blood and fluid, bandages act very beneficially, by the remarkable power they have of exciting the action of the lymphatics (absorbents) by means of the pressure which they produce.

A third object in the treatment in contusions, is to restore the parts to their proper tone. Rubbing the parts with liniments has a good effect in this way. But, notwithstanding such applications, it is often observed that bruised parts continue for a long while weak; and swell, when the patient takes exercise, or allows them to hang down. Pouring cold water two or three times a day, on a part thus circumstanced, is the very best measure which can be adopted. A bandage should also be worn, if the situation of the part will permit. These measures, together with perseverance in the use of liniments, and in exercise, gradually increased, will soon bring everything into its natural state again.

S P R A I N S.

INJURIES of this kind generally affect the wrists, ankles, and knees, being produced by sudden or violent exertions, slipping, or falling, etc. They are followed by violent pain immediately, and then swelling and inflammation. There is generally a rupture of the blood vessels within, and consequently an effusion of blood. The skin is not discolored for some hours after which it generally becomes of a dark bluish or red color increasing or disappearing, from the inflammatory state, in proportion to the extent of injury.

The best remedy for lessening the effects of a sprain, is that nearest at hand—cold water. As soon as the accident happens, the part should be plunged in cold water, or a few pitchers of cold water poured over it. The next remedy is rest, *perfect rest:* the part being kept rather elevated, *never hanging down.* The cold applications stop the effusion of blood, and promote its absorption : the elevation of the part retards the passage of blood to it. It is customary to apply brown paper, (rags are as good), wet with vinegar, or spirits, and water to the part, and continue them wet on the part for several days. Sugar of lead and water, a teaspoonful to the pint is also a good application. Spirits of camphor, or opodeldoc, may with advantage be rubbed over the part, and it should gently be daily rubbed with the hand or a ball of cotton.

Should inflammation come on, you must purge freely, live on low diet, and continue the cool applications of sugar of lead and water, etc. Leeches on the spot most inflamed, will do great good, and so will cupping near it. If the part be much distended and painful, poultices at night of flaxseed, or elm bark, will aid in removing it. But if you will have patience in the first instance to confine yourself and follow the first directions, you will probably never have need of other advice. But from very trivial accidents of this nature, the neglect of necessary precaution has been followed by a loss of the joint, by stiffness, by decay of the bone, and loss of the limb in consequence.

If pains or numbness remain after the sprained part is otherwise relieved, the pouring of water on it from an elevated spout, and frequently rubbing it with camphorated spirits, two or three times a day or more, will be the proper remedies.

WOUNDS.

In cases of wounds, even of a trifling nature, comparatively, there is generally much excitement. This should be overcome by the exercise of good judgment and common sense. Then, the first circumstance, generally, which calls for attention as the consequence of a wound is the effusion of blood, but none of the consequences, perhaps, exhibit greater variation. Sometimes an extensive injury may by inflicted, even the arm torn off at the shoulder, and yet the loss of blood be extreme_y small ; on the other hand, a puncture with a penknife, if it penetrates an artery, may be sufficient to place life in the greatest immediate jeopardy. As a general rule, probably, putting the opening of large vessels out of the question, a greater amount of blood is lost after simple cuts than after any other description of wound. When laceration or bruising takes place, there is usually, by stretching, or otherwise, of the coats of the arteries, a sufficient amount of mechanical imped-iment caused to modify greatly, if not wholly to prevent, any hemorrhage. When a wound is small, the best method of treatment is to tie it up at once with a piece of linen rag ; this is usually sufficient at once to stop the bleeding, particularly if rest and position are attended to ; the small quantity of blood which may exude, quickly dries upon the wound, and forms a kind of glue which effectually excludes the air. As no better dressing can be used, it may be left on till the cut is well ; in some cases, before using the linen, it may be advisa-ble to draw the edges of even a small cut together, by means of adhesive plaster, or material of some kind. Although linen is mentioned in the above directions, of course, should it not be at hand, soft calico may be used, or other soft material.

When a wound is extensive and the bleeding profuse, it will not do to bind it up in this way ; first, because it proba-bly would not be sufficient to arrest the flow, and if it did so ultimately, it would retain a large amount of clotted blood, either in or about the wound, in such a way as to interfere with the healing process. In a large wound, therefore, it is necessary that the bleeding should be almost entirely arrested before it is dressed, that is, closed up, &c.

It should be borne in mind that the first end in view when a wound is dressed, is to get as much of it as possible to heal by the "first intention," or by "adhesive inflammation;" that is, to get the several parts to adhere at once, without formation of matter, and thus with as little pain and trouble as may be. When the wound is a simple cut, this desirable termination may be expected, and often realized. To attain the end, how-ever, in many wounds, considerable care is requisite. In the first place, the wound must not be closed so soon as that a *clot of blood will form between the exposed surfaces ;* if it does, un-less extremely thin, it will *prevent union.* In the second place, when the wound is closed, its surfaces must be placed in as accurate a position as possible, and must be thus kept together till the process is complete. To effect and maintain this con-tact, various agencies are employed, and of these, position is not the least important, that is, the placing of the parts so that the surfaces of the wound may, as far as possible, fall into con-tact, and that, when other dressings are applied, there may be no dragging to get things to meet. Thus, in a wound of the forepart of the neck, it is requisite to fasten the head so as to prevent its being thrown back. At the same time, position must be regulated with a view to prevent hemorrhage. The wounded parts being properly placed, the next object is to draw the surfaces into as close a position as possible ; in some cases, this is sufficiently well done by means of strips of adhesive plaster, placed at such intervals as will permit discharge of matter, should any form. Frequently, however, from the nature, site, or extent of the wound, plaster is not sufficient to keep the edges together, or to counteract the natural tendency of the skin to retract when severed. In such instances, stitches are employed. These consist of a piece of a sufficiently stout silk or linen thread passed through the thickness of the skin, at about the distance of a line from each of the severed edges. The thread is passed by means of a curved surgical needle, if it can be had, and the two ends being tied, bring the edges together, and retain them most effectually in contact ; that is, provided the stitch is not made use of to drag parts into place ; this it should never do. If there is a continued strain upon the stitches, not only do they cause much pain, but they quickly cause ulceration, which, by detaching, renders them perfectly useless. The surfaces of a wound having been brought into contact, a piece of thin linen, soaked in water, should be placed over it, and if possible, a lightly applied bandage. This not only keeps the dressing in place, and assists to exclude air, but gives support, which is always serviceable, and often, in large wounds, absolutely necessary. The bandage may be kept wet with cold or tepid water, as most agreeable to the feelings of the patient. When a wound progresses well toward recovery, when there is no appearance of discharge, or so little that it is neither inconvenient nor offensive, there should be no meddling ; the less the processes of reparation are disturbed the better, and in some cases a week may be allowed to elapse before the dressings are disturbed ; they may of course require it before, especially in warm weather. It ought to be remembered, that in the treatment of all wounds, it is important to exclude the action of the air as far as possible ; and, that rest, simplicity, and cleanliness, are the great promoters of healing ; the last being best attained by the use of water alone. No balsams or similar applications should be employed ; and, except it be a little perfectly sweet fresh lard occasionally, ointments may be entirely dispensed with.

The above observations have been directed specially to simple incised wounds ; when laceration or contusion accompanies the injury, the principle must be to get the wound as much into the condition of a simple incision as possible. To do this—the wound having been thoroughly cleansed from dirt, grit, &c., by means of a soft sponge and water, and any foreign body which *can be easily reached*, removed—all parts not absolutely detached from the body are to be placed as nearly as possible in the natural position, stitches and plasters being used to retain them, and free exit left for the discharge of matter; over these there must be applied either poultice, or cloths dipped in cold water, and a bandage may be necessary or not, according to circumstances, which those in attendance must direct to the best of their judgment.

PUNCTURED WOUNDS.—As a general rule, these require no other treatment than the extraction of any foreign substance which may be left in them, when it can be easily done ; and to lessen the chance of inflammation, by keeping a warm poultice on them, to preserve the surface relaxed, and facili- tate the discharge of any matter which may be formed at the bottom of the puncture. When inflammation is threatened, the means to prevent and lessen it have been pointed out under head of Bruises, &c., and should be pursued.

These wounds sometimes end in convulsions of the muscles, and are most apt to be followed by *lock jaw.* Sometimes it is found necessary to dilate the wound and fully divide any nerve or tendon which may have been punctured ; sometimes a blister over the part has succeeded without the division.

WOUNDS OF THE JOINTS.

IN all cases of wounded joints, it is important to place the limb in such a posture as to favor the union of the sides of the wound, in order to prevent the admission of air, which seldom fails to produce general irritation. Not only *absolute rest* is to be enjoined, but a very low diet, with slight laxatives. The parts should always, when practicable, be brought together, and kept so by slips of adhesive plaster, in *preference to sewing* them up : and when they are to be stitched, the needle should only pass *through the skin*, and never to enter the *cavity of the joint*, where they would increase the inflammation. Treat- ed in this way, they very generally speedily unite without inflammation.

In order more effectually to procure absolute rest of the joints, it is necessary to apply splints to fit the parts, which being lined with soft materials, occasion no inconvenience. In wounds of the knee, ankle or elbow, these splints are indis- pensably necessary. When there is reason to apprehend a stiff joint, it is necessary to choose the position of the limb in which the stiffness will be least inconvenient to the patient, and to preserve that posture during the cure. If, for example, the elbow were to heal with the arm permanently extended, the limb would be almost useless; whereas, an arm bent at the elbow, may be useful.—And the reverse in the knee, as an extended leg would favor walking.

WHITLOW OR FELON.

THIS is a disease with which very many persons are *pain fully* acquainted. It is a painful and distressing inflammation, seated at the end of a finger or thumb, generally terminating in the formation of matter. The inflammation appears in different parts, either at the root or side of the nail, or near the end of the finger, or underneath the whole of the soft parts : or underneath the immediate covering of the bone, and the bone itself. The most distressing kind is that where the tendons are affected, and the inflammation extends along the hand, up the arm—sometimes rendering amputation necessary.

In the treatment of this affection, we should act with an energy proportionate to the degree of disease. In all cases of fever, blood-letting by leeches, from the part, purges and low diet, should be enjoined. In common cases, repeatedly scalding the finger by suddenly dipping it in boiling water, proves sufficient. It is much better to use the strongest lye, than water for this purpose. Coating the entire part with caustic, (nitrate of silver) will also sometimes check the disease. A blister plaster should be applied around the whole finger, in order to excite action on the surface, to relieve that underneath ; and it ought to be kept continually discharging. When matter is formed under the nail, it should be scraped away over it, and a small puncture made for letting it out. Whenever there is reason to believe that matter is formed in any part, by all means, *freely cut down to it,* and give vent to it. Immediate relief from pain will be had, and an end be put to the danger of prolonged, distressing, and dangerous inflammation : for the subsidence of the inflammation and healing of the part are very rapid ; whereas, when the parts burst, as in common boils, these operations are very tedious and painful. Of course, after opening a felon in this manner, poultices should be frequently applied until the healing is nearly completed, then Basilicon ointment is more proper, or some healing ointment twice a day.

C A N C E R .

Tнis disease is not treated of here with the expectation that the unprofessional will at any time attempt the cure of this most dreadful malady, but to point out its symptoms, that it may be guarded against in time, and also to offer such suggestions for the alleviation of the sufferings of the patient as may be employed when a physician can not be had to attend to it.

Cancer is of two kinds, the scirrhous, or hard, and the open or ulcerated; but these may be more properly regarded as different stages of the same disease. By occult or scirrhous cancer, is meant a hard tumor, for the most part accompanied by sharp darting pains, which recur more or less frequently. This tumor, in the course of time, breaks and ulcerates; and then is more strictly denominated cancer. The parts of the body subject to cancer are the following: the female breast and the womb; the lips, especially the lower one, the tongue, the skin, the tonsils, the lower opening of the stomach, and some other parts, chiefly glandular. Chimney-sweepers are subject to a cancerous affection of the scrotum.

In general, cancer begins at a small spot, and extends from thence in all directions. Its progress is more or less quick in different instances. In general, it is too true, that scirrhus is seldom or never dispersed, and that it causes, finally, the neighboring parts, whatever their nature may be, to put on the same diseased action; and thus the skin, the muscle, &c., are all involved in the same destructive process. In consequence of this morbid action, the skin above a cancerous tumor becomes attached to it, and the tumor is also attached to the muscles below. The tendency to this unhealthy action begins in the neighboring parts, even before it can be distinctly seen. As the swelling increases, it becomes knotty and unequal on its surface, and this inequality has been considered as characteristic of the disease; almost in every case a darting pain is experience. The hard swelling which is likely to terminate in cancer, is attended generally by the following assemblage of symptoms: the skin is puckered, and of a dull, livid, color, the part is knotted and uneven, occasional darting pains shoot through it; it is attached to the skin above, or to the muscles beneath; and in some cases there is a peculiar *unhealthy look* about the patient. The skin generally may become the seat of cancer. Of the internal organs, the womb in the female, and the stomach, are the most frequent seats of the disease. Cancer is very rare under thirty years of age. When, from the nature of a tumor, its hardness, situation, age of the patient, and particularly if there be any hereditary bias toward the disease, incipient cancer is suspected, there should be no trifling, no leechings, or rubbing, or fomentings; the advice of a skilful surgeon should be sought at once; and neither time, distance, nor expense, should stand in the way of procuring that assistance which may not only preserve life, but save from a lingering and painful death.

Should the suspicion be unfounded, the mind is restored to peace; should they be correct, the one remedy, excision, or cutting out, cannot be too soon submitted to, before the glands adjacent to the disease, or other textures of the body, *become tainted.* In any stage of the disease, however, the advice of the regular practitioner ought to be taken. *Above all, let the sufferer and the friends beware of being tempted by the advertisements of quack remedies, and of wasting time of which every day is precious.* Those who advertise to cure cancer, you can always set down as impostors; it cannot be done—only by *cutting out.* Many swellings and sores *called* cancer have been healed by simple remedies, these are the *great cures* per-formed by cancer doctors, and cancer remedies.

If cancer has reached the stage at which hope of cure must be given up; when it has become an open, grey-looking ulcer, discharging thin, offensive matter, the seat of shooting and stinging pain, and when the constitution is affected, it only remains to make the situation of the sufferer as comfortable as possible. Opium in its various forms is the great soother, and the other anodynes, hemlock, especially, both internally and as a poultice, are all of service. Codliver-oil in some cases allays the pain and retards the progress of the disease; but the regulation and administration of these remedies must be committed to the care of the medical attendant; the domestic remedies must be the *most perfect cleanliness* and kindest consideration for the comfort and irritabilities of any one who is the victim of cancerous disease. Cancer cannot be said to be propagated by contact; but this should be avoided as much as possible—in the intimate relations of husband and wife especially, whatever the organ or structure affected.

The lower lip is not unfrequently the site of cancer in old people, especially, it is said, in those *who smoke much.* A painful sore in this situation, which will not heal ought not to be neglected, but submitted to medical examination. If there was no other reason why the use of Tobacco should be avoided than this one danger arising from smoking, this should be enough. Beware of the *poisonous* weed which costs millions of money every year, and destroys so many valuable lives without being of any possible advantage to compensate for its destructive properties.

DEAFNESS, AND DISEASES OF THE EAR.

The external ear may be lost by violence, as by cutting off or the bite of an animal, etc. If we see it soon after an accident, and find it much lacerated (torn), we are to attempt its reunion by adhesive plasters, and even by stitches, if necessary. When a bandage is applied it should be only moderately tight, as pressure in this place gives considerable uneasiness. Wounds, and loss of a part, or even the whole of the external ear, do not always occasion deafness. If this occurs from such a cause, an ear-trumpet or similar contrivance must be used. Foreign bodies, as peas, bits of glass, or cherry-stones, may get into the ear, and occasion great pain of the part, as well as impaired hearing. Such bodies have been known to occasion for many years excruciating pain of the head, palsy, convulsions, and other distressing symptoms, all which have speedily ceased when a skilful hand has extracted the offending body. Such bodies should be forced out if possible, by the injection of warm water and the application of a small scoop or bent probe. Worms have been known to produce very violent symptoms, by being hatched in the ear. When there is disease, as ulceration or suppuration in the ear, insects are attracted by it, and deposit their eggs, which in time produce worms. Patients so affected should take care to stop the ear when they go to sleep, in summer and autumn. A slight infusion of tobacco in oil of almonds may be dropped into the ear; and this proves fatal to worms.

A very frequent cause of deafness or impaired hearing, is the obstruction of the passage by thickened or hardened wax. The symptoms arising from this cause are deafness, a sensation as of a *noise or clash when eating*, or of heavy sounds, as of a hammer. This kind of deafness is not very difficult of cure. A little olive oil, or oil of almonds is to be dropped into the ear, and retained there by a piece of cotton; and when the wax is softened, it is to be taken out with a small scooped instrument. Injecting warm water with a little soap, by a syringe, is a method of getting rid of the hardened wax, equally simple and efficacious. A *deficiency* of the wax may occasion a degree of deafness. When this is the case we are to drop in two drops every night of the following mixture Sweet oil, spirits of turpentine, sulphuric ether, of each, equal parts. Shake well before using. The bowels must be kept gently open. When the wax is of bad quality, which is known by its deviation from the healthy color and consistence, it may be improved by frequently washing the passage; and giving once or twice a day, a wine-glassful of the infusion or tea, of quassia, with a teaspoonful of equal parts of rhubarb and magnesia. Discharges of matter take place from the passage in consequence of inflammation going on to suppuration, from scrofulous ulcers, from abscesses after fevers, from small-pox, measles, and other causes. These discharges not unfrequently are attended with the loss of the small bones; and in general, total deafness is the consequence. Exposure to cold frequently produces inflammation about the ear, attended with very acute pain, (commonly termed ear-ache), which continues very troublesome, and even alarming, till the patient is relieved by the discharge of matter. This inflammatory state is to be treated by local bleeding (leeches and cupping), the injection of tepid water, and by fomentations of hops, or hoarhound and vinegar, and the passage should be protected from cold air by the introduction of wool or cotton.

Sometimes there is disease in the drum of the ear, attended with offensive, thick discharge, which makes its appearance at the internal opening, shows that the membrane of the drum is destroyed; and so much disease is in the internal parts, that the small bones are discharged externally. In time, a continual discharge from the ear takes place, and the disorganization is so complete, that a total loss of hearing is the consequence. If this disease be noticed in its early stage, if there is acute pain, followed by a discharge of matter, we know it is from inflammation, and we are to palliate or remove this by topical bleedings (leeches and cupping), purgatives, and small doses of ipecac every three hours; and are on no account to inject *stimulating* spirituous fluids. When the disease threatens to be more chronic, we are to use blisters and setons, as auxiliaries to our cure; to employ laxative medicines, and to foment the part as before noticed; and when there is little active inflammation, to throw in a stringent injection as of sulphate of zinc, a teaspoonful to half a pint of water. If there are fungous growths (proud flesh), they are to be touched with caustic.

Sometimes there is deafness from insensibility of the nerves of hearing, though the structure of the parts may be perfect. If we can ascertain this to be the case, we are advised to put the patient on low diet, and to give saline purgatives, seidlitz powders, salts, etc., once or twice a week, applying blisters occasionally behind the ears. The application of electricity may be tried.

MORTIFICATION.

THE following symptoms will indicate that mortification has taken place. When any part of the body loses all motion, sensibility and natural heat, and becomes of a brown, livid, or black color, it is said to be affected with *sphacelus*, that is, complete death or mortification. As long as any sensibility, motion, and warmth continue, the state of the disorder is termed *gangrene* or mortification.

In inflammations of the external parts which terminate in mortification, the process observed is as follows: the pain ceases, the purulent, thick matter, becomes acrid (irritating) and sanious, (watery), bubbles of air are set at liberty, collecting in small blisters under the skin, or distending the whole organ by swelling. The blood is coagulated (clotted) in the vessels of the gangrened part, and the circulation *can not be restored.* In many cases, a slight delirium comes on, followed either by dejection of spirits or calmness of mind; but in each case attended with a peculiarly *wild expression* of countenance; though sometimes with a very peculiar expression of serenity, and a blackness under the eyes. The pulse is usually quick, low, and often intermitting. In the earliest stages, deep incisions are attended with a discharge of blood, still florid (or red), but the skin, the muscles, etc., soon melt down into a brownish offensive mass. We conclude that similar processes take place in the internal parts when they become mortified. When this occurs in strangulated hernia, (rupture) or in inflammation of the bowels, a remission of the violent pain takes place, and the patient and his friends are deluded with the hope of complete relief; but the experienced physician knows the treacherous symptom, and must not deceive them with false hopes. There is a peculiar kind of mortification called *dry gangrene*, in which the disease begins in one of the toes, particularly in old people, and very often after a person has been *paring a corn or toe-nail.* It sometimes stops spontaneously, and deprives the patient of some of his toes, or even of his foot and leg, as cleanly as if it had been amputated by a surgical operation: at other times it has been successfully treated by giving large doses of opium. In this form of mortification, the parts affected are perfectly *dry, hard,* and not liable to run into putrefaction.

Mortification is brought about by general or local causes. Those which affect the general system, are the violent inflammatory fevers, or the jail and hospital fever; as also scurvy and dropsy, long-continued or intense cold, and some internal changes, which we can not trace nor explain. The local causes of mortification are numerous. Some of them are burns, ex cessive cold, the application of caustics, the strangulation of a part, as in hernia, severe bruises, as gun-shot wounds, bad fractures, violent inflammation, pressure on large blood-vessels, by tumors, &c. Long continuance in one posture, as when a person is confined to bed, gives occasion to gangrene of the parts where the bones have the least flesh upon them, and which are therefore much exposed to pressure ; as the·shoulder blades, the haunch-bones, and the lower part of the spine. Hospital gangrene is produced by some indescribable state of the air in hospitals, jails, and ships. During its prevalence, the small-est scratch or ulcer is apt to turn to a fatal gangrene. In dropsy, occurring in a broken-down and debilitated constitu-tion, if a few punctures be made to let out the effused fluid, or a blister be applied, these are apt to run into gangrene ; it is also not unusual for spontaneous blisters to form and break on such dropsical limbs, and to go on to mortification.

When inflammation is so violent and strong as to give reason to fear that it will end in mortification, it is a call for us to use with great dilligence, purging, low diet, cold applications, and the other means for abating it, taking care that we do not continue them too long, lest we add to the debility and exhaus-tion which are to follow.

When the mortification has fairly begun, our remedies must be very different from those which counteract inflammation. We are now to prevent debility by giving a nourishing diet and tonics. Of the class of tonics, the most efficacious is the Peruvian bark ; and in a great variety of cases, the good effects of the Peruvian bark are very remarkable. Taken in the form of tea, (cold), a gill three times a day ; or one tablespoonful of the tincture. When the weakness is very great, the use of quinine, two grains three or four times a day, or wine may occasionally be required, as also ammonia and other stimulants. We must be careful not to give these remedies when there is much strength of pulse and inflammatory symptoms remaining. When our remedies are successful, and the mortification is about to cease, a separation takes place at the verge of the sound part, caused by a slight degree of inflammation.

Some have advised cold lotions near the verge of the morti-fied part, to check the further progress of inflammation ; bu fomentations and poultices are commonly preferred. To the common poultices, in some cases, are added powdered charcoal or yeast, to correct the offensive odor and to counteract putre-faction. Stale beer grounds, or port wine, with linseed meal, make a good poultice. It is necessary to give vent to putrid matte:, and for this purpose pretty deep incisions are required through the dead parts.

After the mortified parts have completely separated, and a healthy running ulcer is left, the latter is to be treated by common poultices, until healing commences, when the use of strips of adhesive plaster applied over the surface of the ulcer, and proper bandages, will, in general, cause it speedly to heal over. Washing the parts with castile soap and tepid water, is also advisable.

FOREIGN BODIES IN THE GULLET or THROAT.

It is not at all an uncommon occurrence for foreign bodies to stick in the gullet, as pieces of crust, or meat not completely chewed, or small bones, beans, stones, pins, or pieces of money. Some of these would produce a very bad effect if not quickly removed from the gullet; and perhaps still worse, if pushed down into the stomach; but sometimes pretty large bodies have passed downwards into the stomach, and have been discharged by stool in a few days, without any inconvenience.

Pins and other sharp bodies, when they have stuck in the throat, have been returned by swallowing a piece of tough meat tied to a strong thread, and then pulling it up again. If the detained body can be with safety pushed down, the probang, a flexible piece of whalebone, with a piece of sponge secured to its end, is the proper instrument. If the bodies can not be easily moved up or down, endeavors should not be continued long, lest inflammation come on. When endeavors fail, the patient must be treated as if laboring under an inflammatory disease, and the same treatment will be required if an inflammation takes place in the part, after the obstructing body is removed. A proper degree of agitation has sometimes succeeded in removing the body sticking in the gullet, better than instruments. Thus, a blow on the back has often forced up a substance that has stuck in the gullet, or passed into the windpipe. Pins which have stuck in the gullet have been discharged by riding on a horse or in a carriage. Above all things, in cases of this kind, try to "*keep cool*," as the saying is, for in the excitement more harm may be done than good. A knowledge of the nature of the substance in the throat should direct you in extracting it; always being careful in the use of anything in the shape of a *hook*, that the throat may be not injured; the patent must be *firmly held*, to avoid accidents from his sudden movements.

GRAVEL, or STONE IN THE BLADDER.

In this disease it must be observed that the urinary sand or gravel deposited on the sides or bottom of a receiving vessel is of two kinds, *red* and *white ;* and it is of great importance to distinguish the one from the other, as they proceed from different causes, and require a different mode of treatment. The symptoms of *red* gravel are well known. The shade of color may vary from a reddish brown, or pink, to a perfect red. In such cases the urinary secretion is generally *small in quantity,* and *high colored,* and the disease *inflammatory :* the nearer the deposit approaches to a perfect *red,* the more severe in general are the symptoms.

White gravel is less common, but has long been observed to be attended by very distressing symptoms. These consist in great irritability of the system, and derangement of the digestive organs generally. There is often a sallow, haggard expression of countenance; and as the disease proceeds, symptoms somewhat analogous to those of diabetes, (or great flow of urine), begin to appear, such as great languor and depression of spirits, coldness of the legs, and other symptoms of extreme debility. The urine is invariably pale, and voided in greater quantity than usual; and after standing, for a greater or less time, always deposits a most copious precipitate of a white fine powder. In all such cases, the urine is extremely prone to decomposition, and emits a most disgusting odor.

The chief cause seems to be a want of constitutional vigor, and especially in the digestive organs; the periods of life in which this disease occurs most frequently, are from infancy to the age of puberty, and in declining years : while it is rarely met with during the busy and restless term of the prime of life, these complaints being seldom met with in warm climates. The drinking of hard water often influences very sensibly the state of the complaint. White gravel may often be very distinctly traced to an injury of the back.

In a healthy state the urine is always an *acid* secretion, and it is the excess of its acid that holds the earthly salts it contains in solution. If, from any cause, it be deprived of this excess, or, in other words, the secretion of its acid be unduly *diminished,* the earthy parts are no longer held in solution, and a tendancy to form *white* sand or gravel immediately commences. If, on the contrary, the acid be in greater excess *than usual,* instead of deficient, or if the natural secretion of the earthy constituents of the urine be *deficient,* while the acid retain its usual measure, the acid itself has a tendency to form a deposit, and hence the modification of *red* sand or gravel

Gravel.

301

that is so frequently found coating the bottom of chamber utensils.

It is proper to remark that the *red gravel* is by far the most frequent kind of deposit, and the most effectual remedies for it are the *alkalies*, and the alkaline carbonates, such as lime-water, the carbonate of potash or soda, and magnesia. But to be realy useful, they must be conjoined with a proper diet, tonics and mild purgatives; for it ought never to be forgotten, in the treatment of gravel and stone, that they owe their formation chiefly to an irregular and vitiated action of the *digestive organs*, which will invariably require this conjunction, in order to the accomplishment of a permanently beneficial effect.

Half a drachm, or a drachm of carbonate of potash, or soda, may be given dissolved in water, two or three times a day, with an alterative pill of blue mass, five grains, ipecac one grain, rhubarb three grains; the following draught being taken every morning, or every other morning, as a gentle and suitable purgative. Take of Rochelle salt two or three teaspoonfuls, carbonate of soda half-teaspoonful, water three tablespoonfuls—mix, and after adding a tablespoonful of lemon-juice, or thirty grains of tartaric acid, let it be drank immediately.

Magnesia, in this species of gravel, is of considerable use. It may be taken either alone, in doses of ten grains twice a day, or combined with the carbonate of soda, in the proportion of six or eight grains of the former, to ten grains of the latter, twice or three times a day. Or ten grains of magnesia may be dissolved in a glass of soda water, which is an excellent way of administering it.

Ten or fifteen grains of the carbonate of ammonia, twice a day form likewise a useful medicine, especially in cases where great languor, or weakness and coldness of the stomach, is present. The ammonia is a powerful corrector of acidity, and a most valuable cordial.

A very convenient and valuable mode of combining an alkali with an aperient, and gentle bitter tonic, is the following; it is worthy of particular regard when weakness of the stomach, costiveness and *red gravel* are combined : take of carbonate of soda ten grains; Epsom salts half a drachm or a drachm; infusion (tea) of gentian, three tablespoonfuls; tincture of cardamon seed a teaspoonful—mix for a dose, to be taken three times a day. The bowels should be kept gently open by it, and, therefore, the Epsom salt may be either increased or diminished, as needed.

Uva ursi is both tonic and astringent, and has been spoken well of, for its virtues in gravel and stone, by physicans of high authority It may be combined or alternated with the alkalies; and where general debility exists, or there is a discharge of pus-like matter from the bladder, denoting ulceration, or a faulty condition of its secreting vessels, it is at once an appropriate and excellent medicine. The dose is from a half to one teaspoonful of the powder, twice or three times a day; or a strong tea may be made by pouring hot water upon the leaves of the plant; to three tablespoonfuls of which may be added ten grains of bi-carbonate of soda, and drank three times a day. In cases of *white* gravel, it may be given in conjunction with the nitric or muriatic acid, ten drops to the pint of uva ursi tea, to be taken through a quill or straw, to protect the teeth.

The diet of persons troubled with *red* gravel should be moderate in quantity, but of a *nutritious* and wholesome quality, consisting principally of fresh animal food and vegetables. All acids must be *carefully avoided*, and likewise heavy bread, fat meats, hard boiled puddings, and soups.

Active exercise is of great importance in all gravelly disorders; and flannel should be constantly worn next the skin.

Now, in regard to the cases of *white sand or gravel*, an *acid* is the *best medicine*, and all the acids seem to answer the purpose, though the muriatic, nitric, and citric acids, have been in the greatest repute. The citric acid, or lemon-juice, is preferable for children, as being the pleasantest, and that which may be persevered in for the longest time: it may be mixed with water in any proportion that is agreeable. The muriatic acid may be given in doses of from five to twenty drops, twice or three times a day, in a wineglassful of water; and the nitric acid in doses of from five to twelve drops, in the same proportion of fluid, to be sucked through a quill or straw, to prevent injury to the teeth.

The diet should be nutritious, easy of digestion, and moderate in quantity, and be as largely as possible intermixed with acids, salads, fruits, and especially oranges and lemons. Water, saturated with carbonic acid, to be found in most of the drug-stores, in the form of "carbonated water," is the best common beverage in this kind of gravel, and, attention being paid to diet and exercise, will sometimes be alone a sufficient remedy.

When pain attends the gravel, opium or extract of henbane should be occasionally administered, according to the urgency of that symptom. Thirty or forty drops of laudanum, or twenty of the solution of sulphate of morphia, or from five to ten grains of the extract of henbane, may be given alone, or in any drink which the patient may be taking, and repeated until the pain is relieved. Opium seems generally preferable in the *white* gravel; and henbane in the *red*. In *white* gravel, **the** solution of acetate of morphia is particularly indicated as **an** anodyne, since the acid it contains is an appropriate and efficient remedy for the complaint, and, at the same time, counteracts the injurious effects likely to result from the frequent use of opium, when taken in any of its common forms. In case of great pain and irritation about the urinary organs, an opiate injection will be proper, and often of much service; (ten to twenty drops of laudanum, in half pint of tepid water,) or two or three grains of opium may be made into a pill, and inserted within the lower portion of the bowel as a suppository.

A burgundy pitch or galbanum plaster may be applied over the loins with advantage.

Whether the gravel be white or red, when a small stone passes from the kidneys into the bladder there is generally a fit of pain and irritation; to relieve which, the warm bath, or hot fomentations of hops, wormwood, &c., together with forty or fifty drops of laudanum every three hours, will be the most proper and effectual remedies. The passing of a small stone from the kidneys to the bladder, is denoted by a fixed pain in the region of the affected kidney, with a numbness of the thigh on the same side. The pain is sometimes very acute, and accompanied with nausea and fainting, but the pulse is rarely accelerated. During the whole of the passage from the kidneys, the urine is usually high colored, and frequently mixed with blood.

STONE IN THE BLADDER.—The symptoms of stone in bladder are, a sort of itching along the urethra, particularly at the extremity; frequent propensities to make water, and go to stool; great pain in voiding the urine, and difficulty in retaining it; the stream of urine being liable to stop suddenly, while flowing in a full current, although the bladder is not empty, so that the fluid is expelled by fits as it were; and the pain being greatest towards the end of, and just after the evacuation. There is a dull pain about the neck of the bladder, together with a sense of weight, or pressure, at the lower part of the belly; and a large quantity of mucus (or slime) is mixed with the urine, and sometimes the latter is tinged with blood, especially after exercise

The causes of stone in the bladder are the same as thos which give rise to gravel.

The medical treatment to be employed in cases of stone is precisely the same as that for gravel, both in regard to the remedies and diet. There is this difference between gravel and stone, that, in the former, active exercise is highly advisable;

whereas, during the actual presence of stone in the bladder, the patient's exercise ought, for obvious reasons, to be less active and constant.

An injection of castor oil has great effect in relieving the sufferings occasioned by stone in the bladder; the introduction of a lubricating fluid into the bladder, under such circumstances, is productive of ease and advantage. One or two ounces injected when the bladder is *empty*, through a catheter, is about the proper quantity, used once in every two or three days, as may be necessary.

STRANGULATED HERNIA, or RUPTURE.

As this is liable to occur at any time, in those who are ruptured, if not protected by a proper truss, it should be carefully studied, so that it can be detected in·time to send for medical aid. When either an old rupture from some cause has become strangulated, or when some sudden exertion has at once produced rupture and strangulation, the following symptoms occur : there is a swelling at the place of the rupture, painful to the touch, and increased by coughing, sneezing, or by the upright posture. These symptoms are followed by sickness, retching, costiveness, with a frequent hard pulse, and other attendants of fever. The cause of these symptoms is the stricture made on the bowel, by the part through which it protrudes. The object of cure, is therefore to relieve the bowel from this pressure, which is to be effected either by returning the intestine into the belly, by the same aperture through which it came out, or by enlarging the aperture by an operation, which can only be done by a surgeon.

Our first efforts should be to replace the bowel by the hand, if possible ; and various methods are to be put in practice, to produce the relaxation necessary for that purpose : place the patient on his back, with the thighs and knees bent ; and make pressure on the tumor in a direction obliquely *upwards* and *outwards*, if it be an inguinal hernia, (running obliquely a.ong the lower portion of the abdomen,) but the pressure must at first be made *downwards*, towards the thigh, and ther. upwards if the hernia be *femoral*, (running down into the upper part of the thigh.) In a young and strong person, bleeding is very proper, both to induce relaxation, and to prevent inflammatory symptoms. The warm bath may be tried also to induce relaxation. With a view to diminish the bulk of the swelling, and so to render it more easily replaced, cold has been applied to the external parts, by means of ice or of ether. An injection of the infusion of tobacco produces an extreme relaxation of the whole system, and so has conduced to the replacement of protruded bowels. The strength of the infusion is a drachm of the leaves to a pint of boiling water; this is infused for ten minutes; one-half is injected at first, and the other a little afterwards, if no proper effect is produced by the first. *The tobacco injection is, however, a remedy of the greatest danger, and must never be administered, except by an experienced practitioner.* These attempts to reduce the bowel, may be made for a longer or shorter period, according to the symptoms of each case. Much handling will add to the danger of inflammation which is already so great; and too long delay will allow the bowels to get into a state of mortification. Always procure a physician in these cases, if possible.

RUPTURES IN INFANTS.—Ruptures in different parts, especially at the navel, are not unfrequent occurrences in infancy; fortunately, they are not attended with so much danger as similar disorders in grown people. When the disease is confined to the navel, a broad piece of flannel, in the form of a roller, together with pieces of adhesive plaster applied over the part with a ball of cotton, forming what has been termed by surgeons a *graduated compress,* by affording a safe and firm support, prove so useful, that as the infant acquires strength, the rupture commonly disappears. The other varieties of rupture are often cured by the natural increase of size and strength in the body, and require chiefly attention to the due regulation of the bowels, and the daily use of the cold bath. No truss ought to be employed for at least the first two years of life.

ULCERS, or SORES

IT may seem almost unnecessary to state that there are a great many varieties of ulcers, requiring a corresponding variety of treatment. We have first the

SIMPLE PURULENT OR RUNNING ULCER.—Some ulcers are covered with matter of a white color, of a thick consistence, and which readily separates from the surface of the sore. There is a number of little eminences covering the bottom of the ulcer, called granulations, which are small, red, and pointed at the top. As soon as they have risen to the level of the surrounding skin, those next the old skin become smooth, and are covered with a thin film, which afterwards becomes cloudy looking, and forms skin. The principal thing to be done in the treatment of this kind of ulcer, is to keep the surface clean, by putting on a little dry lint, and a pledget (several folds of muslin) over it, covered with simple ointment, made of equal parts of lard, beeswax and tallow. In some patients, ointment irritates and inflames the neighboring skin. Bandages sometimes irritate the sore, and disturb the healing process; but when they do not, they are useful in giving a moderate support to the parts, and in defending those that are newly formed.

ULCERS IN WEAKENED PARTS. Other ulcers are in parts which are too weak to carry on the actions necessary to their recovery. In them, the granulations are larger, more round, and less compact than those formed on ulcers in healthy parts. When they have come up to the level of the healthy parts, they do not readily form skin, but rising still higher, lose altogether the power of forming it. When the parts are still weaker, the granulations sometimes fill up the hollow of the ulcer, and then are suddenly absorbed, leaving the sore as deep as ever. Ulcers are very much under the influence of whatever affects the constitution; and change of weather, emotions of the mind, and some other agents, quickly occasion a change in their condition. Such ulcers as we have been describing, require general as well as local treatment; one grain of quinine three times a day, or a tea of dogwood bark, cold, in wineglassful doses, and nutritious diet, are to be given; and the granulations are to be kept from rising too much, by the prudent application of blue vitriol, lunar caustic and burnt alum, weakened sufficiently by proper admixture of ointment to act as *stimulants,* and not as *caustics.* This will give a proper and healthy action to the granulating surface; whereas the destroying of the rising parts by caustics, seems rather to encourage the growth. Bandages and proper support to the parts, are highly useful. These ulcers, in weak parts, do not seem to be the better of poultices, or other relaxing applications; powders rarely do good, and perhaps the best dressing is the citrine ointment, (see " Medicines, their doses and uses), more or less diluted, if required.

IRRITABLE ULCERS.—There are certain ulcers, which may be called *irritable ulcers.* The margin of the surrounding skin is jagged, and terminating in an edge which is sharp and *undermined.* There is no distinct appearance of granulations, but a whitish spongy substance, covered with a thin watery or milky discharge. Everything that touches the surface gives pain, and commonly makes the ulcer bleed. The pain sometimes comes on in paroxysms, and causes convulsive motions of the limb. Such ulcers seldom do well without a frequent change of treatment. Fomentations with poppy heads, hops, chamomile flowers, or hemlock leaves, are sometimes of use in irritable ulcers. When poultices are prescribed, they should never be allowed to rest or bear weight on the sore limb. Powdered applications are generally too stimulating for irritable ulcers, and bandages also prove hurtful. Frequent washing with cold water, or pouring upon them a stream of cold water, will often be found beneficial in this variety of ulcers. A tea made of oak bark, mixed with *tar water*, will sometimes act like a charm in healing this kind of ulcer.

INDOLENT ULCERS are those which have the edges of the surrounding skin thick, prominent, smooth, and rounded. The surface of the granulations is smooth and glossy; the matter is thin and watery, and the bottom of the ulcer is *nearly level*. A great proportion of the ulcers in hospitals are of the most indolent kind. Indolent ulcers form granulations, but frequently they are all of a sudden absorbed, and in a few hours the sore becomes as much increased in size as it had been diminished for many weeks. The principal applications required for indolent ulcers are those of a stimulating nature, as the basilicon ointment, and occasionally sprinkling with red precipitate. Pressure is to be made by a roller, and by slips of adhesive plaster. Be careful to soften the dressings always before taking them off, by means of warm water or soap suds. The tea of oak bark and tar water is also advisable in this kind of ulcer, or the application of a wash made of a teaspoonful of blue stone in a pint of water once a day, or the lunar caustic, will sometimes be necessary.

WENS.

A DESCRIPTION of these is here given for the benefit of the reader, in detecting tumors of this kind, not expecting any domestic treatment can often be successful.

Tumors on the surface of the body are distinguished by surgeons according to the nature of their contents, and they require treatment varied according to circumstances.

Wen is the common popular name for any fleshy excrescence or tumor growing on any part of the body; most frequently, however, it is applied to tumors about the throat and neck.

Sometimes wens are attached by a narrow neck, and may be removed by the knife, or by ligature (being tied with silk around its neck); at other times they have a broad base, and are so supplied with large blood-vessels that they cannot be removed at all, or cut, without the utmost risk. Sometimes wens are filled with a curdy or cheese-looking matter, and are contained in a cyst or bag, which must be dissected out, along with its contents, and the cut skin will heal and leave very little deformity; in other cases, the tumor is *fungus hæmatodes,* or bloody cancer, which pretty certainly destroys the patient. The bronchocele, or goitre, (on the front part of the neck,) is to be treated with iodine ointment, or tincture of iodine, applied once every day or two, and the tincture of iodine internally (ten drops three times a day in a gill of sweetened water, on an empty stomach). Sometimes very large wens contain a mixed substance, resembling fat or marrow ; they have a firm fleshy feel, and sometimes attain an enormous size.

WRY NECK.

This term is generally applied to a long-continued or permanent turning of the head to one side. It is different from the pain and stiffness which occur from cold and rheumatism, and which prevent the free motion of the head. It arises from various morbid conditions of the part, either from distortions of the spine, from palsy of some of the nerves going to the muscles that move the head, or from some altered structure of the muscles themselves. The removal of this affection, when possible, is accomplished by treatment adapted to the particular cause inducing it. The bones of the neck may be aided by machinery, by which they can be kept in a proper position, if the subject is young ; blistering, friction, and shampooing, long persevered in, have been of service in the paralytic affections of the nerves and muscles ; and at one time it was a favorite practice to cut across the large muscle extending from the ear to the breast-bone, which was generally supposed to be in fault. This severe measure very often was unsuccessful, and is hardly ever to be recommended, even when performed by the best surgeons.

VARICOSE, or ENLARGED VEINS.

This condition of the veins is found mostly in the lower extremities, and is sometimes a troublesome and painful disease. Of course the longer it goes on without being cured the worse it becomes. The affection consists, essentially, in the veins becoming elongated (or stretched longer), so as to permit of their assuming a tortuous, knotted condition, while they are at the same time enlarged. This state is usually associated with *obliteration* or *deficiency*, more or less, of the *valves* within the veins, so that the weight of the entire column of blood bears with distensive force upon the vessels, and upon those parts of them which are most dependent. The most frequent causes of the varicose veins, are such as cause impediment to the upward flow of the blood through the large veins of the abdomen. In this way, pregnancy, if frequent,

t. a most common exciter of the condition: habitual costive ness, diseases of the liver, tumors of any kind within the abdomen, act in a similar manner. The truss worn on account of rupture, or garters too tightly tied, likewise excite the varicose condition, which is usually more common in persons whose occupations require much standing, especially if they are tall.

The veins and limbs generally should be supported by some one of the forms of elastic stocking: these can now be obtained at so moderate a price, that none need be without their valuable aid. It is often surprising how immediately the use of well applied mechanical support, such as the elastic stocking affords, removes the uneasy and painful sensations connected with the condition of the veins in question. Some individuals cannot, however, wear an elastic stocking of any kind; for such cases, an elastic tape or bandage fixed to the foot by a stirrup, and wound spirally round the limb, has been successfully employed. Spaces of about three inches being left between the spirals, each time the band crosses the vein, it acts like a valve.

Frequently a bandage of muslin or calico, wet and rolled moderately tight, beginning at the toes and going up the limb as may be requisite, will answer all purposes. It should be put on when the patient is *lying down*, and reapplied twice a day. Let the limb be plunged into cold water, or laid down and cold water poured over the limb, *beginning at the toes and going along above the knee*, each time before the bandage is applied.

FOREIGN SUBSTANCES IN THE EYES.

This is often the cause of much suffering as well as uneasiness of mind. Persons often go a long ways to a physician on account of some little substance getting into the eye, when the exercise of a little judgment and tact on the part of bystanders would save the patient time, expense, and suffering, by removing it before it had time to produce much local irritation. The membrane, which covers the inside of the lids and white of the eye, is, from its exposed situation, liable to become inflamed from various causes. Minute particles of dust or other substances getting into the eye, and becoming fixed in the lining of the upper eyelid, between it and the globe, cause an amount of pain and irritation which could scarcely be credited from their size, but which is well accounted for by the accurate apposition of the two surfaces between which they lie. A particle so situated may be discov-

ered without much difficulty by a second party examining the
sufferer with the head thrown back, while he slightly everts,
(turns inside out), the upper lid with the thumb and finger
The slightest speck of foreign matter must be removed, and no
better instrument can be employed for the purpose than a
piece of not over-stiff writing-paper twisted, or where it can
be had, a camel's hair pencil. Those who work in metals are
apt to get minute scales imbedded in the forepart of the ball;
they cause much irritation, and are often so extremely difficult
to remove that a surgeon's assistance is required. Pieces of
iron or steel can often be removed by a magnet (loadstone), or
magnetized piece of iron: bringing it in contact with the foreign
substance, it attracts it, and, adhering to the magnet is taken
out. *Never rub the eye* when there is dust or other foreign
substance in it, it may *injure the ball of the eye by friction.*

S T Y E .

This is a *boil* or *abscess* in the lid of the eye, usually caused
by the duct or opening leading from the little glands becoming obstructed. When once they make their appearance they
are troublesome, as one attack after another is apt to follow.
The treatment consists in mild purgatives, a spare diet, and
local applications, such as warm fomentations of hops, etc., or
poultices *in a muslin bag*, lest the particles get into the eye.
When it breaks and gets well, bathe the eye three or four
times a day in cold water.

S O R E   E Y E S .

This being a disease of such frequent occurrence, and by
being *neglected* or *improperly treated*, often occasioning loss
of sight, every person should understand how to treat it.

The first symptom of inflammation of the eye is a sensation
as if a particle of some kind had lodged in the eye, and if an
examination be made there will be seen, not only an enlargement of any small blood-vessels that may be generally visible
on the white of the eye, but a new development of others, the
appearance varying from the slightest apparent increase of
vascularity or fullness, to the most intensely red inflammation.
At the same time there is considerable increase in the mucous
secretion—not in the tears, as is often supposed—and in bad
cases this becomes purulent or mixed with matter. There is,
sometimes, considerable swelling of the surface, usually distinguished as the white of the eye. The above is the most
superficial form of inflammation to which the eye is subject;
if neglected it may extend itself over the cornea or more part

of the eye-ball and produce permanent blindness. It is undis-
tinguishable from the next form, or inflammation of the sclerotic coat, or whites of the eyes, by the size and winding character of the small blood-vessels, and by their being slightly
movable along with the conjunctiva, or membrane, itself, when
the lids are drawn down. It is important that these distinc-
tive characters should be attended to in the first place, that
no error may be committed between this form and a more
serious and deep-seated inflammation of the eye, and also that
proper treatment may be used. A great error is committed
in treating this form of inflammation by means of warm
fomentations, etc., applications tending rather to keep up than
to cure the disease, which is generally quickly removed by
astringents and cold applications. A drop of laudanum mixed
with six of cold water, put in the eye, repeated two or three
times will often cure the disorder; or a lotion of sulphate of zinc,
from one to three grains to the ounce of water, will be found
efficient; but the best of all is the solution of nitrate of silver,
or lunar caustic, of the strength of four grains to the ounce of
water. Of this, a single drop may be introduced into the inflamed eye twice or three times in the twenty-four hours.
Great care should be exercised to get the *exact strength* here
given, as it is a *dangerous remedy if made too strong.* The eye,
of course, should be exercised as little as possible, and if the
bowels are confined or the stomach disordered, five grains of
blue pill every second night, followed the next morning by a
seidlitz powder or castor oil will be found useful. If the disease is obstinate, a blister to the back of the neck may be
applied with advantage.

The disease which has just been treated of, is a comparatively mild disorder, but under certain circumstances it becomes much more virulent. The secretion of matter is very
great, and acquires the power of *propagating* the disease by
contagion from one person to another.

Newly-born and young infants frequently suffer from a
severe form of this disease, which often shows itself within
three days after birth. The inflammation is intense, and the
matter often accumulates largely between the lids, gushing
out when they are separated. In scrofulous children especially,
the affection is often obstinate. The nitrate of silver in solution, is the best application, and small doses of quinine the
best internal remedy. Syringing between the lids with a
solution of alum, four grains to the ounce of water, six or
eight times a day, is also recommended. A little lard should
be used on the edges to prevent them sticking together.

In inflammation of the sclerotic or outer coat of the ball itself (or whites of the eyes) there is more actual pain, it is more deeply seated, and the redness seen on the white of the eye is more of a pink hue than in the other form of the disease just treated of, the vessels appear much smaller and straighter, radiate as it were from the cornea or front of the eye, and are not movable ; the affection is generally a more serious one than the other. Active treatment is necessary ; leeches should be freely applied to the temples, or behind the ears, or cupping on the back of the neck resorted to ; the bowels must be freely purged with ten grains of blue pill mixed with ten of rhubarb, followed next morning, if necessary, by a dose of castor oil or Rochelle salts, in the first instance, and then calomel in two-grain doses, given at intervals of six or eight hours. The diet must be reduced as low as possible, all stimulants avoided, every attempt at exertion even of the unaffected eye forbidden, and the person confined to a darkened room, the only local application being continued hot fomentation to the eye (of hops in a bag, on which hot water has been poured), and a blister between the shoulders. By a continuation of the above treatment, even till the gums get sore with the mercury, much may be done in cases where a physician can not be had, but where possible, a medical man should be called to attend to it.

DEFECTS OF VISION.

LONG-SIGHTEDNESS.—This is a condition of the vision often met with in aged persons. It consists in *near* objects being confusedly seen, while those at a *distance* can be distinguished very clearly. It is thought, usually, to depend on the eye becoming *flattened*, from which results an alteration in the convergency of the rays of light, so that the "*focus*" is formed *behind the retina*, or the expanded nerve of the inner part of the eye.

The defect is to be remedied by the use of *convex glasses*, which must be adapted to the eye by the individual affected.

Short-sightedness, of course, is just the *reverse*, produced by the eyeball being *too round* or *prominent ;* and is to be remedied by the use of *concave glasses*, properly adapted to each individual case.

While on this subject, we would take occasion to warn our readers against *reading in railroad cars*, or any vehicle, *while in motion.* It is very injurious to the sight, by the constant strain or effort to fix the " focus." Railroad conductors from this cause have become blind.

We would also warn the reader against the use of what has

been lately brought into use, by extensive advertising, called "*eye sharpeners,*" to produce a greater convexity of the eye ball, by applying the instrument to the front of the eyeball. Such things are very injurious to the eyes, in unprofessional hands.

GROWING IN OF THE TOE-NAIL.

THIS is a frequent and troublesome complaint, caused by the nail pressing down into the soft parts, and kept up by its continuance. It is generally the result of *tight,* or *misshapen* boots and shoes.

There are many palliative methods of treating this affection, such as the use of caustics, scraping the nail away, the application of poultices, etc. ; but perhaps there is no certain mode of treatment but the thorough removal of the entire half of the nail up to the root, on the offending side. This, of course, ought to be done by a physician. Wearing an "old shoe," or loose slipper, or going barefooted for a while, will afford great relief, and sometimes effect a cure, especially if the foot is often soaked in water, and the nail *scraped thin in the middle.*

INFLAMMATION OF THE TONSILS, or Sore Throat

SORE throat of this kind usually commences with chilliness, and often flushes of heat; the tonsils and back part of the throat soon become red, swollen, and painful. The pain is acute and darting, and usually extends to one or both ears. It is increased by every attempt to swallow, and by external pressure. These local symptoms are generally attended with some degree of fever. Swallowing is greatly impeded as the disease increases, and speaking, and even breathing, are endered difficult. In a few cases, small white spots are to be observed upon the tonsils. When the inflammation is *very violent,* the eyes become red, swollen and watery; the cheeks flushed and swollen, and the patient is unable to open his mouth. Externally, large tumors can be felt, or even sometimes seen by the eye, on each side the jaws. The sense of suffocation is intolerable, and the patient is obliged to be supported in an erect posture.

When inflammation of the tonsil occurs repeatedly in the same individual, within a short space of time, a peculiar susceptibility to the disease is established, so that it is produced by the slightest causes afterwards.

The active symptoms in this disease may either rapidly decline or produce suppuration and an abscess in the throat; or remaining a long time in a chronic state, cause an enlarged and hardened condition of the tonsils, by which sometimes breathing, swallowing and speech are so much affected as to require the removal of the organs by a surgical operation.

At the very commencement of the attack, before the inflammation of the throat is of any considerable extent, an emetic of ipecac, ten to fifteen grains, given in warm water, will frequently be found to remove it at once. When, however, the disease has run some time, or is from the first of a violent grade, bleeding, by leeches to the throat, will be demanded. This should be succeeded by a dose of Epsom salts or castor oil, or when the act of swallowing is attended with great difficulty, a tablespoonful of the following mixture may be taken every two or three hours: Epsom salts, one ounce; nitre, (saltpetre,) one drachm; tartar emetic, two grains; and boiling water, twelve ounces. This, with the occasional use of the warm foot-bath, an injection of warm water, will have the effect of opening the bowels, producing a gentle perspiration, and reducing the inflammation.

A variety of acid and astringent gargles have been proposed in this disease, but there are few cases in which the patient can make use of gargles in such a manner as to derive much advantage from them. In general, more benefit will be derived from inhaling the vapor of warm water or vinegar and water, or of bitter herbs, &c., as ordered in catarrh in the head, which may readily be directed to the throat by means of a common funnel.

If the inflammation should not be reduced by these means, a large blister is to be applied around the throat, or the throat may be enveloped for five or ten minutes with a cloth wet with spirits of turpentine. At the same time, the mixture directed above, with the inhalations, should be continued.

Volatile and other liniments to the throat, which is so frequently resorted to in this disease, is productive of little good, and in some cases is even injurious, poultices seeming to afford more relief.

The patient should be allowed nothing in the form of food or drink, during the disease, excepting barley, beef tea, soup, or gum water, rendered slightly acid by the addition of lemon-juice.

When we discover that the swellings in the throat appear evidently inclined to suppurate, or come to a head, this should be encouraged by the frequent inhalation of the steam of hot water, (as before recommended,) and in certain instances by poultices externally. The moment they become soft they should be punctured with a lancet, to allow of a discharge of the contained matter. After this, a gargle of sage tea, alum and l. i. g;, several times in the course of the day, will complete the cure.

We should bear in mind the great liability to a recurrence of the disease, which will point out the importance of the patient being on his guard for a considerable time subsequently to his recovery, against exposure to cold or damp, to sudden transitions of temperature, &c.

Bathing the throat night and morning in cold water is a good means of preventing a return of the disease, in those who have been subject to it.

INFLAMMATION OF THE LARYNX;

(*Or top of the Windpipe.*)

This commences with the usual symptoms of fever, from irritation or local inflammation. The voice very quickly becomes hoarse and indistinct, sometimes entirely extinct; the breathing laborious, with a painful sense of constriction in the throat; on examining the back part of the throat, we now find that every portion of it is of an intense, dark red color, and considerably swollen. The face soon becomes red and bloated, the eyes red, swollen and often protuberant, as in cases of strangulation. The pulse is very quick and frequent, and the tongue coated. Every attempt to swallow is attended with intolerable distress; the muscles of the throat and chest being thrown into violent spasmodic action, threatening the patient with instant suffocation, and causing him to cry out for the admission of more air into the room.

This disease is extremely acute and rapid in its progress, often destroying life, by suffocation, in a day or two, or even in less time, unless attacked in its very commencement by the most active remedies.

In many of its symptoms it bears a close resemblance to croup, and to distinguish them from each other is not always very easy. This, however, is not of much importance, as the treatment of the two diseases does not differ in any important particular.

This is an affection which calls for the exercise of the most energetic and best-directed medical treatment as soon as it can be procured. But it is of the highest importance that no time should be lost, even while waiting for that aid, and that some properly directed means should be at once resorted to. First, from half a dozen to two dozen of leeches, according to the strength of the patient, should be applied to the throat and upper part of the chest; or, if leeches cannot be obtained

from six to twelve ounces of blood are to be taken from the
back of the neck by cupping. Tarter emetic, in eighth of a
grain doses, or twenty drops of antimonial wine, is to be repeated at intervals of from one to two hours, *at first*, and
calomel given in four grain does every four hours, with a
quarter of a grain of opium in every, or every second dose,
should purging ensue. Hot bran poultices are to be kept
constantly to the throat, the feet put in hot water, and advantage may be derived from breathing the steam of hot water,
and bitter herbs, &c., as recommended under the head of Catarrh in the Head; the patient, of course, being kept perfectly
quiet in bed. These measures will do all that can be done until
the arrival of a physician.

COLD IN THE HEAD.

It is unnecessary to give the symptoms of this disease, as they
are too well known to all by frequent and unpleasant experience. However familiar we may be with this disease, yet
we should not look upon it as a trivial affair. From its tendency to recur, and also to produce and keep up irritation of
the lungs, it is not only *not to be neglected*, but should be
hecked at first, if possible, and for this purpose various methods of treatment are recommended. The injection of a
solution of sulphate of zinc, five grains to the ounce, into the
nostrils, at the *very commencement* of the disorder, has been
said to stop it without fail. A teaspoonful of paregoric, or
six or eight grains of Dover's powder, when taken at bedtime
repeated for two or three nights, will often check a cold in the
head at once ; and the usual system of hot foot-baths, confinement to bed, low diet, and diluent drinks, along with boneset
tea, is certainly calculated to mitigate the disorder, and may
be followed with advantage. It is the common practice to
drink copiously of tea, gruel, or some other diluent during a
cold ; as long as this promotes perspiration it is of some utility,
and although it augments the flow from the nose, it has the
effect of diminishing its acrimony or irritating qualities, by
dilution. It is the acrimony of this discharge, which reacting
on the membrane, keeps up the inflammation, and its accompanying disagreeable symptoms. On this circumstance depends the efficacy of a measure directly opposed to that just
noticed — we mean *a total abstinence from liquids* in any
shape, water, tea, coffee, milk, beer, &c. To those who have
the resolution to bear the feelings of thirst for thirty-six or
forty-eight hours. we can promise a pretty certain and complete riddance of their colds, and what is, perhaps, more im

portant, a prevention of those coughs which commonly succeed to them. Nor is the suffering from thirst nearly so great as might be expected, especially when a piece of orange or lemon peel, sassafras bark, or something of the kind is kept in the mouth. This method of **cure** operates by *diminishing the mass of fluid* in the body to such a degree that it will no longer supply the diseased secretion. Anything that **will** contribute to reduce the quantity of fluid in the **body** will assist in the plan of **cure**, and shorten the time necessary for it to take **effect**. It is therefore expedient to begin the treatment with a purgative of salts, followed by a sweat at bed-time, as is usual, and this is the more necessary when any fever attends; but beyond this no further care **need** be taken, and the individual can devote himself to his **usual** employments with much **greater** impunity than under the ordinary treatment. The coryza, or running from the nose, begins to be dried up about twelve hours after leaving off liquids; from that time the flowing to the **eyes and** fulness in the head become less and less troublesome; the secretion becomes gelatinous, and between the thirtieth and the thirty-sixth hour ceases altogether: the whole period of abstinence need scarcely ever to exceed forty-eight hours. It is then as **well** to return to the *moderate* use of liquids, **as the** first indulgence is apt to be excessive. It is not necessary to limit the **solid** food any more than to that which is plain and simple, except **where there** is an acceleration of the pulse, or irritation of the stomach, in which cases animal food should be avoided. **For the** sake of comfort in mastication, the food should **not be of the** driest kind. Thick puddings and **vegetables, with** or without meat, will be the best dinner; and toasted **bread or biscuit** *merely moistened* with tea or other liquid for **other meals.**

A *single cup of tea* is sufficient to bring back the coryza, or watery discharge from the nose, immediately, after twelve hours' abstinence has removed it. We doubt not that it will be said that this plan of **cure** is worse than the disease, and so it may be in some instances. It may be **called** always a choice of evils; but we do not believe that **any one** who is liable to severe colds, after once experiencing **the** amount of good and evil resulting from this method, would **hesitate** between them. *Moderation in liquid food*, is one of the best preventatives against the bad effects of exposure to cold. When there is a large quantity of liquid in the system, there must be *increased perspiration*, and therefore greater risk from the effects of cold. Nature seems, to some extent, to provide for these changes in the atmosphere, as in cold weather we drink but little, and in warm weather drink more, as by the evaporation brought about by perspiration, the temperature of the body is reduced or kept down.

INFLAMMATION OF THE LUNGS, (*Pneumonia.*)

INFLAMMATION of the lungs commences with the usual symptoms of fever: a feeling of coldness or shivering, succeeded by increased heat and dryness of the skin, thirst, flushed face, furred tongue, and increased frequency of the pulse. In some cases, however, there is little or no increase of heat, and the pulse is not more frequent than natural. Very soon a difficulty of breathing is experienced, and a pain, more or less acute, in some part of the chest, increased upon inspiration, (taking air into the lungs,) and in particular positions of the body. Occasionally the pain is dull, or rather there is a feeling of weight and oppression in the chest, rather than of pain. The pain is commonly fixed, but sometimes shoots towards the shoulder, or upper part of the breast. It is invariably accompanied by a short, dry, distressing cough, which greatly aggravates the disease. In the beginning of the disease, the cough is seldom accompanied with much expectoration; a little frothy mucus is generally, however, brought up by it, which, in the course of the disease, is often streaked with blood. Subsequently, however, an expectoration of yellowish thick matter takes place, which becomes whiter, softer and more easily brought up as the disease progresses. The foregoing symptoms are of greater or less violence, according to the extent and intensity of the inflammation.

A favorable termination of the disease is indicated by a gradual subsidence of all the symptoms. The respiration becomes more free, the expectoration more copious, the cough less frequent and distressing; the fever disappears, and the pulse becomes softer and less frequent. The disease is sometimes suddenly arrested by a spontaneous discharge of blood from the nose, or a very copious expectoration of a thick, yellow-colored mucus, brought up without much cough; and sometimes, but more rarely, by the appearance of an eruption on the skin.

We should fear a termination of the disease by abscess, by the obstinacy and but little violence of the symptoms, and their not yielding to an appropriate treatment within the first four or five days, and if there be but little expectoration, or especially if delirium, with a soft, undulating pulse, supervenes. Where suppuration, or abscess, has actually taken place, the symptoms are, frequent, slight shiverings, a mitigation or cessation

of the acute pain, with a continuance of the cough and a difficulty of breathing ; the pulse being soft, fuller, and either
slower or more frequent ; by a redness of the cheeks and lips,
an increase of thirst, and other symptoms of fever towards
evening. An abscess being formed in the lungs, the breathing becomes very short and laborious, and attended with
rattling in the chest; the cough short, dry and obstinate ; the
patient is able to lie only on the affected side ; the urine is
muddy, the countenance pale, the body becomes quickly
emaciated and enfeebled, and night-sweats and diarrhœa make
their appearance. When the abscess is situated on the external surface of the lung, immediately beneath the ribs, a soft,
indistinct swelling may be sometimes felt externally, with an
evident fluctuation of matter. In such cases, an opening may
be made into the abscess, between the ribs, the matter discharged, and the life of the patient frequently preserved. This,
of course, can only be done safely by a physician. When the
abscess is deeper seated in the substance of the lungs, in may
burst into the air-cells of the lungs, and if it do not immediately cause the death of the patient by suffocation, the matter
may be discharged by expectoration, and the patient be
finally restored to health.

This disease requires prompt treatment, and of course if
possible, a physician should be called at the earliest moment.
When one is not to be had conveniently, let no time be lost,
but pursue the course here marked out, which in a great many
cases will be the means of curing the disease, or checking it
while medical aid is being procured. Open the bowels by
means of an injection, and also giving some mild purgative,
such as castor oil. Epsom or Rochelle salts, or rhubarb. Apply leeches, ten to twenty to the side affected, if they can be
procured ; if not, scarify and apply the cups, after which a
warm poultice of bran, Indian meal, or linseed meal or slippery elm, etc., to be sprinkled over with a little laudanum or
paregoric ; to be applied frequently. Small doses of ipecac
either in powder or the syrup, should be given every three
hours, just so as to produce slight *nausea but not vomiting.*
When this has been continued for about twelve hours, then
use the following mixture : water, eight ounces (about one
gill) ; syrup of ipecac, one table-spoonful, or five grains of the
powder ; chlorate of potash, one drachm, or about one teaspoonful ; spirits of nitre, two tablespoonfuls. Dose : a
teaspoonful every three hours ; if much sickness of the stomach is produced, not so often. Let the patient have plenty
of cooling drinks, such as flax-seed tea, gum arabic, or slip

pery elm water, toast water, etc. The bowels to be moved occasionally by a dose of castor oil.

PLEURISY.

THE symptoms of this disease are very similar to those of inflammation of the lungs, at least, requiring usually a physician to tell the difference. However, the treatment recommended for that disease, is adapted in every respect for pleurisy

INFLAMMATION OF THE STOMACH.

THOUGH not a very frequent disease, this is a very painful and dangerous one. The symptoms are marked by a more or less acute pain and feeling of burning in the region of the stomach ; these symptoms are aggravated by every thing taken into the stomach, by the motions of the body, and by pressure.

Inflammation may come on very gradually, or be suddenly developed, according to the causes by which it has been produced. When the disease is fully developed, the pulse is very small, hard, and frequent ; there exist great anxiety, oppression, and a greater prostration of strength than in most other acute inflammatory affections. Every thing taken into the stomach occasions vomiting with painful reachings. Hiccup is also an early symptom. The features of the face are contracted, shrunk, and altered from their natural expression. There is distressing thirst, a continual tossing of the body, constant wakefulness, and in general a costive state of the bowels. In violent cases there is difficulty of breathing, with increase of pain on a deep inspiration. In the course of the disease, fever, with intense heat of the skin is sometimes developed, and at others, delirium, convulsions, and stupor.

Inflammation of the stomach is produced by large draughts of *cold fluids* taken when the patient is in a *profuse perspiration,* or over-fatigued by exercise ; cold applied externally, under similar circumstances ; contusions or blows upon the abdomen, intemperance in eating ; hard or irritating articles taken into the stomach, the violent operation of emetics ; or the excessive *use of ardent spirits.*

Inflammation of the stomach is always a dangerous complaint, terminating fatally, when violent, in the course of a few hours, or when less acute, producing a long series of distressing and painful symptoms, and causing incurable disorganization of the stomach, or by sympathy, producing disease of the skin, joints or brain ; therefore it is all-important that it be treated in every case with promptness and energy

The region of the stomach should be covered with leeches, and the flow of blood encouraged after they have fallen off, by warm fomentations. This treatment is the one adapted to the more acute cases, the important object being to reduce the inflammation with as little delay as possible. In all cases of inflammation of the stomach leeches are an important and indispensable remedy, and they should be repeated again and again, until the local symptoms are entirely removed; their number and the intervals of their application being adapted always to the urgency of the disease, and when some degree of inflammation still remains, the region of the stomach may be covered with a blister. In regard to internal remedies, there is none, with the exception of minute portions, frequently repeated, of cold gum arabic water, or iced water, but what would have the effect of irritating the stomach and increasing the inflammation. The vomiting and sense of burning by which the patient is often so much distressed, can be relieved only by the leeches, and by the cold fluids just alluded to. In some cases, advantage has been derived from a teaspoonful of powdered ice slowly swallowed, and occasionally repeated.

Bathing the feet in warm water, and the application of mustard to the ankles, will, in some cases, be found beneficial. The bowels should be kept regularly open by injections of warm milk and water, or soap suds, etc.

Chronic inflammation of the stomach is to be treated by the judicious application of leeches, by a very spare diet, of barley-water, gum arabic water, tapioca or panado; by blisters to the region of the stomach, by the warm bath and frictions of the skin, and by gentle daily exercise in the pure open air, friction over the region of the stomach, etc.

INFLAMMATION OF THE BOWELS.

This disease, in its symptoms, does not differ widely from the foregoing, and perhaps in a majority of cases, at least in a partial degree, accompanies it. Inflammation of the bowels usually commences with a slight chill, and a sense of uneasiness in some portion of the abdomen, at first intermittent, but gradually becoming permanent, and finally changing to a fixed pain which spreads over the whole abdomen. The latter is somewhat swollen, and sore to the touch. Obstinate costiveness generally attends the disease, and sometimes severe vomiting. The pulse is *very small, hard* and *frequent,* and the tongue *dry* and *furred.* The thirst is extreme, the urine high-colored, small in quantity, and most commonly discharged with difficulty. The breathing is short and laborious, and the patient generally lies upon his back, with his knees drawn up towards his breast. If the disease be allowed to proceed, these symptoms augment in violence. The abdomen becomes greatly distended with air, small mucous discharges take place from the bowels, with considerable straining ; the action of the bowels sometimes becomes inverted, and the contents of the bowels are discharged by the mouth. Suddenly the gony of the patient ceases, he appears to have obtained relief from his disease, but his intermittent and scarcely perceptible pulse, the paleness and livid hue of his face, the icy coldness of his extremities, and other alarming symptoms, indicate that mortification has taken place, which is quickly succeeded by death.

Inflammation of the intestines may be distinguished from colic by the presence of more or less fever, by the fixed and *continued pain*, increased upon pressure, and by the hard, frequent pulse. In colic, there is *no fever*, the pain comes on in paroxysms, with distinct intervals of rest, and is diminished rather than increased by pressure.

In the treatment of this disease leeches are to be applied over the seat of the pain, and repeated so long as the local symptoms remain, and the strength of the patient will permit, followed by a blister. The same cold fluids internally, will be proper, as those ordered in inflammation of the stomach.

After the violence of the inflammation has been reduced, four grains of calomel, every two hours, should be given, combined with mucilage of gum arabic, until a copious evacuation from the bowels is obtained; the operation of the calomel being aided by laxative injections (See *Medicines, their uses and doses*). In some cases, fifteen grains of calomel with two of opium, will produce a very prompt evacuation, and relieve greatly the remaining symptoms of the case.

During convalescence from inflammation of both stomach and bowels, the greatest caution must be observed, by a mild, well-regulated diet; abstinence from all stimulating drinks; by guarding against exposure to cold, and over exertion of the body ; by keeping the bowels regular, and using gentle daily exercise in the open air, wearing flannel next the skin, sponging the body with water once a day, followed by friction with a rough towel, etc.

DISEASES OF THE LIVER.

ACUTE INFLAMMATION.—As in other forms of inflammatory diseases this is generally ushered in by a chill, succeeded by all the symptoms of fever. To these are soon added pain in the region of the liver, sometimes acute and shooting, with a sense of fullness in the right side; at others, fixed and severe, or deep-seated. The pain commonly extends to the breast, collar-bone, and shoulder of the right side. The pain in the side is increased by pressure, especially when the patient lies upon his left side. The pain is often increased during breathing, and it is in consequence often impeded, more especially when the portion of the liver in contact with the diaphragm is inflamed. A severe cough is then also generally present, and in the course of the disease, hiccup commonly occurs. The cough in this disease is usually dry, short and frequent.

From the cough and difficulty of breathing, which so often attend inflammation of the liver, it is often mistaken for inflammation of the lungs; happily, the treatment of the two diseases does not *materially differ*. The skin, eyes and urine have the same deep yellow tint as in jaundice. The pulse is various, being sometimes small and feeble, at others, full and strong; but most commonly hard. The urine is ordinarily high-colored, the heat of the skin and the thirst considerable; the mouth dry, and the tongue coated with a yellowish mucus, which, in the course of the disease, becomes often dark brown or even black. There is likewise wakefulness, restlessness, and in a few instances, delirium.

A favorable termination is often preceded by a discharge of blood from the nose, or from piles; sometimes by a copious perspiration, or increased discharge of mucus from the lungs. A copious flow of deep-colored urine, occurring about the fourth day, and depositing, after standing, a red or whitish sediment, is also a favorable symptom; the same is true of free bilious discharges from the bowels.

From the obstinacy of the symptoms, we are to fear an abscess. As soon as it is formed, the acute pain in the side is changed into a feeling of weight and pulsation; the former being increased when the patient lies upon the left side. There are also frequent irregular shiverings, and finally, all the symptoms of hectic fever. When the abscess is seated on the external surface of the liver, a tumor and fluctuation of a fluid can be detected just below the ribs, on the right side, and by an incision, the matter may be evacuated, by which the chance of the patient's recovery will be greatly increased. When the

abscess is more deeply seated, an adhesion taking place between the liver and intestines, the matter may find its way
into the cavity of the latter, and be discharged by stool. In
this case, the patient often recovers.

Its most common causes are the action of excessive heat
upon the skin; sudden changes of weather; the sudden application of cold or damp to the body when heated; contusions or violent blows upon the head or other parts of the
body; the excessive use of wines and spiritous liquors; high
living, and intemperance generally; violent passions of the
mind, particularly anger and rage; the suppression of various
habitual discharges, and irritations of the stomach generally.

In the treatment of this form of liver disease, which is an
inflammation, the treatment is the same as for inflammation
of the lungs, bowels, &c., excepting that calomel or blue pill
should be made use of freely in this disease, and the leeches,
&c., applied to the region of the liver. The other general
treatment is the same, as far as can be undertaken, without
the advice of the attending physician.

CHRONIC INFLAMMATION OF THE LIVER.—This is attended by
the same symptoms as the acute, but assuming a more obscure
and insidious character, and are more slow in their progress.
In conjunction, also, with the peculiar symptoms produced by
the disease of the liver, we have also those of ordinary dyspepsia—wasting; defective or variable appetite; acidity; flatulence; feeling of fullness or uneasiness about the stomach,
dry, harsh, and discolored skin; disturbed sleep; great depression of spirits, despondency, irritability of temper; irregular bowels; disinclination to exertion, whether mental or
bodily; indeed, all that train of symptoms to which the inde-
finite term *nervous* is so generally applied. Ordinarily,
chronic inflammation of the liver is attended with considerable difficulty of breathing, and a short, dry, teasing cough;
sometimes, however, the cough is attended with expectoration.
There is frequently a decided paroxysm of fever towards evening; more or less yellowness of the skin, and when the disease
has been of long continuance, night-sweats, great emaciation,
and a wasting diarrhœa. Chronic inflammation of the liver
may either produce a great enlargement and hardening of the
liver, perceptible to the eye and feel externally, or it may terminate in suppuration, (coming to a head,) the matter being
discharged in the same manner as in the acute form of the
disease. In most cases, chronic disease of the liver is attended
with dropsy, either externally or of the abdomen, or both; a
species of chronic dysentery, with ulceration of the bowels,

sometimes takes place; and again, in other cases, from the operation of various causes, but particularly indulgence in ardent spirits, chronic inflammation of the liver may be rendered acute, when it is rapid in its course, and generally fata..

In its early stage, local bleeding by cups or leeches, applied over the liver and stomach, and repeated according to circumstances, followed by blisters to the right side, in conjunction with a very light vegetable diet, the warm bath, and friction to the surface: with a pill, every night and morning of th blue mass, five grains: soap, three grains; ipecac, one grain and a. ns, two grains, aided in its operation upon the bowels by an occasional dose of castor oil, or laxative injections, wil very speedily remove the disease, provided the patient, at the same time, take gentle exercise, when the weather wil! permit, daily, in the open air—his body being defended from any sudden diminution of temperature by appropriate clothing, especially flannel next the skin. A permanent drain from the side, by inserting an issue or seton over the liver, has occasionally been found beneficial.

In cases of chronic affection of the liver, the dandelion has been strongly recommended by various practitioners; it may be given in the form of extract, five to ten grains three times a day, or a gill of the strong decoction, or tea, *cold.*

A bath of the nitro-muriatic acid is strongly recommended by some physicians. It may be applied either to the legs and feet, or by sponging, with the acid diluted with water, the whole surface of the body. The use of the remedy should be persevered in for a length of time.

In this form of disease the food should be very light but nourishing, easy of digestion, and taken in *small quantities* at a time. Milk, rice, potatoes, and especially a free use of *tomatoes*, cooked or raw, twice a day, should be allowed. Avoid cold and dampness, and be careful to keep the *bowels open* at all times. A sea voyage is sometimes of great service in this disease. Out-door exercise should be taken daily.

INFLAMMATION OF THE KIDNEYS.

THE symptoms of this disease are a feeling of heat, uneasiness, and a dull, or sharp pain about the loins, and often a dull pain in the thigh, of one side, and great stupor or heaviness. The urine is at first *clear*, and afterwards of a *reddish* color, often bloody, and voided frequently, and in small quantities at a time. The urine generally coagulates, or becomes lumpy by heat. The disease is often attended with vomiting, costiveness, difficulty of breathing, and cold extremities. There is a painful feeling of uneasiness when the patient is sitting upright, or standing; the easiest position being that of lying on the side affected. If the inflammation of the kidney be severe, or occur in a broken-down constitution, it most commonly gives rise to more or less dropsy, either externally, or of the abdomen.

Inflammation of the kidneys may be induced by cold; by habits of intemperance; by the use of powerful diuretics, as spirits of turpentine, cantharides, &c.; by bruises or sprains of the back or loins; gravel; violent or long-continued riding, &c.

When the local symptoms are severe, the patient possessed of a considerable degree of strength, cups or leeches should be applied over the kidneys, and repeated until the disease is broken up. The use of the warm bath, or fomentations of hops, or horehound and vinegar, to the loins, constitute an important remedy, after cupping or leeches, and should be repeated daily. A mild purgative, as an ounce of castor oil, or a dose of salts, with the addition of one drachm of lemon-juice, in the dose of a table-spoonful every two hours, should be given, so as to keep the bowels regularly open, with frequent emollient injections of flaxseed-tea, infusion of slippery elm or thin starch, &c.

The patient should make use plentifully of thin gum arabic, or barley water, or flaxseed tea, and abstain from all solid and irritating food, and stimulating drinks. In case of violent pain continuing after the cupping or leeches, a grain of opium, or two of hyoscyamus may be administered at bed-time. A decoction or tea of the dried leaves of the peach tree has been said, when taken to the amount of a pint a day, in many cases, to produce considerable relief. They are best given mixed with the uva ursi leaves.

If the disease become chronic, the insertion of an issue or seton at the loins, and internally twenty to thirty drops of balsam copaiba, or one teaspoonful of powdered uva ursi, or a gill of the tea (cold) made of the same, will have sometimes a most happy effect.

INFLAMMATION OF THE BLADDER.

This disease is characterised by a feeling of fullness, uneasiness, and pain in the region of the bladder, and a frequent desire and great difficulty in discharging the urine, often a total suppression, with frequent ineffectual efforts to evacuate the bowels, occasioned by the irritation extending to the lower part of the bowel. These symptoms are generally accompa nied with fever, sickness and vomiting; great anxiety and restlessness: sometimes delirium, coldness of the extremities, and clammy perspiration ensue.

The mucus or lining membrane of the bladder, is likewise affected with a chronic inflammation; in these cases, there is a dull, uneasy sensation in the part, frequent desire to pass urine, which is generally thick, from being loaded with mucus; sometimes bloody, or, if ulceration of the bladder has taken place, mixed with matter. The coats of the bladder become often, when effected with chronic inflammation, thickened, or hardened, and otherwise disorganized, giving to the patient great uneasiness, and causing a constant inclination to urinate, or a total suppression of the urine.

In the acute variety of the disease, the early employment of leeches over the region of the bladder, followed by the warm bath and fomentations, will be required. The bowels should be kept gently open by mild laxatives and emollient injections, as directed in inflammation of the kidneys. The leeching should be repeated until the pain and uneasiness of the bladder are subdued. During the disease, the patient should observe a very low, unirritating diet, and make use of some mucilaginous fluid for drink. Any of those mentioned in the preceding article may be employed. If the urine be retained so as to cause distension of the bladder, the catheter will have to be used to draw it off. The patient must keep perfectly quiet, avoiding all manner of exertions of mind or body, during this disease.

When the case is chronic, leeches or cups, applied occasionally to the region of the bladder, will be advisable; the bowels should be kept open, and a warm bath taken twice a week at bed-time. Opiates, as ten drops laudanum, half teaspoonful of paregoric, or ten grains of Dover's powders, at bed-time, will be necessary to ease pain and procure rest.

M U M P S.

This is a very common affection: it is a painful inflammation of the parotid glands, situated behind the angle of the jaw, and often extending lower down. Though sometimes confined to one side of the jaw, it usually occurs on both sides; it is at first clearly defined and moveable, but soon become fixed, and spreads to a considerable extent. It increases in size till the fourth day, and often involves the neighboring glands in the inflammation; it is supposed to be contagious, and often prevails as an epidemic. After the fourth day, the swelling gradually declines; and, for the most part, it is unattended throughout with fever, and scarcely ever calls for medical aid. As the swelling of the throat subsides, it not unfrequently happens, that a swelling takes place in the privates of males, and in the breasts of females.

Generally, all that is requisite in this affection is, to keep the head and face moderately warm, to avoid exposure to cold to observe a mild diet, and to open the bowels by a very gentle purgative, as a dose of magnesia, rhubarb, or salts.

When the privates and breasts simply enlarge, they ought not to be interfered with; but should they be painful, and tend to suppurate or break, a purgative of Epsom salts, or Rochelle salts should be given, a few leeches applied, and afterwards a warm poultice.

Should high fever occur, with other alarming symptoms, the usual means of reducing inflammation must be applied—purging, small doses of ipecac, every two hours, with cooling washes, such as green tea (cold), tea of mullien leaves, &c.

ANGINA PECTORIS, or BREAST PANG;
(Pain in the Breast.)

This disease has been called by several names, as will be seen above. It is a disease characterized by a sharp pain and oppression, seated at the lower end of the breast bone, inclining to the left side; it comes on in paroxysms, and is accompanied with great uneasiness, difficulty of breathing, violent palpitation of the heart, and a sense of impending suffocation.

In the first stage of the disease, the pain is felt chiefly after some exertion, as going up stairs, or up a hill, or walking quickly, particularly when the stomach is full; but in the more advanced stages, slighter exertions are sufficient to cause a paroxysm of pain, as walking, riding, coughing, sneezing, or speaking; passions of the mind also have the same tendency. In the first stage, the uneasy and threatening symptoms soon go off; but afterwards they continue longer and are more distressing, causing the patient to fear immediate dissolution. During the paroxysm, the pulse is feeble and irregular, the face pale, and covered with a cold sweat, and the patient appears as if in a fit of apoplexy, without the power of sense or motion. The disease makes occasional attacks, at longer or shorter intervals, for years, and at last suddenly puts a period to the patient's life.

This affection appears to be of a spasmodic or convulsive nature, as is shown by the manner of treatment, which is most successful. It is found to attack chiefly those who are of the make which has been supposed most liable to apoplexy, viz., those with large heads and short necks, and who lead a sedentary, luxurious and inactive life, who are disposed to be corpulent, and especially such as are of a gouty habit. It seldom attacks persons under fifty years of age.

Our treatment should consist in *preventing*, as well as curing, by diminishing the quantity of blood in the system, by purgatives sufficient to keep the bowels open, spare diet, and avoiding everything that would quicken the circulation. During the paroxysm, stimulants must be very cautiously employed; the head and temples are to be bathed with cold water and vinegar ; a slight bleeding may be necessary to relieve the overloaded heart, gentle pressure should also be employed on the *left side, to empty the heart ;* mustard should be applied to the extremities, and on the appearance of returning breathing, hartshorn is to be applied to the nose at intervals. The disease has been mitigated by forming issues in some part of the body. Also by blisters, or tartar emetic ointment applied to the chest. Mustard to the ankles, and active purgatives, are almost always beneficial.

Persons who are subject to this disease should be careful to shun all mental irritation, and every outburst of passion or excitement that would hurry the circulation. Moderate exercise should be daily taken in the open air, but no violent exertion indulged in ; and all attempts at going up a rising ground should be avoided, or if made, should be with the utmost care. The food should be plain, moderate in quantity, and easily digestible ; such as is not liable to occasion flatulence. Fermented and distilled liquors are altogether improper. On any appearance of fullness of blood, animal food should be entirely abandoned, and mild purgatives (epsom salts, rhubarb, castor oil, rochelle salts, &c.) frequently taken. A perpetual blister or other irritation in the region of the heart is sometimes useful, and warm bathing to the feet and legs will help still further to prevent the undue flow of blood to the other parts of the body. Mustard or cayenne pepper in the warm water, will render the bath more stimulating and of greater service.

DISEASES OF THE HEART.

OWING to the use of tobacco and intoxicating drinks, which prevail to such a fearful extent at the present day, diseases of the heart are greatly on the increase. Each issue of the morning paper, almost, records the "sudden death" of some loved one, who has left home and friends in apparent good health "in the morning," but is brought home cold in the embrace of death. What a pity that intelligent beings should use such deadly poisons! However, diseases of the heart are brought about by other causes; still, where there is any tendency to the disease, or it naturally exists, though unknown to the individual, rum and tobacco render it doubly dangerous.

If from any cause one or more of the valves of the heart should become deficient, it is evident that each time they close, a small portion of blood will pass back, or " regurgitate," into the heart— and this actually occurs in cases of disease. And the consequence of the long-continued and constantly-repeated disordered action is to cause stretching, or " dilatation," of the cavity which receives the regurgitated blood. This one instance will explain how one slight derangement in the nicely-balanced machinery of this important organ gives rise to another. From somewhat analogous causes, the blood may regurgitate into, or be dammed up in the lungs, or in other parts of the body, causing hemorrhage, dropsy, &c.

Sometimes there may be impediments to the circulation; the valves above alluded to, or others, may not yield as they should do, or there may be other causes which render it difficult for the heart to propel the blood through the body. In such a case, the heart, like any other muscle under similar circumstances, acquires increase of substance, in consequence of the continued increased exertion demanded of it, to maintain the proper circulation; and thus we have a cause and effect, producing enlargement of the heart—an evil certainly, but a lesser evil to prevent a greater, for in this very enlargement— this strengthening, as it were, of the heart to do its extra work —the patient's safety lies.

Of course there are many other forms of heart disease, but the above instances will convey some rational idea of the nature and peculiarities of the disorders of the organ in general. It would be quite profitless in this work to enter into anything like detail respecting diseases of the heart or their treatment; disorders so varied in their nature and symptoms, can only be properly investigated and managed by a medical man, conversant with the mechanism and the functions of the human frame at large, and *in their relations of mutual dependence.*

Affections of the heart are manifested by pain in the chest difficult breathing, cough, palpitation, &c.; and at other time by faintings, giddiness, irregular pulse, &c.; but there is not one of these symptoms, or any combination of them, which may not be developed under certain bodily conditions, al though the heart is perfectly sound. None, therefore, need alarm themselves merely because such symptoms occur ; they happen at times more or less to all ; still they ought not to be neglected : if they continue to recur, a medical man should be consulted. If there is no disease the mind is set at rest, and any general disorder which may have caused the symptoms will probably be rectified. The above cautions are given, because there is no class of diseases of which people are so apt to fancy themselves the subjects as those of the heart; and the more they think of the symptoms, the more likely are they to continue or increase, from an organ so intimately connected with the emotions of the mind as the heart. Even if the heart be unaffected, it is by no means advisable to permit it to continue to be functionally disordered, (irregular in its actions,) either by mental emotion or by sympathy with other organs, for the functional disorder may end in organic disease; that it does so sometimes is evident from the fact that there is no more fertile source of heart disease than those convulsions, either commercial or political, which occasionally agitate society.

When disease of the heart, either insipient or confirmed, does exist, it cannot too soon be discovered by examination, nor the necessary precautions and regulated mode of life too soon adopted ; for with these precautions, a large majority of persons who are the subjects of heart affection may not only continue to live for years, sometimes many years, but to enjoy life. True, the knowledge to any one that he is himself the subject of heart disease may be uncomfortable, but it cannot be unprofitable. He may be aware that heart diseases are sometimes apt to have a sudden termination, and that *his* life *may* be somewhat more in jeopardy than that of an unaffected person ; but surely to every right-thinking man, this fact would rather be an argument why he should know his real condition. The possibility of his being called away from the affairs of this life without warning, should be a reason for his keeping them well arranged ; and still more important, should it be a reason that in conducting his earthly stewardship, he should do it, not only with reference to this world, but to give account of it in another. And when the many chances and contingencies of life are considered, the consciousness of being the subject of heart disease amounts to little more than such contingencies assuming a more prominent position in the mind ; and to the individual it may be a merciful dispensation of providence, as, being reminded more frequently of the uncertainty of human life, he may be constrained to look for a

realization of *true* happiness only beyond tl.is vale of tears,—
the Christian's home in heaven !

Besides the two already mentioned, there are other causes
of disease of the heart: mental disturbance and agitation is a
most frequent one ; also mental depression and grief, which,
if long continued, appear to exert much influence over the
organ, and to make the phrase " a broken heart" not altogether
a poetical fiction. Violent passion strongly affects the heart
—its indulgence may lay the foundation of disease, which its
repetition strengthens, and may bring to a fatal termination.
Rheumatism, or rather rheumatic fever, is probably another
of the most fertile sources of heart affection. In this disease,
inflammation of some portion of the membranes covering or
lining the heart, is apt to occur, and to be followed by such
effects as induce permanent change. Violent physical exertions, and *dissipation of all kinds,* are causes of disease of the
heart.   It has been said that persons with heart affections may
continue to live and enjoy life, but it must be under a more
regulated and restricted system of living than is imperative on
persons in health. Every thing which may be a cause of
heart affection must also be a source of aggravation ; all mental or physical excitement especially so. When these are
guarded against, the rest may be summed up in—strict attention to the general health. Whenever an old symptom becomes aggravated, or a new one, such as swelling of the legs,
&c., appears, medical advice should always be obtained.

PALPITATION OF THE HEART.—This is often caused by the
various emotions of the mind, as fear, joy, anger, or excitement of the mind from any cause. When it arises, however,
on every trivial occasion either of mental emotion or of physical exertion, or without occasion at all, as it often does, even
during rest in bed, then it requires attention, not solely on
account of the discomfort it gives rise to, but because it may
lay the foundation of disease of the organ which is so constantly subject to over-excitement. Affections of the heart
become more common after seasons of much public excitement
of any kind—an effect traceable only to the frequent disturbance of the organ by the passions or emotions of the
mind.

Independent of disease, palpitation of the heart is liable to
occur in the young of both sexes, and in females particularly,
soon after the age of puberty--in the latter being very generally associated with hysterical tendencies ; in such cases, it is
met with in its most aggravated forms, and often of such
violence as to prove alarming. In any case the tendency to

palpitation is more common in the nervous temperament, and is increased by whatever gives undue predominance to that temperament, such as indolence, luxurious habits, and the indulgence of feelings and imagination artificially excited; and having once begun, it is kept up and aggravated by the continued attention with which the mind is apt to dwell upon the ailment. The individuals subject to it easily imagine themselves as subjects of heart disease, watch every motion almost of the heart, and thus under the influence of their own imaginary fears, produce the very symptoms they dread.

This nervous condition (for it is generally nothing else) is only to be got rid of by those measures which give a more vigorous and healthy tone both to mind and body. The false excitement of imaginative literature, I mean reading trashy novels, etc., if it has been indulged in, must be exchanged for something which calls for some healthy mental interest. Where it can be adopted, the pursuit of some branch of natural history, botany, geology, or any other out-door occupation, are the best pursuits; they occupy the mind, and draw it away from its own morbid fancies, even in the time of exercise, which is rendered doubly invigorating by the mental excitement which accompanies it. Along with these means, a system of diet calculated to give good nourishment should be adopted; heated and ill-ventilated rooms, above all things, are to be avoided, *early hours observed*, and if a feather bed has been habitually lain upon, a firm hair or husk mattrass should be substituted. One article of diet requires especial mention, as being peculiarly injurious in such cases; *tea* of any kind is better avoided, but *green tea is absolute poison;* coffee is scarcely allowable, and cocoa or milk should invariably be substituted for either of the above more stimulant beverages; wine or malt liquor may be injurious, or the reverse, according to the previous habits of the patient and the nature of the case; if depression or debility follow their withdrawal, the tendency to palpitation is certain to be increased. In addition to these measures, regulation of the bowels, the use of the shower-bath, or better, of the *douche*, or water poured down the spine, and occasional mustard plasters on the chest or between the shoulders, are all useful, especially if, as frequently happens in cases of aggravated palpitation, any tenderness of the spine is found to exist. In cases of nervous palpitation, medicine is not much called for, unless to remedy other disorders, such as indigestion. Some patients derive much benefit from a teaspoonful of a mixture of equal parts of tincture of valerian, tincture of henbane, and spirits lavender: taken twice or three times a day in water.

Hartshorn in teaspoonful doses with cold water, is often useful, especially if there is much flatulence; or ether may be taken in ten or fifteen drop doses, either alone or with the above-mentioned remedies given in water. The ether, however, is more generally serviceable as a remedy, *during an attack* of palpitation than when taken regularly. In most cases of this kind it will be found necessary to change the remedies occasionally, or they will lose their effect.

S C U R V Y.

By medical writers, and by well-informed non-professional men, the term scurvy is now applied to that disease which is produced by a long abstinence from fresh vegetable food, exposure to damp, and the influence of the depressing passions, and which is therefore frequently observed in long voyages, in camps, and in besieged towns. It comes on gradually, with heaviness and aversion to motion, with dejection of spirits, anxiety, and great debility. The countenance becomes sallow and bloated, the breathing is easily hurried, the teeth become loose, the gums are spongy, and bleed when slightly touched; and livid (or purple) spots appear on different parts of the body. A very curious circumstance sometimes occurs in scurvy: old wounds, which have been long healed, break out afresh. If the disease is not checked, the joints become swelled and stiff, the tendons (sinews) of the legs stiff and contracted; a dark colored blood issues from the nose, the ears, and other parts; offensive stools are discharged, and the patient dies with symptoms of diarrhœa or dysentery.

The cause of scurvy is long confinement to a diet destitute of a due admixture of fresh vegetable substances. The frequent occurrence of scurvy in long voyages, during which the ship's crew are compelled to live much on salt provisions, and in besieged towns, where the provisions are scanty and bad, and in cold, damp and poor situations, where human life is with difficulty supported, is thus accounted for. Among other very exciting causes, we are to reckon want of *cleanliness* and *ventilation*, a damp and cold atmosphere, and, above all, depressing passions. It has been ascertained that by keeping the ship perfectly clean and dry, by allowing the men as much recreation and exercise as possible, and by taking to sea a proper supply of lime *or lemon juice*, and distributing to the ship's company a portion of it every day, when their fresh provisions begin to fail, they may be kept from scurvy as effectually as any number of persons living on shore, and using fresh vegetables every day. This has, for a considerable number of years, been ascertained on a very extensive scale of experience. From one to two ounces a day are a sufficient quantity for this salutary purpose; and it may be given diluted with water, or made into lemonade, with sugar and water, so as to form a healthful and refreshing beverage. When in any individuals the tendency to scurvy appears stronger than in others, as indicated by the spongy and easy bleeding gums, by stiffness of the hamstring, by inertness and dejection of spirits, it will be proper to give an ounce of the juice three or four times a day till the tendency is diminished. Lime juice is apt to ferment, and not to keep properly, owing to the quantity of pulp and mucilage squeezed out along with it. To prevent this fermentation, it is proper to mix with it a little alcohol or carbonate of lime. It has been ascertained by experiment that the curative properties of the lemon juice depends very much on the *potash* which exists in combination with the acid. So that cream tartar and the carbonate of potash given in small and repeated doses, say one teaspoonful three times a day in a glass of water, will act well where the lemon juice can not be had. And so far as our experience goes, the same powerful yet simple agent which prevents scurvy, is also fully sufficient to cure it. When circumstances admit of it, and we are able to procure for a ship's company an abundant supply of fresh meat and vegetables, this is the natural and appropriate remedy; but in the last stages of the disease, when the debility is great, it is a matter of much danger to take the sick on shore, as they not unfrequently die in the boat that carries them thither; and instances have even occurred of the land air being too oppressive for the lungs of those, who did not previously show marks of so great feebleness. It is by far the safest way to attempt the cure of the men *on board their ships;* and when the disease abates a little, and the strength is beginning to return, it may then be accelerated by a removal on shore, and by the usual diet and exercise to be found there. Many auxiliary circumstances are to be called into action, both in the cure and prevention of scurvy; the greatest attention to ventilation and cleanliness, frequently washing the ship in fine weather, fumigation between decks by the vapors of the nitric or muriatic acid, by pouring either of them over a small quantity of common brown sugar, or pieces of zinc, iron, etc. attention to increase the real comfort of the men, and to check all intemperance; to promote regularity of discipline and cheerfulness of mind. Various articles are to be used in diet which counteract the pernicious tendency of a long continued use of salt provisions, as spruce or treacle beer sourcrout, preparations of oatmeal, and occasionally parboiled fresh vegetables, such as water cresses, radishes, scurvy grass, lettuce and the like, which may be eaten raw ; or cabbage, turnips, spinage, cauliflowers, boiled ; or ripe fruits, as oranges, melons, pine-apples, plantain, etc.

CONSUMPTION.

This disease has been truly termed the " great destroyer of our race." It enters the cottage and the castle ; it kills the rich and the poor, the old and the young, and regards not sex or condition, but rushes on, slaying its tens of thousands every year. Consumption is only a *portion*, strictly speaking, of a constitutional malady which very frequently develops its intensity in the organs of respiration, (the lungs), but may do so in other modes and in other organs of the body. Its *constitutional* nature requires to be impressed upon the mind of people in general ; for, regarded only as a disease of the lungs, alarm is not taken, nor are remedies generally resorted to until its effects upon these organs become manifest ; the antecedent period in which the constitution is giving way is overlooked, and that time is lost in which the first indications of disease might have been successfully attended to.

We may state that the causes of pulmonary consumption are all those which occasion debility generally, not excepting the most frequent of all, hereditary predisposition, or that tendency to the disease which exists so strongly in some families, that no care or precaution can ward it off, nor prevent it seizing in succession member after member of a household. Fortunately, this intensity of hereditary transmission is not so very frequent, but there are few families in which the tendency does not more or less exist ; there are few which cannot number amid their deceased relatives some victim of consumption With a susceptibility so widely diffused, it becomes a serious consideration with all by what this tendency is encouraged, and how it may be diminished. The first consideration that presents is marriage. There can be no question, that from errors in the contraction of this great engagement of life, much of the hereditary tendency to consumption is developed, and especially when the union is between parties nearly related by blood ; doubly so if the predisposition already exists in the family. Delicacy or debility of either parent, particularly of the *father*, is very apt to entail consumptive tendencies upon the children ; and the same follows if the parents are either *too young*, or if the father *be advanced in life*. The mistake is a very common one, that marriage and child-bearing act as a check upon the progress of consumption, and the step is often advised even to the comparatively young with this view. The error is a serious one; nothing can be more trying even to a healthy female than having a family before the constitution is formed; and most certainly it is so to the weak. It is true, apparent *temporary* amendment of consumptive symptoms sometimes occurs, but the powers of life are sapped by the too early call on their exertions.

Great care is required near the approach of the age of puberty, especially in the case of those who have displayed any scrofulous or consumptive tendency. The development of the body which is going on requires a full supply of the most nutritious food, *animal food* particularly. The secretions should, if possible, be kept in healthy activity, and, more especially, all sources of exhaustion most strictly avoided. Youths especially must be warned against the evil of *prolonged physical exertion;* and not less so against the mental efforts, which those especially, who partake of the nervous and excitable constitution of the hereditary consumptive, are apt to give way to, in competitions at school or college.

Along with the consumptive tendency, two very different conditions of physical development are found. In the one, there is the fair fine skin and bright red complexion, the fair hair, the light eye, with its pearly looking white, and the tapering fingers; in the other, the dark hair and skin, the latter almost dirty-looking, and the swollen-looking upper lip. Consumption varies much in its initiatory stage; sometimes it steals upon the patient most slowly and imperceptibly; at others, developed probably by some acute attack, it appears to start at once into activity. Generally, for a considerable period before marked symptoms—or at least symptoms which attract general attention—show themselves, the person has felt weak, languid, and *complained much of cold*, probably has sunk in flesh, and a short dry cough has come on, apparently without cause, or there has been continued dyspepsia. If the patient is a female, the monthly discharge has become irregular, or stopped. It may be that these symptoms have been aggravated during winter, and disappeared partially or entirely with the advent of warm weather. Such symptoms may go on for a longer or shorter period, ebbing and flowing, but still gaining ground, or they may progress more unremittingly, though still slowly, or become suddenly aggravated by taking cold, some unusual fatigue, or the like. The emaciation becomes too evident to escape notice, the cough is unabated and becomes troublesome, the voice assumes a peculiar hollow sound, the breathing is quickened, and it may be that spitting of blood, profuse night perspirations, or even diarrhœa, have set in before the patient's condition excites either alarm in their own mind or in that of their friends. Indeed it very frequently happens that the patient is the last to take the alarm, the last to entertain the idea of the fatality of the disease, of which this hopefulness of recovery is a well-marked symptom.

As regards the prospect of recovery from consumption—for recovery does undoubtedly take place—much depends upon the original and existing constitution and the habits, past or present, of the individual, and the worldly means within his power.

When once the disease has become established in an individual of strong hereditary tendency to it, or in one who has broken down his constitution by dissipation or intemperance, or who is the subject of some other debilitating disease, hope of amendment can be but small. If, on the contrary, the affection is more probably induced, and rather accidental than the result of original constitutional tendency, the probability is that under proper management, and with the aid of the great curative powers of cod-liver oil, and medicated inhalation, (see Catarrh in the Head in another part of this work) not only amendment, but permanent recovery may be obtained. When consumption has advanced beyond its first stage, all the symptoms already mentioned are increased, the cough and perspiration particularly become more distressing, and the tendency to diarrhœa, (frequently with severe spasmodic pain in the bowels,) notably increased; expectoration is often difficult, either from weakness or from the thick matter expectorated; the hair falls off, and emaciation continues.

The *prevention* of so fatal a disease as consumption is a more important subject in a work like the present, than its treatment; and in those predisposed, the preventive or "prophylactic" system must be continued life through, even into old age. *It is a popular error, that by the time middle life is reached the liability to consumption is over.* Such is not the case, for even the "three score and ten" is sometimes terminated by the disease.

In any condition of life, the question of tendency or not to consumptive disease should influence the choice of field for exertion, and not only of field, but also of the nature of the business of life. Any occupation which renders the inhalation of irritating substances unavoidable, is to be avoided by the consumptively inclined man; and not less so, that which

involves confinement in a constrained position or in a close
room.

The most eligible employments are those which require
muscular exertion of not *too exhausting a kind*, and without too
great exposure to the weather ; the gardener, the carpenter,
butcher, the farm-servant, are all less likely to be the victims
of the disease. In whatever situation or grade of life, however,
a person may be placed who is predisposed to consumption,
much may be done to keep up the powers of resistance by
keeping up the general health to the highest possible standard, by diet, early hours, attention to the skin, etc. (See " *How
to preserve Health*," in the first part of this work.)

The chest and shoulders should be bathed every morning
with cold salt water, and rubbed afterward to promote reaction.
Cheerfulness of mind and moderate mental exertion are important, while perfect temperance in the use of alcoholic
stimuli is indispensable ; but any change to their total disuse
cannot be made suddenly, if the patient has been accustomed
thereto without danger. All the usual sources from which
" cold is taken" are to be shunned, particularly wet feet,
sitting in damp clothes, crowded ball-rooms, and public assem-
blies ; and, lastly, when exposure to cold air, especially to east
winds, or to the foggy atmosphere of night, is unavoidable,
the protection of a respirator to go over the mouth and nostrils—a silk handkerchief being about the handiest—should
be resorted to.

Medical men are often asked their opinion as to the commu-
nicability of consumption from one person to another. That
it is not generally communicable is certain ; that it has been
thought to be so under circumstances of predisposition, and
when there has been close communication between two persons, should be sufficient to caution other members of a consumptive family from hanging too much over one affected with
the disease ; and certainly in any case forbid *the occupation
of the same bed*.

This disease has thus been brought under consideration
more with a view of pointing out the *preventive remedies*, and
the general course to be pursued by those predisposed to the
disease, than any curative treatment after the disease has
advanced too far to be treated by any course of domestic
medication. A physician should be called to the patient at
as early a period as possible, and above all things avoid the
thousand and one remedies that are daily advertised by men
whose chief talent consists in knowing how " to make
money."

DYSPEPSIA or INDIGESTION.

This distressing complaint has been said truly to be " no respecter of persons; " its symptoms are usually well known; it originates or is brought about by a great variety of causes; among which it is often found associated with a diseased state of the *liver*. Persons who have used spirits of any kind to *excess*, or stimulants of any description, such as spices or highly seasoned food, and those also who have used *tobacco* to great excess, by which the coats and functions of the stomach have been impaired and debilitated, are liable to indigestion. A costive habit, acquired by permitting the bowels to remain too long without evacuation, will bring on this disease, and persons who are long confined to any stationary or sedentary business, without taking the necessary exercise, are often subject to it. When the complaint is firmly seated in the stomach, it is marked by belchings of wind, gnawing and disagreeable sensations at the pit of the stomach, risings of sour and bitter acid into the throat, occasioned by the food not being properly digested; great irregularity of appetite, which is sometimes voracious and at other times greatly deficient, and a sinking and oppressive debility or weakness of the stomach. In addition to these symptoms of indigestion, on gratifying the appetite at any time, the stomach in a short time afterwards becomes oppressed with sensations of *weight* and *fullness*, the head becomes confused, the sleep very much disturbed, the bowels very irregular and costive, the urine high-colored, and the poor victim commences taking medicines for relief, and brooding in dejected silence over thousands of unhappy retrospections of his past life, and countless melancholy anticipations of the future.

Should disease of the liver be connected with this disease, a dead and heavy pain will be felt in the right side; the water deposited in the chamber will have, on cooling and settling, a brick-dust colored sediment, which, if permitted to remain any length of time, will adhere in rings of a reddish hue to the inner sides of the urinal; a pain will be felt in the top of the shoulder and back of the neck; the feet and hands will frequently become benumbed, or what is called " getting to sleep," from want of regular circulation; the complexion will become of a yellowish hue or tinge, and general uneasiness of the whole system will be felt.

The treatment of this disease consists in *keeping the bowels open regularly*. If they are not rendered so by the use of stewed fruit and ripe fruit uncooked, with injections of half pint of cold water up the bowels every morning, a dose of some mild purgative medicine should be taken. (See "*Medicines, their doses and uses,*") so that the bowels are *opened every day.* Friction and percussion, by striking with the open hand, gently over the region of the stomach for a few minutes, half an hour before each meal, should be attended to, and be sure to *never eat as much as the stomach seems to crave,* that is, always "*get up hungry,*" and never eat any article of food known to disagree with the stomach. Eat *slow* and *at regular hours.* Avoid the use of intoxicating drinks, and do not eat meats for supper; sleep on a hard bed and take a sponge bath every morning, followed by the use of a dry, rough towel and the flesh-brush; take plenty of active outdoor exercise, and partake of a *mixed diet;* and above all avoid the poison called *tobacco,* in every form.

HEARTBURN, or SOURNESS OF THE STOMACH.

HEARTBURN is found to exist under very different conditions of the system. It is prevalent, frequently, with females, during pregnancy, the causes of which were explained under the head of "Pregnancy and its Diseases." It is also of the greatest consequence to the dyspeptic; and may be caused by any thing which interrupts the process of digestion, especially *costiveness* of the bowels. The symptoms of this disease consist of a burning sensation, felt either at the pit of the stomach or top of the throat, and occasioned by undue acidity, or by irritating substances in the stomach. It is generally relieved by soda, magnesia, or chalk, ten grains in a little water, or twenty drops of hartshorn in a wineglassful of cold water—which neutralizes the acid. It is not, however, advisable to have too frequent recurrence to these palliative remedies, for they are *only* palliatives,—they cannot be taken habitually without weakening, not only the stomach, but the system generally. Heartburn cannot continue to recur without there being an error somewhere; either the diet is badly regulated, or the digestive organs require something more than simple neutralization of the superabundant acid. This acid is a badly-formed gastric juice, and if it is neutralized, whatever digestive power it might possess is destroyed; consequently the stomach is called upon to secrete *another supply* before the food can be digested—a call upon its powers which cannot fail to be injurious. Moreover, persons finding how quickly a dose of alkaline medicine removes the uncomfortable sensation of heartburn, are very apt to trust to the palliative, and continue their indulgences, rather than to practice the self-denial requisite to effect a cure of the cause.

Prevention, therefore, or removal of the *cause*, is what is of most importance in this disease. The bowels must be kept open by mild purgatives, so that there is an operation regularly *every day*. Every thing known to *disagree* with the stomach must be avoided, exercise taken in the open air, and all the means pointed out under the head of " How to preserve Health," strictly adhered to.

WATERBRASH.

MEDICAL men have strange names for diseases, as well as those who are not medical men. *Pyrosis* is the medical name for this disease, but usually called *Waterbrash*. It is a peculiar affection of the stomach, in which the patient brings up frequently a considerable quantity of thin watery liquid, sometimes insipid, at others intensely acid. Before the fluid is brought up, often there is more or less pain experienced at the pit of the stomach. This complaint attacks, mostly, persons past the middle age, particularly females, and the fit comes on generally in the morning and afternoon. It usually begins with a severe pain in the pit of the stomach, attended with a feeling of constriction, or oppression, and soon after a quantity of thin watery fluid is thrown up, which is sometimes insipid, at other times it has a highly acid or burning taste. The causes of this complaint are various, but whatever disorders the stomach may give rise to it. It appears to be owing to a peculiar state of irritation of the stomach; and is most certainly relieved by the use of the white oxyde of bismuth, from two to three grains made into pills with extract of gentian, three times a day. This medicine will often perfectly cure *waterbrash ;* but attention to the diet, as laid down under *dyspepsia*, is of much consequence, and will be absolutely necessary in order to render the cure permanent. A diet of plain animal food may be allowed, with which may be united the use of biscuits, home-made bread, and preparations of rice and milk. Daily exercise must also be taken, and frictions, with the flesh-brush, over the region of the stomach and bowels, are of no small service. The bowels must of course be kept open by purgatives, when necessary, even when making use of other curative means.

JAUNDICE.

THIS disease is often improperly called "janders :" it is characterized by yellowness of the eyes and skin, whitish or clay-colored stools, and saffron-colored urine, which communicates to substances immersed in it a saffron dye. Jaundice is caused by the duct or tube leading from the gall-bladder to the stomach, becoming *stopped up,*—that is, something in the first place *stops the flow of bile* from the liver; and jaundice, which consists in an absorption of bile into the blood, is the result. Probably, gall-stones, or thickened bile, are the most common obstructions; but tumors which press upon the duct, or spasm, may also stop the bile and induce jaundice. Jaundice has sometimes been caused by violent mental emotions, (uneasiness of mind, from various causes,) also, an irregular or sedentary mode of living will produce it as well as intemperance, especially continued indulgence in spirituous liquors. The presence of bile in the blood is quickly manifested by the color of the skin, and also more particularly of the white of the eye; the shade of color varying from the slightest perceptible tinge, to deep golden yellow, or even brown. At the same time, the stools become white and chalky-looking, and the urine—sometimes the perspiration—is deeply tinged with bile; the constitutional symptoms are generally those of disordered digestion, (see *Dyspepsia,*) headache, languor, &c.

The principal objects of treatment are, to allay irritation in the stomach and bowels, and to remove the obstruction existing to the free passage of the bile, through the biliary ducts, and along the bowels. It will be frequently found, that the best means of allaying irritation in these parts will be the most effectual in removing the jaundice. In the young and robust, bleeding by leeches applied over the region of the liver, will often be demanded, especially in recent cases; cupping over the stomach and region of the liver is generally an important remedy.

The patient should also take a warm bath at ninety-six or seven degrees, every other morning about eleven o'clock, with the following pills during the day: Castile soap, a drachm and a half; rhubarb, in powder, eight grains; ipecac, in powder, ten grains; oil of juniper, ten drops; syrup of orange peel or lemon, a sufficient quantity to make the whole into twenty-four pills. Three to be taken twice or three times a day. Or when there is a good deal of pain in the bowels, the following may be used instead: compound extract of colocynth, and extract of henbane, of each, a drachm; divide into twenty-four pills; one, two, or three to be taken as above directed.

Gentle purging is perhaps the most beneficial mode in most of cases one or two tablespoonfuls of both the Epsom and Glauber's salt may be dissolved in half a pint of lukewarm water, and taken every morning for a fortnight; and resumed for another fortnight or three weeks after being laid aside for a week.

Milk Sickness—Colic.

If there is much pain in the affected side, mustard plasters or poultices, applied frequently, will be of much service. Sometimes giving an emetic of ipecac will cause the gall-stone or thick bile to pass into the stomach, and thus relieve the difficulty at once.

Many persons who have resided long in a hot climate, contract a sallow, yellowish complexion, which hue often pervades the whole skin ; this is generally regarded as a mild sort of jaundice, or as arising from the absorption of the bile, but it is usually of a different nature. In most instances, it is not owing to this cause, but to a peculiar alteration in the circulation, on the external surface of the body, in consequence of the skin's sympathizing with a weakened and irritated condition of the digestive organs, more especially of the stomach and bowels.

MILK SICKNESS.

This is a peculiar affection which occurs in many of the Southern and Western States, in the autumn. It has been called by various names, such as "swamp sickness," "tires," "slows," "stiff joints," "river sickness," "puking fever," &c. It affects both the cattle that eat of the herb which causes the disease, and also the persons who use the milk taken from the cow, or who eat the flesh of animals affected with the disease.

It has not as yet been satisfactorily settled what kind of an herb or plant causes this sickness ; however, the symptoms show it to be a *narcotic*, or *poison*. There is vomiting, purging, extreme nervous agitation, great prostration, trembling, &c. The treatment consists in giving plenty of warm sweetened water every few minutes, until the stomach has been emptied by vomiting ; then a mustard plaster over the stomach and bowels, and injections up the bowels of lukewarm water, until the bowels operate, after which a mild purge of castor oil. The patient must be kept perfectly quiet in bed ; should symptoms of sinking show themselves, give stimulants —brandy, whisky, wine, hartshorn, or whatever is handy To allay the thirst, give strong coffee (cold) without sugar or milk.

COLIC.

This disease is produced by so many different causes, is so varied in its symptoms, under different circumstances, and requiring treatment suited to its different causes and symptoms, that we have thought it best to treat it under the following divisions :

COLIC FROM INDIGESTION.—Among the most common symptoms occasioned by an excess in eating, whereby the stomach is loaded beyond what is compatible with the regular and healthy performance of its functions, is a violent pain or colic, accompanied with nausea, headache, and dizziness, preceding the ejection of the contents of the stomach by vomiting; and terminating subsequently in a griping looseness of the bowels. But it is not only by excess of food that this species of colic is produced; it is occasioned also by the *quality* of the food · various high seasoned and made dishes; certain articles of a highly indigestible nature; malt liquors; cider and wines of a bad quality; the stones, kernels, husks and enveloping membrane of various fruits, swallowed when the latter have been eaten, &c., frequently give rise to it. We find it, however, in many instances originating after meals, from causes more obscure, and accompanied by various additional symptoms of a much more violent and distressing nature, as though the food itself had proved poisonous, or some poisonous substance had been intermixed with it. Occasionally these additional symptoms consist of an intolerable feeling of suffocation; a feeling of constriction in the throat; the face and eyes become swollen; with excessive thirst; a burning heat over the whole surface; a feeling of itching or prickling in the skin, and an eruption, sometimes in the form of minute raised points, at others in that of larger elevations; the skin peeling off on the subsidence of the attack: in addition to which we sometimes have a species of delirium, with twitching of the muscles. At other times the symptoms consist of great anxiety; difficulty of breathing; dejection of spirits; spasms of the limbs, as well as of the bowels and muscles of the abdomen; coldness of the extremities; loss of sight and hearing; convulsions, &c.

The first of these two last varieties of colic from indigestion arises in general from a peculiarity of the individual, the condition of the stomach at the time of eating, or from some peculiar principle connected with animal food of a particular description.

The articles of animal food which in general give rise to the species of colic under consideration, are various shell-fish, mussels, crabs, lobsters, &c.

Animal substances, in the process of cooking, or in the different processes to which they are subjected with the view of preserving them for future use, or from their being *improperly or too long kept*, may undergo a change, rendering them improper articles of food.

The second variety mentioned of the colic from indigestion is produced generally by eating deleterious vegetable substances, either mixed with our food or eaten in mistake for healthy articles.

The treatment in all the varieties of this species of colic is to be commenced with an emetic, in order to unload the stomach of the offending matter. Where the disease has been induced merely by a surfeit, or by too rich food, a dose of ipecac will be proper; in some cases, even warm water will be sufficient; but in the two other varieties, a more prompt and powerful emetic is demanded, and it has been recommended to prescribe immediately a full dose of the sulphate of zinc (twenty grains, in a gill of warm water.) The emetic is to be followed by a brisk purge, which may be aided in its operation by injections up the bowels. In the second variety, the vital powers of the system are in general rapidly, and to a most alarming extent, exhausted; it therefore becomes necessary, as soon as possible after the evacuation of the stomach, to rouse the system, by the administration of the most diffusible stimulants and cordials, such as sulphuric ether, ginger tea or cayenne pepper tea; and vinegar, diluted with water and sweetened, should be drank in abundance. These remedies may be aided by stimulant applications to the extremities, and all the other means of rousing the vital powers from their state of exhaustion. In the second variety, the treatment differs but little from that already laid down. The great indication is to *get rid of the offensive matter as quickly as possible,* by active emetics and purgatives; afterwards to rouse the system by external stimulants, mustard, friction, &c.; and to subdue the irritation and general convulsions by opiates. A mixture of ether and laudanum, ten drops of each, in water, is an excellent internal remedy; and in many cases, much advantage will be derived from the plentiful use of water and vinegar, sweetened with sugar.

FLATULENT COLIC.—In addition to the general symptoms of colic, in this species we meet with a considerable and unequal distension of the abdomen, occurring suddenly. There is a rumbling of wind in the bowels, and a frequent expulsion of it both up and down. These discharges of wind in general occasion some relief to the patient, and the pain is also diminished by pressure upon the abdomen, bending the body forward, &c. Flatulent colic is produced by every thing which occasions derangement in the stomach and bowels, and it is frequently complicated with dyspepsia. In many cases of the disease, the affection appears to be induced by a morbid production of air by the bowels themselves, but in general it is dependent upon the use of fermentable substances as articles of diet, particularly vegetables. The fruits of the season, cabbage, beans and peas, new cider, wines, beer and porter, honey, onions, various kinds of nuts, &c., &c., being introduced into the stomach when that organ is in a state of *debility* frequently undergo a rapid fermentation, and give rise to an enormous development of **gas,** producing the symptoms peculiar to this species of colic.

In the flatulent form of colic, our indications are to relieve the spasm, expel the wind distending the bowels, and afterwards, by a proper regulation of the diet and the judicious administration of tonic remedies, to restore the healthy action of the stomach and bowels. Immediately upon an attack of flatulent colic, if we are convinced that no degree of inflammation exists in any portion of the bowels, we should administer ten drops of laudanum in combination with some aromatic or diffusible stimulant; (strong essence of peppermint being about the best, ten to twenty drops,) and at the same time apply mustard plasters or stimulating fomentations externally to the abdomen. Vinegar and hops will be found very good, or large bran poultices, frequently renewed, will answer. The best internal remedy is probably a combination of ether and laudanum, in proportions suited to the age of the patient and the violence of the case, exhibited in a draught of aniseed or mint water, or the compound tincture of lavender; at the same time we may administer injections, composed of some aromatic, a little cinnamon tea, mixed with soap suds may answer, but the one most to be depended on, is composed of a teaspoonful or two of turpentine and a sufficient quantity of peppermint rubbed up with a proper portion of some thin mucilage, or thin gruel; tincture of assafœtida and laudanum, ten to twenty drops of each; assafœtida, a few drops in gruel, has also been recommended, in the form of injection. In many cases, hartshorn, in doses of twenty drops in water every half hour, will give prompt relief. After the pain has somewhat subsided, it will be proper to administer an active but mild purgative; the best is probably magnesia, combined with calomel, to which should be added a few drops of essence of peppermint.

There is another form of colic caused by *costiveness.* The disease will be known by the fact that nothing in the way of diet having been taken to produce the symptoms, and is more *gradual in its attack;* a feeling of uncomfortable fullness in the bowels being felt for some days before the attack; also loss of appetite, headache, restlessness, &c. The treatment must consist of giving an emetic, injections of soap suds up the bowels, and a brisk purgative and mustard plasters to the bowels, until relief is obtained, then being careful afterwards to keep the bowels regular. (*See Costiveness.*)

BILIOUS COLIC.—This form of the disease is marked by a violent and intolerable pain of the bowels, which in some cases seem to be, as it were, tied together, and in others closely puckered up, and with a sensation as though they were bored through with a sharp pointed instrument; the pain occasionally abates, but quickly returns. In the beginning, the pain is not fixed to one particular spot, as it is in the progress of the disorder, while vomiting also is less frequent, and the bowels more easily yield to the action of purgatives; but, as the pain increases, it becomes obstinately fixed to one place. Frequent vomiting succeeds; the bowels become more and more costive, until at length the symptoms, rapidly increasing in violence, unless the patient be speedily relieved, a total inversion of the action of the bowels takes place. Every thing administered by the mouth, or injected into the bowels, is then thrown up violently by *vomiting;* the matters discharged from the stomach are various in appearance; sometimes of a green, yellow or dark color.

In this disease there is violent irritation of the bowels, in consequence of which they are thrown into a state of spasmodic contraction, by the vitiated secretions poured into them from the liver and surrounding glands.

Though this is a formidable disease, if taken in time it can be successfully treated in most of cases.

In every case where the symptoms are of any considerable violence, particularly if the patient be of a robust habit, it will be proper to commence the treatment of bilious colic by applying leeches to the bowels, and afterwards a warm poultice to encourage the bleeding from the parts; or if leeches are not handy, scarifying, and the application of cups will be advisable.

Immediately after the leeches, or cupping, ten grains of calomel and one of opium in combination, should be administered by the mouth, and the patient may at the same time be immersed in a warm bath; or if this be not practicable, warm fomentations of hops and vinegar, should be applied over the whole of the abdomen, and continued for some length of time.

If, after a reasonable time (two or three hours), the patient is not better, the calomel and opium should be repeated. This will usually relax the spasm of the bowels, and at the same time excite the healthy action of the liver, and unload the bowels of the diseased secretions by which the irritation is kept up The action of the calomel may be assisted by injections of a laxative nature thrown into the bowels, and frequently repeated or, where the stomach will receive and retain it, we may fol low the calomel by Rochelle salts or castor oil, in repeated doses.

When the stomach is very irritable, and frequent vomiting present, these symptoms will in general be relieved by the calomel and opium; we may at the same time, however, exhibit the effervescing mixture (soda powders), and a large blister or mustard plaster should be applied over the stomach. In every case where the symptoms are violent and obstinate, the application of a blister or mustard plaster, after the employment of warm fomentations has been continued for some length of time, will be of advantage, and should not be neglected. After the violence of the disease has been removed, we should next direct our attention to restore to the liver, the stomach, and bowels their healthy action. This is to be done by giving small doses, say five to ten grains of rhubarb, half grain of opium, and half grain of ipecac every four hours.

Of course the diet of the patient should be light, nourishing, easy of digestion, and taken in small quantities at a time. He should particularly avoid all fat, coarse and irritating articles of food, all crude vegetables, and all stimulating liquors; he should make use daily of moderate exercise, but above all, riding on horseback.

Bilious Colic is a disease which is easily reproduced by any impropriety of diet or regimen, or by exposure to cold or damp, and all these exciting causes of the disease should be carefully guarded against by the patient for some considerable time. He should be particularly guarded against over-heating himself, either by exposure to the sun or by over-exertion, and particularly when such has been the case, should he be cautious not to expose himself to cold, either by throwing off any portion of his usual clothing, sitting in a draught of air, going out into the night air, or drinking cold fluids. His bowels should be kept moderately open either by the use of the ripe fruits of the season or some gentle laxative, and intoxicating drinks should be by all means avoided.

COSTIVENESS OF THE BOWELS.

THE bowels, generally speaking, ought to operate thoroughly once in the twenty-four hours. In persons who do not experience such relief it may be called a case of costiveness of the bowels. There are, however, exceptions to this rule; with some individuals, a single evacuation of the bowels once every three or four days, and even less often, seems to be sufficient, and perfectly compatible with their enjoyment of perfect health; and when such is the case it is of course superfluous to endeavor to correct it, and it is better to let well alone. If, however, in conjunction with this condition of the bowels, the person suffers from headache, from languor, from distention of the abdomen, if the breath is disagreeable, and the tongue furred, the state is *not* compatible with health, and should be corrected.

There are numerous causes producing this disease, a few of which will be enumerated. The nature of the food, as might be expected, exerts considerable influence ; bread badly made, and especially if alum be mixed with it, cheese, milk with some persons, farinacious articles, such as arrow-root or ground rice, and food of too concentrated a character, all tend to produce costiveness. Deficient exercise, particularly if combined with much exertion of mind ; any drain upon the system, as in suckling, abundant perspiration, loss of nervous power, and old age have the same effects.

The large bowel is very frequently the seat of the costiveness ; it loses tone, allows itself to be distended, sometimes to an enormous extent, or contracts to a very narrow calibre in some portion of its course. Lastly, a very common inducing cause of costiveness, particularly in females, is *inattention* to the intimation of the laws of nature to relieve the bowels.

Every effort should be made to correct the disorder without the aid of medicine. In the food, all these articles which have been enumerated, or which are known to produce costiveness, must be avoided. The bread used should be made of unbolted flour (called " Graham flour " sometimes); if vegetables and fruits agree in other respects, they may be freely consumed, and cocoa substituted for tea or coffee : food is not to be taken in a state of too great concentration, but so that by the *bulk* of its refuse it may afford substance to *stimulate* the action of the bowels. In addition, there are various articles of diet which exert an aperient or laxative effect, and which may be used or not, according to the taste of the person : such as Scotch oatmeal in the form of porridge, honey, prunes, etc. Exercise, whether on foot or on horseback, is another valuable aid in the removal of the costive state ; it not only quickens all the functions, but it assists the action of the bowels by the mechanical motion communicated to them. Friction by rubbing with the hand or coarse towel over the bowels, has a good effect.

Another very important point is *regularity* in the time of evacuating the bowels ; not waiting for the urgent sensation, but retiring for the purpose at one set period of the day, when time can be given, directly after breakfast, being the best

Persons who are liable to costiveness should give themselves at least a quarter of an hour, or even longer, for the daily evacuation of the bowels.

In cases which are not benefitted by a diet of fruits, &c., as often recommended in this book, of course appropriate medicines must be used.

If there is simple costiveness, without disorder of the digestive organs, the best remedy will be the regular use of injection of cold water every morning; if, on the other hand, furred tongue, with acidity of stomach, flatulence, pain between the shoulders, headache, &c., betoken deranged digestion, medicine will be required, at all events in the first instance: the liver is probably at fault, and five or six grains of blue pill, followed in the morning by castor-oil, will be requisite to commence the treatment; or seidlitz powders, or Rochelle salts, may be used instead of the oil occasionally.

If the stomach, liver, and upper bowels have been well cleared by the above medicines, it is requisite to *keep* the bowels open; otherwise a few days will see all the symptoms returned—and, in fact, such is too often the case. Persons are content with taking a dose of strong opening medicine every few days, or once a week, as the case may be, and rest content with thus having a good clearing out. The practice is one incompatible with sound health, and is most injurious to the stomach and bowels themselves: many cases of obstruction, and even inflammation of the bowels are produced by it. The principle to be proceeded upon in the treatment of costiveness is, that it is more easy to keep the bowels in action than to excite them to it when they have become thoroughly torpid, and therefore the individual should not rest content without the daily evacuation.

If simple costiveness exists, caused by torpor or inaction of the lower bowel, injections of cold water up the bowels every morning will often cure it; but medicine may be required, perhaps daily, for some time, or it may be used alternately with the injection. Some medicines are better adapted than others to the treatment of habitual costiveness, and of these castor-oil, aloes alone, or in its combination alternately with senna, and ipecac are the principal; their great advantage is not losing their effect by continued use. When castor-oil can be taken regularly, in most cases it answers extremely well; and if taken regularly, the dose requires rather diminution than increase. It is a medicine, moreover, which never seems to injure the tone or the mucous coat of the bowels. Aloes is peculiarly well adapted to relieve certain forms of costiveness, particularly that of sedentary persons, and may be taken in the form of pill, in combination with soap, five grains of each at bed time.

Senna, in the form of the ordinary infusion, (tea) or powdered, and taken with syrup, is a safe, and at some time effective purgative, as it does not lose its effect, like many other medicines. Or a quarter of pound of Epsom salts, dissolved in a pint of water, of which a wineglassful or two is taken at bed time, often operates freely next morning.

DIARRHŒA, or LOOSENESS.

This complaint consists in a too frequent discharge from the bowels, in consequence of their increased secretion and motion. The appearance of the stools is various: sometimes being thick, thin, slimy, whitish, yellow, green, dark brown, &c. Each discharge is preceded by a feeling of weight in the lower part of the belly, which, for the time, is relieved by the evacuation. The causes of diarrhœa are very numerous, but may, perhaps, be classed under three heads—nervous causes, causes which act upon the surface of the body, and irritating causes which act directly upon the bowels themselves.

Influences affecting the surface of the body, particularly cold, and especially cold feet, often produce diarrhœa. Cold nights succeeding hot days are often said to occasion the disease; but it is also remarkable, that diarrhœa is apt to occur at the breaking up of a long frost—indeed to be epidemic, that is, of general prevalence.

But it has been noticed that the most frequent cause of diarrhœa, however, is irritation in the bowels themselves, caused either by undigested or indigestible food, by acid, by acrid, morbid bile, or by the deficiency of that fluid permitting the digested food to become unduly changed. It may also be caused by costiveness, or by the lodgment of such matters as the skins of dried peas or beans, or of raisins, in the folds of the large bowel. Some persons habitually have a relaxed condition of the bowels, which is natural, and at the same time essential for health, and then to check it is dangerous. It is evident, that in a disease depending upon so many and various causes, a due discrimination of these is requisite for proper treatment. It must not, either, be lost sight of, that diarrhœa is in many cases salutary—an effort of nature to free the constitution from such morbid matter which, if retained, would produce disorder or disease. On this account, the simpler forms of diarrhœa are better left to right themselves, so long as they keep within moderate hounds. *This caution is particularly to be observed with regard to that which occurs in the teething of children*, which, when moderate, is a safeguard; but when it becomes so frequent that the child is evidently weakened by it, and especially if the evacuations appear to be losing their healthy character and become like shreds of skin, or streaked with blood—in such cases, a warm bath for six or eight minutes should be used, for two or three evenings in succession; isinglass or gelatine given in the milk-food, and a teaspoonful of the castor-oil emulsion with the yolk of egg beat up together, in equal proportions, with a little mucilage of gum Arabic, given three or four times a day, each dose containing a drop of paregoric.

In diarrhœa, resulting from exposure to cold, the best plan of treatment is to moderately re-excite the skin by giving small doses of ipecac every two hours; taking in the meantime some ginger-tea, mixed with a little blackberry brandy.

If diarrhœa is caused by irritating matters in the bowels, one thing is evident—it cannot be properly relieved unless the bowels are freed from the irritating matters. It may, it is true, be stopped under these circumstances, but it will recur, unless the irritating substances have been removed by purging, previous to the use of the astringent medicine, and the continuance of the diarrhœa is merely the consequence of the previous irritation. In many cases in which the diarrhœa is owing to irritating matters in the bowels, particularly to bile, all that is requisite is to diminish the acridity by means of drinks of slippery elm, or gum arabic water, or flax-seed tea, largely used, to which, if there is acid in the stomach, a little carbonate of soda or prepared chalk is to be added. In other cases, when the action of the bowels is constant, painful, and exhausting, it is absolutely necessary to check these symptoms in the first place, and to soothe the bowels, before means are resorted to for freeing them from the irritant cause. For the former purpose, a dose of paregoric, or a mixture of one teaspoonful of paregoric, four tablespoonfuls of water, and one of prepared chalk, repeated if necessary, will answer as well as any; or a strong tea made of allspice and cinnamon, with a little chalk, given every hour, will answer. Mustard to the bowels will ease pain, and often helps the disease very quickly. If the active diarrhœa does not, from its comparative mildness, require these remedies at first, or when it is sufficiently moderated, the bowels should be thoroughly cleared out with a tablespoonful dose of castor oil, to which ten drops of laudanum may be added; this will probably clear away the irritating matters, if they consist of indigestible substances. When castor oil cannot be, or is not taken, the best substitute is twenty grains of rhubarb and fifteen of calcined magnesia, with some aromatic, as cinnamon or allspice, or a teaspoonful of tincture of rhubarb, and, if there is much pain, five to ten drops of laudanum, the dose being repeated, if requisite. After the action of the opening medicine, one or two doses of astringent medicine, as mentioned above, may again be required, as the bowels are apt to keep up acting simply from irritability.

If the disease has been permitted to pass into the stage of irritation, when there is tendency to fever, the belly tender, the tongue red, and the motions resemble shreds of skin, or pieces of jelly, and are mixed with blood, the case is of that serious nature that medical assistance should at once be obtained, if it has not been so before. In the meantime, the mixture of castor oil with yolk of egg, in small doses, will be found the safest and most effectual medicine; two tablespoonfuls, with five drops of laudanum, being given every four hours, and starch and laudanum injections, the diet being as unirritating as possible, and containing abundance of gelatine. A most excellent drink in these cases is rice-water, in each pint of which from a quarter to a whole ounce of gelatine or isinglass is dissolved, with a piece of toasted bread introduced to flavor it, or a little cinnamon. Also, a blister or mustard plaster will be of very great service in some cases. Flannel should be worn next to the skin in all these cases.

CHOLERA.

As we do not propose making this a "historical" work, only a short chapter will be devoted to the subject of cholera, simply giving its symptoms, its manner of propagation, the means of prevention, and its treatment.

The true nature of this terrible disease is not known. However, such is not the case with respect to those conditions of body, and still more with respect to the external circumstances, which favor the attacks of the scourge and foster it into activity. Dirt, bad air, bad water, bad food, insufficient clothing, excitement of any kind, and irregular and vicious habits all favor the production of cholera when the disease is prevailing as an epidemic; and were it not for these, singly or combined, it would, in all probability, become a comparatively manageable disease.

The symptoms of this disease are pretty generally known, even to the unprofessional reader. However, a reference to the principal or more prominent symptoms, will be in place here.

The violence of its symptoms, and the fearful rapidity with which it often terminates life, render cholera one of our most alarming diseases. Occasionally, but more frequently in hot climates than in temperate ones, persons are, as it were, prostrated at once by the cholera poison, and die, perhaps within an hour of the first attack, without any other symptoms than total collapse of the powers of life. More generally the seizure is not so sudden : probably there has been slight diarrhœa, or rumbling movements of the bowels, with sinking sensation at the stomach, for some days previously ; or, at all events, the person has felt unwell. When the disease sets in earnestly, which in the larger proportion of cases it does during the night, the patient vomits, and is purged with more or less frequency and violence, the evacuations quickly coming to resemble thin gruel or rice-water ; cramps of the limbs succeed, the surface becomes cold, blue, bathed in sweat, and has, particularly the fingers, a peculiar shrunken appearance ; the tongue is cold, the pulse imperceptible ; the urine is suppressed, and the voice acquires a peculiar pitch of tone. Many die in this, the collapse stage of the disease ; but if it is passed through, reaction comes on, the surface gets warm, the thirst continues, the quick pulse becomes perceptible, the tongue is dry and brown, and delirium is present ; in short, fever is established, and may end either in recovery or death.

Many persons do wrong, during the prevalence of cholera, by making material changes in their *ordinary modes of living*, and, by so disordering the regularity of the functions, lay themselves open to attacks of the disease. Of course, if a man is aware that he is habitually indulging in practices injurious to health, such as intemperance, debauchery, &c. &c., he only acts wisely as regards his physical safety in changing those habits ; but it is hazardous to alter regular modes of living, which have hitherto been found compatible with good health —it being understood that whatever tends to lower the standard of health favors the attack of the disease. There is, however, one important precaution which ought to be observed, at all times, but more particularly during the epidemic of cholera : the perfect *purity of the drinking water* should be ascertained, and its freedom from all *decomposing organic matters* made certain.

Care is also to be observed *not to take active purgatives*, especially *salines*, such as Epsom or Rochelle salts, seidlitz powders, etc., which produce watery evacuations ; if aperient medicine is required, it ought to be of a warm character, such as magnesia and rhubarb, with some aromatic, (cinnamon ot allspice), for whatever produces free action of the bowels apparently increases the susceptibility to attack.   For this reason, too, *the slightest tendency to diarrhœa should at once be arrested* by a dose of paregoric, or laudanam, or what is preferable, a mixture of prepared chalk, one tablespoonful, cinnamon or allspice powdered, one tablespoonful, white sugar and flour, one tablespoonful each, water, one wine-glass; paregoric, two tablespoonfuls; Cayenne pepper, half a teaspoonful. Mix, and take a teaspoonful every half hour, or as may be needed, and the use of milk, and farinaceous preparations (corn starch, farina, flour, etc.,) containing gelatine, for food. The speedy adoption of these measures, in places distant from medical assistance, might do much to check the disease. Should the astringents above recommended fail, use the remedies recommended below.

As to the actual treatment of the disease itself, when fully established, many different methods have been proposed and practised, and few of them, perhaps without apparent advantage in some cases, but as yet no treatment which can be called decidedly successful (a cure), has been discovered.

The treatment which would be safe in the hands of others than medical men would be about the following : When vomiting and purging have set in, with cramps, give the following mixture : Tincture of Cayenne pepper, laudanum, spirits of camphor, of each one ounce; spirits of hartshorn, half an ounce; mix together and take one tablespoonful every hour or half hour according to the symptoms. Or give one grain of opium, one of camphor, one of Cayenne pepper, (made into a pill with a little flour and water) every hour, or as may be needed.

The patient should be wrapped at once in a blanket, or flannels next the skin. For the cramps use the following as a liniment : Tincture of Cayenne pepper, spirits of hartshorn, chloroform, turpentine, or kerosine oil, two ounces of each. Mix and rub over the affected parts with a woolen cloth. Be *careful to remove the contents of the chamber from the room immediately and bury it in the ground*. Also mix with the discharges from the stomach and bowels, as soon as voided, some sulphate of iron (common green vitriol), also dissolve some of the green vitriol in hot water, and set the same in vessels around the room and in the different parts of the house; and then throw some down the sinks, privy, cellar, and such places, once every day. Keep the sick chamber well aired, and by all means try to cheer and comfort the patient, so as to keep up his spirits. A mixture of mustard and Cayenne pepper moistened with strong vinegar, applied to the stomach and bowels is good to check the vomiting and purging, or applied to the limbs for cramps.

During the prevalence of this disease the greatest care is necessary in regard to cleanliness, ventilation, etc. (See " *How to Preserve Health*," in first part of this work). It may be mentioned also that warm bricks or warm stones, irons, or hot salt should be applied to the limbs or body where there is coldness or cramps. An injection up the bowels of half a teaspoonful of laudanum, four or five tablespoonfuls of brandy or whisky, with a little thin starch, is often very beneficial in the active stage of this disease, to be repeated if necessary.

DYSENTERY or BLOODY FLUX.

This differs from *diarrhœa*, as may always be known by the symptoms. In dysentery there is an inflammation of the lining or mucus membrane of the large bowels; the stools are frequent and often bloody, attended with griping and bearing down; the ordinary excrement being seldom discharged, and when it is, the quantity is small, and voided in the form of hard lumps. Fever very generally attends the acute form of the disease.

Sometimes the disease comes on with shivering, succeeded by heat and thirst, and other symptoms of fever; at others, the affection of the intestines is the first symptom observed. There exists unusual flatulence in the bowels, severe griping, frequent inclination to go to stool, loss of appetite, nausea, vomiting, frequency of pulse, and a frequent discharge of a small quantity of mucus streaked with blood, pure blood, or of a peculiarly offensive matter by stool. The matter discharged from the bowels in dysentery, varies very much in appearance, in different cases; being sometimes, as we have just stated, pure mucus, or mucus mixed with blood; pure unmixed blood, and in other instances, pus, a jelly-like dark colored matter, or a putrid watery mass, which contains films of a membranous appearance, or small fatty masses, floating in a large quantity of liquid matter. Hardened excrement is likewise sometimes passed. There is great emaciation and debility, a quick and weak pulse, a feeling of burning heat, and intolerable bearing down of the bowels.

There are two forms of this disease, the acute or active, and the chronic. In the *acute* form, the symptoms are urgent and clearly inflammatory, the natural faeces (contents of the bowels) very rarely appearing, the pain and bearing down great, and blood often passing in large quantities.

Usually the *chronic* species is a consequence of the acute, and is, as its name imports, of a less inflammatory and more protracted character than the latter: here the stools are often frequent, loose, and have much the appearance of the natural excrement, but mixed with blood and mucus, and passed with severe bearing down.

The principal causes of dysentery, are suppressed perspiration, a damp atmosphere succeeding to a high temperature, and exposure to noxious exhalations and vapors. The indications of treatment in *acute* dysentery are to subdue the local inflammation, to allay irritation, and to restore a healthy secretion from the skin, and these objects are most certainly secured by leeches, calomel, and anodynes, especially opium, with ipecac. In the majority of cases, the application of leeches to the lower part of the belly will be very useful. In many instances a large blister or mustard plaster may be laid over the abdomen with advantage.

When severe bilious symptoms are present, a dose of calomel or rhubarb, ten grains of each, may be advantageously employed, and followed in the course of a few hours by a dose of castor oil.

A combination of calomel or blue mass and ipecac will often be found of inestimable service in this complaint. A grain of calomel or three of blue mass, a grain of powdered ipecac with a third or fourth of a grain of powdered opium, may be made into a pill, and given three times a day ; or, a grain of calomel and four grains of Dover's powder, made into a pill in the same manner, may be administered three times a day; the other measures above prescribed being previously employed. These combinations of calomel with an anodyne are sometimes of the greatest advantage.

In the *commencement* of acute attacks of dysentery, more especially if the inflammatory symptoms run very high, opium, however, should not be given, either in a liquid or solid state until a dose of rhubarb and calomel, followed by castor oil, has been given, and the bowels cleared of their contents. In such a condition, our chief means must be directed to lessen the existing inflammation by leeches and injections of thin starch or flax-seed tea. As an anodyne in this complaint, there appears none so suitable and efficacious for general use as Dover's powder. If therefore, it is not exhibited through the day, as one of the principal remedies, a dose of eight or ten grains may be given in the form of pills, every night.

The diet should be mucilaginous and fluid, and consist chiefly of barley, rice or gum-arabic water, during the inflammatory stage of the disease, and this only in *very small quan tities.* Subsequently preparations of sago, rice, arrow-root, milk and the like may be allowed, and to them the patient should be confined for some time after recovery.

Sometimes when the patient begins to recover, his appetite being vigorous, he is apt to eat too much; and care must be exercised not to exceed a very moderate quantity of food, even where the appetite is keen; for if too much be indulged in, the bowels will suffer increased irritation, and a severe relapse invariably follows.

Laying down in bed is the best position for the patient, and perfect rest must be constantly observed during the active stage of dysentery, and the greater the irritation the more requisite they are. The patient ought not to *give way to the frequent inclinations to stool* by which he is harassed, *but stifle them as much as possible.* The stools must be immediately removed from the patient's chamber, which should be freely ventilated at all times, and kept perfectly clean.

For the constant griping and bearing down which attend this disease, the best remedy is frequent injections of thin starch, combined with sweet oil, to each of which, after the violence of the disease has been somewhat reduced, thirty to forty drops of laudanum may be added.

Flannel should be constantly worn next the skin during and after the disease, and be careful to avoid dews, damp night air, and sudden atmospherical vicissitudes, more especially in hot or unhealthy climates.

If pain and irritation are still occasionally felt, four or five grains of Dover's powders may be given at bed time.

In the protracted species of the disease (chronic dysentery,) in which the acute inflammatory symptoms have subsided, or been subdued, our objects are nearly the same as in the acute variety, only we are called upon constantly to remember, that *debility* is invariably associated with this form of the complaint, and therefore every means of preserving and increasing the general strength must be employed. Local bleeding by leeches or cups will, in most cases, be demanded if the strength is sufficient; in conjunction with which, one of the best remedies yet discovered for chronic dysentery is calomel or the blue mass, combined with ipecac and opium, with frequent frictions of the skin, rest, and properly regulated diet. Many patients, tormented for a long time by this painful malady, have been speedily relieved and ultimately completely cured by this plan of treatment. It is equally adapted to the protracted dysentery so often met with in hot climates, and to that of colder regions.

Three grains of calomel, two of ipecac, and one of opium, given three times in the 24 hours, followed next morning by a teaspoonful of oil, is a good proportion of the remedies.

The patient must constantly wear a flannel bandage round the bowels, and keep the feet and legs warm by wearing woollen stockings and drawers.

CHOLERA MORBUS, or VOMITING AND PURGING.

This disease is occasionally ushered in by chills, pains of the head, giddiness, propensity to sleep, and a sense of numbness in the limbs. Sometimes the disease commences gradually; at others, it attacks suddenly. At first the patient is troubled with some belchings and pains in the stomach; these symptoms are soon followed by vomiting, which is almost constant. At first the contents of the stomach are discharged; afterwards a fluid, sometimes green, whitish or colorless, and at others, dark colored, or even black. Discharges from the bowels of a similar character occur simultaneously with the vomiting. The patient, at the same time, experiences great thirst, pains in the stomach and bowels, and tension or fulness of the abdomen. If the disease be *violent* and *protracted*, the limbs are affected with spasm, the strength is greatly prostrated, the surfaces of the body and limbs become cold, the pulse small, frequent and often imperceptible, a cold clammy sweat breaks out, and is succeeded by continual hiccup, delirium and death.

In the ordinary cases of cholera morbus, particularly when the discharges are green or tinged with bile, the best practice is to give the patient, plentifully of some mild diluent drink, as toast, gum, barley, slippery elm, or rice water; to p'ace his feet in warm water, and subsequently administer an injection of a pint of thin starch, a teaspoonful of sweet oil, and forty to sixty drops of laudanum.

If the case is violent, especially when the pain of the bowels is constant and severe, the free application of leeches or cups to the abdomen, mustard poultices to the extremities, and the administration by the mouth of a grain or two of opium, in a pill, or a teaspoonful of paregoric, with twenty drops of essence of peppermint, to be repeated in an hour or so if necessary, will be demanded, and will often arrest the disease almost instantly. The same injection as recommended above, will also be beneficial when the discharges from the bowels are frequent and copious. After the vomiting and purging are suspended, it will be prudent to administer a dose of calomel, or a pill composed of blue mass ten grains, and opium one grain, which may require probably to be repeated on the ensuing day; the patient at the same time confining himself strictly to thin gruel or panado, encouraging the healthy functions of the skin by the warm bath and frictions of the skin.

When the powers of life appear to be sinking, the skin becoming cold, the pulse small and feeble, and a constant hiccup taking place after the vomiting, the patient should have stimulants, as brandy or whisky, with small pieces of ice, and also injections of thin starch and whisky up the bowels occasionally, with warm bricks, &c., to the extremities, and brisk frictions, and mustard poultices applied to the inside of his legs and arms, and over the stomach. As it is all-important in these cases to put as early a stop as possible to the discharges from the bowels, which often continue after the vomiting has ceased, injections composed of a pint of water, in which has been dissolved twenty grains of sugar of lead, and two grains of opium, may be administered every three or four hours; or injections of allspice tea, *cold*, in same quantity, is often a good remedy.

DIABETES, or GREAT FLOW OF URINE.

In this disease there is discharged an unusual large quantity of urine; sometimes the amount is enormous, and the urine contains a great deal of saccharine (or sweet) matter. The attack of this complaint is generally slow and gradual; the urine is clear and transparent as spring water, and accompanied by a faint smell, as if mixed with rosemary leaves. These symptoms generally occur without pain, and are usually attended with a voracious or greedy appetite. The serious nature of the affection renders it one of those which should be trusted for treatment only to medical hands. The same reason renders it important that its first symptoms should be known, that they may not be neglected. It may creep on a person insidiously, or be suddenly developed. The first and most prominent symptom which usually awakens attention, is the frequent call to pass urine abundantly, at the same time the thirst is extreme.

In addition to the symptoms already mentioned, the mouth is dry, and the tongue clammy and sticky, often very red; there is flatulence and indigestion, and the bowels are generally costive. Emaciation and general debility also occur; pain and weakness in the loins, and feebleness of the limbs. The leading symptom, however, is the discharge of urine, which has been known to exceed *forty pints in the twenty-four hours*

At the commencement of the disease the urine may still retain the urinous properties of the diluted secretion; but this passes into the saccharine (sweet) condition. If yeast be added to the urine, it ferments, and alcohol is formed, the sugar partaking more of the character of grape, or fermentable sugar, than of the cane. Diabetes is often accompanied by other diseases, especially by pulmonary consumption. The dietetic treatment of diabetes is probably of more importance than the medicinal; the chief precaution being the avoidance of whatever—either sugar of any kind, or vegetable starchy matter—is capable of being converted into grape or fermentable sugar This, of course, involves the prohibition of bread made from ordinary flour, which contains all the starchy matter of the grain. This privation is always much felt and complained of, and various substitutes have been proposed. The following, by Dr. Percy, is probably the best:—" Take sixteen pounds of potatoes, washed free from starch; three-quarters of a pound of mutton suet, half a pound of fresh butter, twelve eggs, half an ounce of carbonate of soda, and two ounces of dilute (half water and half acid) hydrochloric, (muriatic) acid. This quantity to be divided into eight cakes, and baked in a quick oven until nicely browned.

" It is, as must be obvious, an expensive article, but with many diabetic patients this will not be an object of consideration. It is somewhat improved in taste by being slightly toasted."

Animal diet is principally to be depended upon for nourishment, and some of the green garden vegetables, such as spinach, are permitted. Distilled water, or *boiled water*, but *not toast-water*, may be used for drink. Dr. Prout, in some cases, found porter beneficial; and, in France, claret has been given with advantage.

Persons who suffer from any suspected tendency to diabetes cannot attend too strictly to the state of the digestive organs. Wet feet must be particularly avoided, and flannel should be worn next to the skin; while all sources of debility must be guarded against.

All that can be done by others than medical men, will be to follow out the directions given above, as regards diet, &c.; also keeping the bowels open by mild purgatives, as castor oil, &c., and if there is restlessness at night, twenty drops of laudanum, or a teaspoonful of paregoric, at bed time. It has been stated by some medical men of eminence that emetics o ipecac, given once a day for a few days, has checked the disease and the patient has recovered. It is worthy of a trial where you can not have the attendance of a medical man.

BLOODY URINE.

This is produced by external violence, as blows, bruises, &c. or it may be the consequence of violent exercise, as in riding or jumping; or it may be occasioned by the irritation of a stone in the kidney or bladder; it may also take place without any cause that we are able to ascertain. In some cases, the quantity of blood voided is very large, and the debility induced is very alarming. In the treatment of the disease, we are to be guided by the cause by which it has been produced. When it is occasioned by external violence, we are to diminish inflammatory symptoms by topical bleedings, (leeches, and cupping over the lower part of the abdomen and small of the back,) by giving mild purgatives, and directing the patient to drink largely of diluent drinks, such as flaxseed tea, lemonade, &c., to which a little spirits nitre may be added, in order to dilute the contents of the bladder. If the symptoms lead us to believe that stone or a gravelly complaint is the cause of the disease, that must be attended to, for the treatment of which, see " Gravel." The spontaneous voiding of blood is to be checked by the application of cold to the region of the bladder, and by injecting cold water into the bowels. Small doses of opium, laudanum, or paregoric, may be given to allay irritation. Blood is sometimes discharged with the urine, mixed with matter. In all of these cases, a tea made of equal parts of uva ursi and buchu leaves, taken in doses of a wine-glassful, three or four times a day, are the best remedies.

RETENTION OF URINE.

By this is meant that the urine, though *secreted* by the kidneys and conveyed into the bladder, it is not discharged in the usual way.

The distinguishing symptom is a swelling at the lower part of the belly, occasioned by the distended bladder, and accompanied by pain on pressure; fever, and deficiency of urine, either total or partial. Sometimes the bladder may be distended, although there may be a *partial* flow of urine, and without great care patients may be deceived by this circumstance. If violent efforts at expulsion take place, some portions of urine may be discharged, and the patient may be supposed merely to labor under a strangury, (or difficulty in voiding urine.) By examination of the lower part of the belly, and the introduction of the catheter, the disease may, however, almost always be ascertained. Retention of the urine may arise from palsy of the bladder, which is not an unusual occurrence in advanced life.

Palsy of the bladder may be owing to a person acquiring

the habit of not evacuating the bladder *when nature prompts him to do so.* Retention of urine occurs also in the malignant forms of typhus fever. It comes on sometimes gradually, with a degree of debility which hinders the patient from completely emptying the bladder, so that he still feels a desire to do so. The inconvenience increases; at length, the patient is unable to discharge any urine, and the bladder becomes distended. In cases of retention, the urine is to be drawn off by the catheter, and when relief is given by this means, it is not unusual for the bladder to recover its tone; pretty speedily, when the complaint has come suddenly on, and more slowly, when it has been gradual in its progress. In addition to the regular emptying of the bladder by the catheter, we are to try the effect of cold applications to the parts, and of blisters to the lower part of the back.

Regularity in the times of passing the urine will have a beneficial effect also. Sometimes ten or twenty drops of tincture of cantharides (Spanish fly,) taken three or four times a day, will excite the bladder to contraction, and the urine be thus thrown off.

SUPPRESSION OF URINE.

THIS is a condition of the system in which the urine is *not secreted* in the same quantities as usual; it is in one respect just the opposite of *Diabetes*, or too great a secretion of urine. It is mostly a disease of old persons. The symptoms are: the patient passes no water, and if the catheter should be introduced, still none passes; the patient complains of feeling " unwell ;" sometimes there is nausea and restlessness at night; if there is much perspiration, it has the smell of urine.

The treatment consists in putting the patient into a warm bath for fifteen or twenty minutes; the application of cups over the kidneys, followed by mustard plasters, or warm fomentations of hops, flaxseed meal and slippery elm. Give a dose of Epsom salts or castor oil, and also twenty drops of spirits nitre in a little gum-arabic water, every hour. Sometimes a tea (given cold) made of peach leaves, uva ursi or bu chu, and taken freely, will have the most happy effect. Or a strong tea made of Virginia snake-root, given hot, every hour, to produce perspiration, will start the secretion of the kidneys.

STRANGURY.

This is a frequent desire of making water, attended with much difficulty and pain in voiding it. It arises from various causes, as an inflammation of the urethra, of the neck of the bladder, or of other neighboring parts; the application of a blister when the matter of the cantharides is taken into the body, the internal use of cantharides in powder or in tincture; excess in drinking wines or spirituous liquors, or from gravelly particles in the passage. It is sometimes a symptom of gout, and very often arises from disease of the prostate gland, (situated at the neck of the bladder.) When strangury is owing to the application of a blister, the patient should take plentifully of diluent drinks, as barley-water or thin gruel, slippery elm or gum-arabic water; to which a little spirits nitre may be added. In severe cases, fomentations or poultices to the urethra and neighboring parts may be required; and it will be proper to use injections to evacuate the bowels, as the accumulation in them will increase the strangury, from whatever cause it originates. If the strangury is an attendant on inflammation, it must be treated by local blood-letting, by leeches, by cooling purgatives, by fomentations, the warm bath, &c.; and if from spasm, an opiate (laudanum or paregoric) by the mouth or by injections, according to age, &c., will be proper and beneficial.

DIPHTHERIA.

This is a peculiar affection, met with mostly in children, and is characterized by a feeling of great prostration and want of appetite; paleness, dryness, or a feeling of roughness in the throat; swelling of the tongue and at first, perhaps, a little redness about the palate, &c. This feeling may continue a day or two, or more, before more active symptoms set in. Then, on examining the throat, palate, &c., it will be observed that they exhibit small, white, or yellowish patches, having a lardy or curdy appearance; there is also at this time swelling of the glands of the throat, some difficulty in swallowing, &c. The patches are at first small and distinct from each other, but may soon spread until the inside of the mouth and throat are covered.

In a short time a bloody fluid oozes from the affected parts, coloring the spittle. There are also at this time red and inflamed spots or patches on some part of the throat; sometimes it is of a greyish color, and looks as if it had been cauterized or burnt with caustic. There is also, in bad cases, a discharge of a thin, offensive fluid, sometimes tinged with

blood from the nostrils and the mouth; and also at times
bleeding from the nose. This false membrane, or patches, is
thrown off from time to time, but is speedily reproduced,
though *thinner* and of a more whitish appearance. Sometimes
these patches of false membrane become softened and *mixed
with the spittle*, or *absorbed*, instead of being thrown off.

This disease, when it terminates fatally, generally does so
by the peculiar inflammation *extending into the stomach and
bowels, or the lungs. Whitish patches* sometimes appear in
ordinary sore throat, so that you must not take *this* as a *sure
symptom* of diphtheria; the latter can be told by the fact that
the patches do not *extend*, and being of only *transient duration*.

Diphtheria is usually regarded as a contagious disease, and,
of course, other children should be kept out of the apartment;
also let brown sugar be burnt on live coals or a hot iron, once
a day, and chloride of lime kept in open vessels in the different
rooms, especially in the sick chamber; or *green vitriol* dissolved
in hot water may be set in open vessels about the apartments
and the place—*kept well ventilated.* The attendants on the
sick should also carry a mixture of gum camphor and green
vitriol in a little bag about their persons.

By all means have a physician at once in this disease, if
possible; but when you can not, the following treatment is
advised :—Dissolve one heaping tablespoonful of common salt
in a pint of lukewarm water, and with this let the throat be
gargled every hour; also let the patient take a little salt in the
mouth occasionally, and when dissolved, swallow it. Give
plenty of flaxseed tea (cold,) or lemonade, mixed with slippery
elm, or gum arabic water, thin gruel, &c. Apply poultices
sprinkled with Peruvian bark (not too warm) to the neck, and
give a mild purge of castor oil or rhubarb. Then follow by
giving a tea made of Peruvian bark (cold), every three hours,
or one or two grains of quinine three times a day. Inhalations
of bitter herbs (to which add some Peruvian bark,) as men
tioned under head of catarrh in the head, where the patient is
old enough to use it, three or four times a day, should also be
tried; or, if too young, let the vapor of *burnt tar* be inhaled,
which can be done by burning the tar in the room. Gargles
of oak bark or tar water may be necessary also. A solution
of nitrate of silver, fifteen grains to the ounce, is applied by
physicians, with a sponge or mop, to the affected parts.

Polypodium Vulgare, or Common Polypody.

(THE ROOT AND TOPS.)

This is found in mountains, on rocks, throughout the United States and Canada ; the root has a sweet, mucilaginous taste. This plant is good for colds, coughs, influenza, and _worms._ Some of the plant stewed in syrup and a tablespoonful given every hour or two, is an excellent thing in colds or chronic cough. United with an equal quantity of liverwort, and made into a syrup, it makes an excellent remedy in diseases of the lung and throat ; to be taken freely every two hours. The root powdered, and mixed with powdered rhubarb, in equal parts, and the same quantity of syrup, given once or twice a day ; from ten to twenty grains at a dose, to children, will expel worms after many other remedies fail. I may be used also as a tea, in colds, etc.

368

Inula Helenium, or Elecampane.

(THE ROOT.)

This plant is possessed of pretty energetic tonic properties. It acts likewise as an excitant, owing to the camphorated oil which it contains. It is an excellent article, in combination with others, in colds and coughs, in pulmonary irritation, (as in consumption, etc), and in some forms of indigestion, when it proceeds from a debility of the digestive organs. It may be given in powder, decoction or infusion. Dose of the powder, from half to one drachm ; decoction or infusion (tea), from half an ounce to one ounce, every three or four hours.

Hedcoma Pulegioides, or American Pennyroyal.

(THE TOPS.)

This is found in all parts of the United States and Canada, in dry woods, plains, etc. The fresh or dried plant chewed and the juice swallowed, is good to expel wind from the stomach; made into a tea, and taken warm, it produces perspiration very freely. In case of suppressed menstruation from cold or fright, a tea made of this plant, given warm at bed time, often assists nature to restore it; taken as a tea, mixed with a little spirits, it is good to cure pains and cramps caused by drinking cold water, suppressed perspiration, etc. The tea should be taken strong, warm, and freely.

Sanguinaria Canadensis, or Blood Root.

(THE ROOT.)

THIS is an emetic (producing vomiting), narcotic (reducing the pulse and quieting the nerves), expectorant (for cough), etc. It must be used with care, to avoid taking an over dose. It is recommended in rheumatism, diseases of the liver, typhoid—pneumonia, coughs, colds, etc. Dose, from one to five grains of the root powdered, and given in the form of a pill, every three or four hours, according to the symptoms, disease, etc.

Eupatorium Perfoliatum, or Boneset.

(THE LEAVES AND FLOWERS.)

THIS is a plant found in most parts of the United States. The whole plant is exceedingly bitter; taken as a tea, *warm*, it produces vomiting; taken cold, it acts as a gentle purgative. A wine glass or about a gill of the tea taken cold, half an hour before each meal, acts as an excellent tonic, and will act almost like a charm on some persons who can not take quinine or other tonics. The leaves may be powdered also, and given in mucilage; ten or fifteen grains three times a day, as a tonic and alterative.

Berberis Vulgaris, or Barberry.

(THE BUSH.)

THIS shrub blossoms in April and May; the berries ripen in June. The whole shrub, even the root, is *acid* or *sour ;* the bark is yellow and bitter. The berries contain a red and very sour juice, that is beneficial in chronic dysentry, or diarrhœa; also, as a cooling drink mixed with water, sugar, and orange peel, or cinnamon bark ; it is useful in fevers for abating heat and quenching thirst. A syrup may also be made from the berries or bark ; or a tea may be made of the bark, mixing cinnamon or allspice to suit the taste, and give a wine glass, cold, every three hours. The bark or berries added to hard cider, and used freely three times a day, are recommended in diseases of the liver.

Aralia Racemosa, or Common Spikenard.
(THE ROOT.)

This grows in deep woods and good soils, from New England
to the far South and West. The root is a *healing pectoral*, stimulant cordial, and causing gentle perspiration. It is much used
by the Indians; the roots bruised, chewed, or pulverized is used
by them in all kinds of sores, bruises, and ulcers. In coughs
and colds the root may be used freely, boiled in syrup; or i
may be used as a tea, cold, mixed with flaxseed-tea, lemonade,
or toast-water, when the cough is troublesome. A little piece
of the root chewed and the juice swallowed is also beneficial in
allaying a tickling cough.

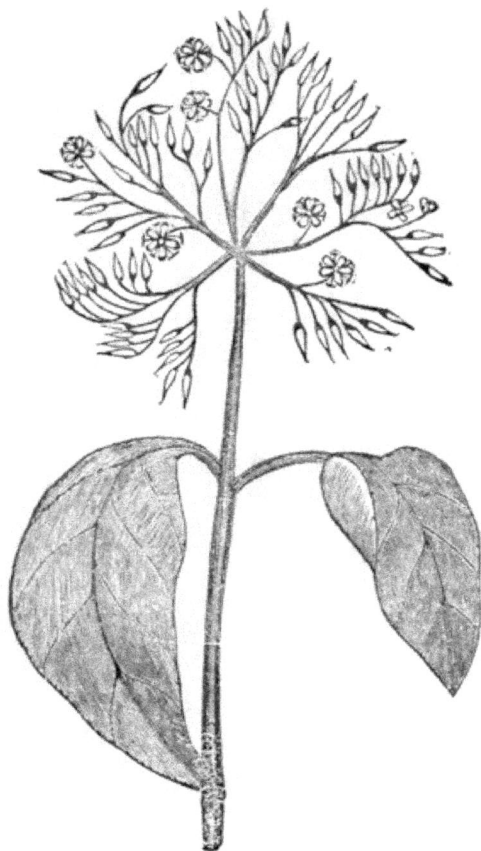

Cornus Sericea, or Rose Willow.

(THE BARK.)

THIS grows near brooks, along the banks of rivers, and on upland meadows; it is known throughout the United States by the name of red rose-willow, which distinguishes it from the *black willow*, or the *puss willow*, which grows in swamps, and along the sides of moist meadows. It is a powerful *astringent* and tonic—preferred by some to the Peruvian bark or Columbo Root, and is much employed in the Northern States, in substance or otherwise, in diarrhœa and dyspepsia. In vomiting, this is an excellent remedy, given in the form of an infusion; in the vomiting particularly arising from pregnancy. This is a valuable article. It is mostly administered in the form of infusion, or tea, given *cold*.

Asclepias Tuberosa, or Pleurisy Root.

(THE ROOT.)

This is a valuable popular remedy, and a mild sudorific (causing sweating) acting safely, without producing any stimulating effect upon the body. Its action is specifically upon the lungs, to assist suppressed expectoration, and to relieve the difficult breathing of patients laboring under pleurisy. It relieves difficulty of breathing and pains in the chest. It sometimes acts as a mild purgative, and is suitable to the complaints of children. In low stages of typhus fever, and other diseases of a like nature, it has been known to excite perspiration when other medicines have failed.

From twenty grains to a drachm of the root, in powder, may be given several times a day ; but as a diaphoretic (to produce sweating), it is best given in decoction or infusion (tea), made in the proportion of an ounce to the quart of water, and given in the dose of a teacupful every two or three hours till it operates.

Liatris, or Devil's Bit.   *Tanacetum Vulgare, or Tansey.*

(THE ROOT.)   (THE LEAVES.)

LIATRIS; or, *Devil's Bit.* (The Root.) The root of this plant has a bitterish, pungent, spicy taste, and smells like turpentine or juniper. It is a powerful diuretic (acting on the kidneys) yet *acting mi'dly*, and may be used freely without danger; it also acts as a diaphoretic, producing perspiration; and is also a tonic, strengthening the system; it is also good bruised, and applied to ulcers, sores, etc. It is a useful medicine in dropsy, sore throat, gravel, scrofula, etc. A wine-glass full of the tea, or infusion (cold), given every three hours, is good to operate on the kidneys, or taken warm to produce perspiration; or three times a day (cold) before meals as a tonic; or ten grains of the powder may be given three times a day in syrup.

TANACETUM VULGARE; or *Tansey.* (The Leaves.) This grows in moist pastures, edge of cornfields, and is cultivated in gardens. Given in the form of tea, warm, but not too strong, it produces perspiration freely, and is good in debilitated persons; a wine-glass may be taken, *cold*, three times a day in dropsy, hysterics, and disease of the kidneys. It is good taken warm to expel wind from the stomach.

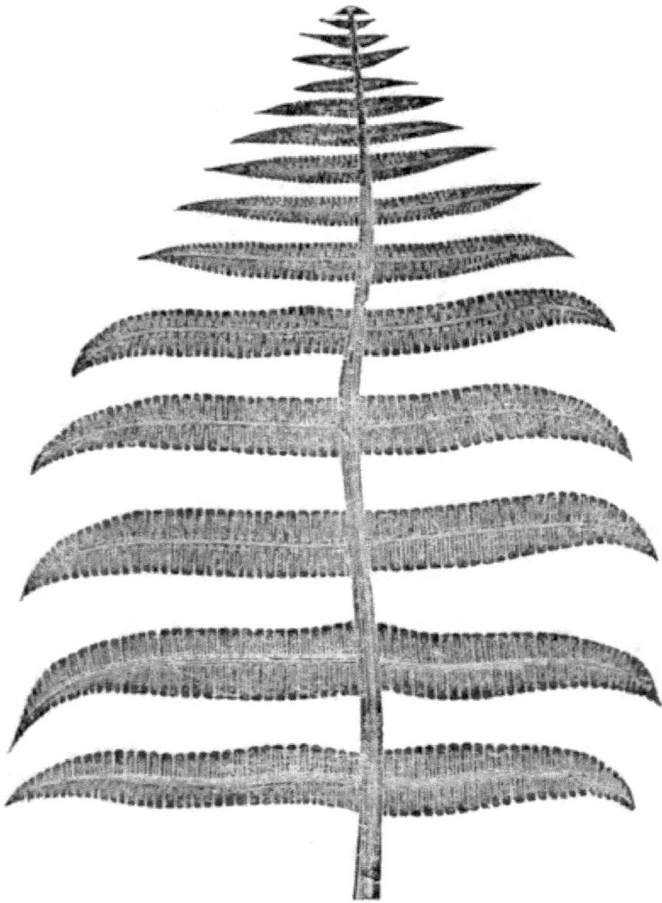

Aspidium Filix Mass, or Male Fern.

(THE ROOT.)

THIS plant grows in shady pine forests from New Jersey to Virginia ; It is likewise a native of Europe, Asia and North of Africa. It is used as a remedy against tape worm. Dose of the powdered root from one to two teaspoonfuls, given with powdered white sugar, white of egg, beat up, or the thick juice or water of slippery elm ; this dose to be re peated night and morning for two days ; then give a brisk purgative of castor oil, and should that not operate in three hours, give a dose of Epsom or Rochelle salts

Cassia Senna, or Alexandria Senna.

(THE LEAVES.)

This, is a very useful cathartic, operating effectually and mildly. It is necessary to combine this article with other ingredients, such as manna, aniseed, etc., to prevent its griping effects. It is often administered, principally in the form of infusion. INFUSION OF SENNA is made thus: Take of Senna leaves, an ounce and a half; ginger root, sliced, a drachm; aniseeds or caraway seeds, a drachm; boiling water, a pint. Macerate (keep warm) for an hour in a covered vessel, and strain the liquor. Take one half for a dose, the balance to be taken in three hours if necessary.

Arislotochia Serpentaria,
or Virginia Snake Root.

Asclepias Syriaca, or
Common Silk Weed.

ARISLOTOCHIA SERPENTARIA; or, *Virginia Snake Root.* (The Root.) This was first introduced as a remedy against snake bites, from which it derives its name, and was used by the Indians for that purpose. It possesses powerful and lasting stimulant virtues; but besides this general action, it acts also on the skin, producing perspiration. It is very useful in all cases where there is not active inflammation, in promoting perspiration, especially in typhoid fevers. etc. Dose of the powder, ten to twenty grains, in syrup or flaxseed-tea: or it may be used as an infusion (or tea), which is preferable, putting half an ounce to a pint of boiling water; of which four or five tablespoonfuls may be given every three or four hours.

ASCLEPIAS SYRIACA; or *Common Silk Weed.* (The Root.) This plant, growing plentifully throughout the United States, along roadsides and sandy grounds, is a powerful *diuretic* (operating on the kidneys), and is useful in dropsy. Boil eight ounces of the root in six quarts of rain water down to *three* quarts; strain before using. For dropsy, take a gill of this decoction four times a day, increasing the dose, or otherwise, according to the symptoms. It is used the same way for suppression of urine. It may be taken in powder, twenty to thirty grains three times a day.

Crocus Sativus, or Garden Saffron.

In small doses, saffron is employed as a diaphoretic, soon causing perspiration; in large doses, it acts upon the whole animal economy in the same way as a stimulus. It extends its action considerably to the uterus, (womb.) It is useful to allay the lumbar pains (in the back) which accompany menstruation in some females. It is useful also in chlorosis or green sickness, hysterics, &c. It may be employed likewise as a stomachic (or cordial) and antispasmodic (for spasms, &c.)

Dose—In powder, twelve grains. To make an Infusion, put half to one drachm, in one pint of boiling water; let stand half an hour; give a wine glass full every two hours. It is very valuable in all eruptive diseases, such as measles, small pox, &c.

Hyoscyamus Niger, or Black Henbane.

(THE PLANT.)

THIS is an annual plant, native of Europe, but grows plentifully in this country, along road-sides and among rubbish, flowering in July. This plant applied externally, made into a poultice or fomentation, is useful to allay pain in all cases of obstinate and painful inflammations, such as boils, fistulas, core throat, and swelling of the breast. The dose and manner of administration, internally, will be found under the head of Medicines, their doses and uses; and also in the treatment of various forms of diseases treated of in this work.

Scutillaria Laterifolia, or Scull Cap.

(THE PLANT.)

THIS is found in all parts of the United States, in meadows, woods, near water, &c., flowering in the summer. It is highly recommended for St. Vitus' Dance, given in the form of tea, (cold,) a wine glass full, or more, three times a day on an empty stomach. Given in the same way, it is highly recommended also for worms, to be followed the third day by a purge of castor oil.

The plant or leaves may also be powdered, and given in doses of from ten to fifteen grains three times a day.

LIST OF MEDICINES.

Their Doses and Uses.

PROPER TIME TO GIVE MEDICINES.—This is a matter of considerable importance, and should not be overlooked. There are certain times of the day more convenient than others for giving some medicines. Purgative medicines should, in general, be given late at night, or early in the morning; the bowels not being so easily acted upon during the time of sleep; for this reason pills and other medicines which do not act speedily, when given in the evening, have time to dissolve fully, and to produce their due effect on the bowels. Saline purgatives, such as salts, &c., are best given in the day time, that the cooler state of the surface may determine their action to the kidneys. Emetics are best given in the evening, as they produce tendency to sleep and perspiration, which are best encouraged by retiring to bed. Medicines for *perspiration* should not be given during the process of *digestion.*

QUALIFICATIONS OF ATTENDANTS ON THE SICK.

THERE is very little doubt but that recovery from sickness depends materially on the nurse, or attendants upon patients, as well upon refraining from officious interference, as giving timely attention. All the necessary qualifications can seldom be found in any *one ;* but the nearer they can be got to follow the following directions, the better :

Great attention to cleanliness of the mouth, the body, the bed, and the room ; often washing the mouth, and speedily removing all filth ; changing the clothes with as little labor to the sick as possible, being careful not to give them cold—the greater the perspiration, the more frequently changing will be necessary.

Keep the room always of a moderate degree of warmth, regulated by the season, with that all-important article, more wanted in sickness than in health, *fresh air*, to be gently admitted without a *current*, and no bed-curtains being allowed.

Keep quiet ; disturb the sick as little as possible, by talking or making any kind of noise ; never communicate any bad news,—remembering that perfect rest to the patient is of great importance.

Administer with faithfulness, and in the most palatable state, the medicines prescribed, and observe their effects: which report to the prescriber. Unpalatable pills may safely be surrounded by a thin piece of paper, jelly, or gold leaf; the great disgust to medicine being frequently caused by the nauseating manner in which it is given.

Have in readiness a bed-pan, and never suffer the patient hen very weak, to sit up on it long, as in that state they may expire from exhaustion. A cheap and ready mode at all times, of making a proper pan, would be to saw down a pail or bucket to a depth of three inches, on which a top can be placed, with a hole in it like that of a privy. This of course applies to situations in which no better facilities are at hand.

Keep constantly a supply of various articles for drink, in a proper state of the weak kind of teas, in addition to such medicinal drinks as are prescribed, which, when solely enforced, prove disgusting; remember to give but *small quantities at a time;* not *very cold,* as they *increase thirst.* Drinks can be made of any of the garden herbs generally used for teas; of toasted bread, barley, of apples, cut up in water of gruel, elm-bark, flax-seed, of lemonade, of chicken, or lean fresh meats, of tamarinds, vinegar, or cream of tartar, of currant jelly; in short, of any thing used in families, possessing no stimulating powers. Spirits, in any state of combination, wine, porter, cider, and the like stimulating drinks, unless particularly directed or called for from fainting or the disease of the patient, must be carefully avoided.

Diet is a subject too, to which the nurse should pay the greatest attention. As this must vary with each varying state of the system, it is impossible to give any further directions in this place, than to state that in high fevers, it should be very moderate, entirely of *small quantities* of vegetables, and that in low fevers, it should be of well-seasoned, palatable food; the more like that the sick person *had been accustomed to,* the better; but take care to give it often, and little at a time, as a full meal, in such states of system, often produces death.

INJECTIONS.

ADMINISTER an injection when required, of whatever article directed. These often operate only by their bulk; and, unless given as medicine, their component parts are not material; they are generally made of warm water, with salt, or soap, or sugar, or oil, and the common syringe is the instrument generally used; but the best kind is what is called the Rubber Extension Syringe, to be found now in most drug stores. In order to give it, the patient should be laid on the side at the edge of the bed, a little over the edge, the knees drawn up near the belly, and then the pipe, with the finger before it, is to be applied to the fundament; and on pushing t m, the finger is to be taken away. It is gently to be pushed a little backwards, or towards the back bone, and then the contents is to be forced out with one hand while with the other the syringe is firmly held. It only requires that the injection should be made stronger, to irritate and excite the lower part of the bowel, which brings on the action of the rest of the bowels by sympathy.

CUPPING.

This is an operation, so easily performed, and often so important, that all ought to learn how to do it. If blood is to be drawn, the part ought to be cut in many places, each cut about the length and distance of a finger-nail apart; the cut only deep enough for a little blood to flow; then a glass, or mug, or a gourd, of suitable mouth, is to be taken from a basin of hot water, and fitted to the part; then a bit of paper about as large as a dollar, dipt in spirit, is to be held near the cupping-glass, and set on fire by a candle, when, as soon as it blazes, the cupping-glass being leant on one side, the burning paper to be quickly thrown into it, and then the glass applied close to the skin, as when first fitted. As the paper burns, the air will consume, and thereby cause the extraction of the blood. Paper burnt brown, or any inflammable article is often made to answer; but the use of spirit is to be preferred. A small quantity of spirit put in the cupping-glass and set on fire will answer without paper: and its blaze is less apt to burn. There are articles made for the purpose of cupping, to be had in drug stores, in cities, much better than the rude apparatus here referred to, but in country places they are not likely to have them.

DRESSING BLISTERS.

If it be desirable not to keep the blister running, then make a very small opening with a needle or scissors, and let out the water very slowly, holding cloths to absorb it; and apply to the part cabbage leaves, freed of their stem, warmed and rubbed before the fire, or paper moistened with lard. Plasters of tallow, or suet, or of hog's lard, will answer. When the blister is to be kept running, the skin should be *cut off* with scissors, and such stimulating ointment applied, as shall be directed. Basilicon ointment is often used, as also a very weak blistering plaster.

BANDAGES.

THAT most commonly used is a long piece of cotton, linen, or flannel, about three inches wide, rolled up smoothly; its application is to be equal, compressing no one part more than another; and to insure this, you begin at the extremity of the toes or fingers, and wind it around, making each edge lap an inch over the other; and when the part over which it goes is irregular, the bandage is to be so turned, edge for edge, that it shall become suited to the bulging or irregular parts it has to pass over. This properly applied, is a powerful remedy in sores, but does great harm when put on so as to produce *unequal pressure*, interrupting the circulation and thereby increasing the disease.

There is another kind of bandage called the eighteen tailed bandage—because generally made of that number of slips of cotton or linen, of length sufficient to go once and a half around the part to be bound up. This is used when it is improper to move the limb. These detached pieces, in number sufficient to bandage the part needing it, are put under the limb at its first dressing—one to overlap the other about an inch; then the ends are to be wound smoothly around the limb, the one end over the other; and they of course will lap half around the part and each end will bind the other. When they are to be removed for dressing, and are found filthy, each can separately be pulled out, with a clean one attached to its end, and in that manner carried to replace that removed, without moving the limb.

POULTICES.

THE chief object of poultices is to relax the skin over which they are placed and allay irritation. When made of Indian corn meal, or bread and milk, they should be soft, and the part going in contact should be greased. The best is made from flax-seed meal, made by pounding it or grinding in a coffee-mill. The powder is gradually to be added to hot water and stirred until it is of proper consistence. A poultice made of slippery elm bark cut small, and boiled with a little Indian meal, is very soothing, as also one of thick jelly of water-melon seed, obtained by well boiling the seed in a little water. In most cases the chief good is derived from the moisture or warm water, which can be fully had by simply dipping cloths in hot water and applying them, to be removed on becoming cold and soaked in hot water again.

Mode of Applying Bandages.

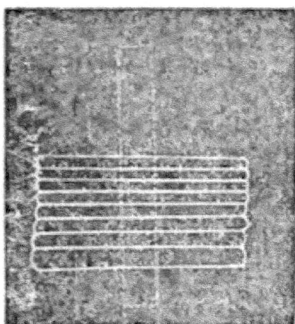

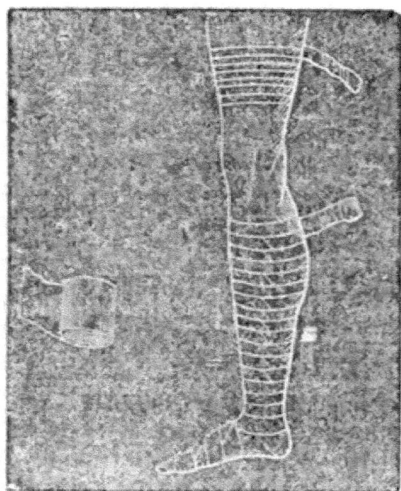

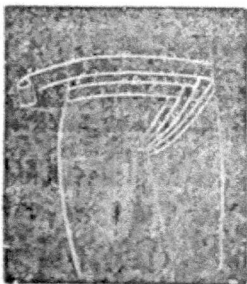

(See next page.)

How to Apply Bandages.—The art of applying a bandage well, that is, both neatly and efficiently, requires some practice and attention, but it is often a most useful accomplishment; for a bandage, if required at all, must be properly applied, otherwise it is worse than useless. If, therefore, none but the surgeon can undertake the task, it necessitates a much more frequent attendance on his part, than might otherwise be requisite. In general, the first few applications of a bandage will be made by the medical attendant himself, and ought to be in the presence of the individual to whom the duty may be afterward deputed. By careful attention on the one hand, and kind explanation on the other, much may be learned and taught, but not all, as the inexperienced bandager will discover on the first attempt. By all means, therefore, let the first attempt be made *on some one in health*, before the call is made to the invalid. Attention to the following directions will facilitate the application of the previous practical lesson, or in some measure supply its place, if from circumstances it has been wanting. Whatever the material, the width of the bandage or roller must be proportioned, in some degree, to the size of the part to which it is to be applied. If too narrow, it is apt to be stringy, and to cut; if too broad, it does not adapt itself readily to the inequalities, and the pressure is unequal. For an ordinary sized adult male leg, a bandage of 2½ inches broad is a good proportion; for the arm of the same person, one of two inches ought to be sufficiently well adapted. The material for bandages must neither be too strong nor **too weak**; ordinary "shirting" or "calico" is a very convenient texture. The length, of course, must vary according to what is required, but rollers are usually put up in six or eight yard lengths; they are better *torn* in one continuous strip, free from joinings, and without selvage edge. The strip, when prepared for use, must be rolled up as firmly as possible, into a single head. If the bandage is a new one, of calico or linen, the loose threads of the roll at each end must be roved off, otherwise they become troublesome when the roller is applied. Bandages may be applied in simple circles (B), in spiral, etc., or in reverses (C). They are also applied in various other forms, to suit the different portions of the body. In applying a bandage, the rolled up strip being held in the right hand, the end which is commenced with is secured by the first turn. If it be the simple circular bandage, round the trunk of the body, or round a limb of nearly equal girth throughout, either naturally or from swelling, the roller is carried round and round, each succeeding turn slightly overlapping the one before it. If the spiral bandage be required, the rolls are carried up very obliquely; but if, as most likely, it is the reversed bandage, then wherever the inequality of the parts prevents its being laid on flatly and evenly, the bandage must be turned upon itself (C), so as to become reversed, the surface of the cloth which was next the skin being turned outward, and *vice versa.* It is difficult to describe the manœuvre, and it is a little difficult at first to execute it neatly and well ; but when practised, it becomes perfectly simple. This is by far the most useful form of bandage, and a person who can put it on well, will have but little difficulty in accomplishing the other varieties. *For the purpose of retaining dressings upon the head*, nothing answers better than a close-fitting calico cap ; a handkerchief will often serve every purpose, or the split cloth may be used ; applied as seen in the engraving, by the upper tails being brought beneath the under ones, and fastened under the chin, the under tails being carried to the back of the head. *When it is desirable to retain the head in one position*, it may be done by bands attached to a cap, and fastened as required to a band going round the chest. When for this purpose, or to fix a broken rib, such a band is required, it ought to be from eight to ten inches wide, made of tolerably strong double calico, and sewed firmly round the body.

To retain a pad or poultice in the armpit, a good sized handkerchief answers better than any bandage, the middle being placed at the armpit, the ends crossed, at the side of the neck opposite, carried under the corresponding armpit, crossed and brought and tied on the shoulder. For the groin and parts adjacent, the spica, or figure of S bandage, is also used. A roller eight yards long is taken, the end secured by one or two turns round the pelvis (hips), and then the bandage is brought down across the front of the thigh, carried evenly between the legs, and again brought up and carried round the pelvis : this being repeated at each turn till the roller is exhausted. As a general rule, leg bandages, habitually worn, ought to be put on before the individual gets out of bed in the morning. A bandage which gives pain after its application, without obvious cause, ought to be taken off, and be reapplied. If there is reason to suspect inflammatory swelling beneath, it will be well to try the use of cold water before disturbing matters.

There is some little management required in taking off a roller, as well as putting it on. As each successive turn is unrolled, it should be gathered in a bunch in the hand, and not, *as is often done*, three or four yards of bandage at full length pulled round the limb every time.

CLASSIFICATION OF MEDICINES.

---•◦•---

The medicines wanted for common use are *very few :* although such an immense variety is to be found in apothecary shops. They are all arranged under different heads, according to their most conspicuous effects upon the system. Under each head will be mentioned those deemed most important in common use.

Remember that in taking medicine of any particular kind, the **system** becomes *habituated to it,* and requires an *increased dose ;* therefore, medicines of *similar nature* ought often, when practicable, be substituted for the one previously taken.

The doses stated are for adults—the ratio of doses **for children will be** found under Tabular list of Medicines.

PURGATIVES.—Medicines which Open the Bowels.

Saline Purgatives, (or *Salts.*)—Epsom salts, Glauber salts, Rochelle salts. Dose of each one ounce, (about one tablespoonful) given dissolved in a glass or cup of cold water, on an empty stomach, the morning being preferable. Dose for children, about half the above quantities, or less, according to age.

Seidlitz Powders.—The contents of one *blue* and one *white* paper, taken on an empty stomach, or the contents of two blue and two white papers may be taken ; directions go with the seidlitz powders ; for children about half doses.

Phosphate of Soda.—This is a *tasteless* salt, and on that account is well adapted to children, as it can be given in *soup,* or *beef tea ;* dose from one to two teaspoonfuls.

Castor Oil.—This is a standard article, the world over, almost. Dose from one to three tablespoonfuls for a grown person, and children in proportion to the age. To prevent castor oil from *griping,* let it be taken with a few drops of essence of peppermint or cinnamon and some loaf sugar ; or for grown persons it may be given in some cold coffee or lemonade.

Senna Leaves.—A small handful (three drachms) steeped in half pint of water for an hour, (like tea,) a few caraway seeds, some orange or lemon peel, tamarinds or sassafras bark, added to the senna, prevents it, to a great extent, from griping. This is not a suitable purge for *children.*

Scammony—In powder ; dose from three to ten grains, for grown persons.

Jalap—In powder ; dose from ten to twenty grains.

Gamboge—Dose from two to five grains, in powder.

Rhubarb—The powdered root ; dose from ten to twenty grains.

Cream of Tartar and Sulphur—Mix one tablespoonful of sulphur with two of cream of tartar, to be taken in a glass of cold water on an empty stomach ; or for children, take the above quantities of each, mix thoroughly with syrup or molasses, and give one half in the course of the day ; the balance the next day, if needed.

Aloes—Dose in powder, five to fifteen grains, in tincture, one to two tablespoonfuls ; for children, half the dose.

Calomel, is not often given alone as a purgative, except in cases of great irritability of the stomach, five to ten grains, in a pill, will sometimes remain on the stomach when nothing else will. Some other purge must always be given in twelve hours to carry off the calomel, to prevent *salivation.*

Active or Brisk Purgative—Jalap, ten to twenty grains; cream of tartar, thirty grains, for a grown person.

Manna, is a good purgative for children; dose from one half to two drachms, according to age.

Castor Oil Mixture—Powdered gum-arabic, a heaping teaspoonful, powdered white sugar, two teaspoonfuls; peppermint or cinnamon water, two ounces; mix these together, and add a tablespoonful or two of castor oil, or less, of course, for children. (The peppermint and cinnamon water is made by dropping a few drops of the *essence* into cold water.)

Oil of Turpentine and Castor Oil—Take two drachms of turpentine, and six or eight drachms of castor oil; mix them well together, for a dose. This combination is proper, when a prompt action on the bowels is required, as in affections of the brain, &c.

Purging Draught—Take of Epsom salt, Glauber's salt, each two drachms; mint-water, an ounce and a half; antimonial wine, forty or fifty drops; tincture of senna, two drachms. Mix.

This is a very valuable and effectual purgative for all acute diseases, and most common purposes. If a purgative which will operate quickly and actively be required, the following will be found useful:

Compound Senna Tea—Take of senna leaves, one ounce; manna, half an ounce; cream of tartar, five drachms; cinnamon bark, half an ounce; boiling water, a pint and a half. Infuse for two hours. The dose is a wineglassful every two or three hours.

Pills of Rhubarb and Soap.—Take of powdered rhubarb, one drachm; white soap, ten grains; with water enough to mix into a soft mass. Divide into fifteen pills; dose, two to four at bed-time, for a gentle purgative.

Laxative Powder.—Take of powdered rhubarb, twenty grains; calcined magnesia, ten grains; essence of cinnamon, ten drops. Mix. A good mixture in acidity of the stomach or bowels, when a gentle purgative is needed.

EMETICS, or Medicines that produce Vomiting.

Ipecac, the powder; dose, from fifteen to twenty grains. Put the ipecac in half pint of warm water, and let the patient take about four tablespoonfuls every five minutes, until vomiting is produced. Drinking freely of warm water, during the intervals, assists the effects of the medicines For children, one-fourth to two-thirds the dose, according to the age.

Syrup of Ipecac, for children, is the best: half a teaspoonful every ten or twenty minutes, or oftener, if necessary, will have the desired effect.

Wine of Ipecac, is given in doses from one to two teaspoonfuls, every half hour.

Tarter Emetic.—Dose, dissolve two grains in four ounces of hot water; a tablespoonful every ten minutes. It is not safe for children.

Antimonial Wine—Two teaspoonfuls every ten minutes.

White Vitriol, (Sulphate of Zinc.)—Thirty grains dissolved in water, in cases where a speedy emetic is wanted, as in poisoning.

Tickling the throat with a feather, will often produce vomiting at once, in cases of poisoning, &c.

Ground Mustard—A teaspoonful or two mixed in a gill of water, is an active emetic in cases requiring prompt vomiting.

LAXATIVE AND PURGATIVE INJECTIONS.

For Grown Persons.

TAKE of common salt, a dessert-spoonful; tepid water, or water-gruel, a pint; add a tablespoonful of sweet oil, or melted butter.

A more active injection is made as follows:

Take a strong infusion (or tea) of senna, a pint; Glauber's salt, or Epsom salt, an ounce and a half. Sometimes, to increase the purgative effect, a spoonful of oil of turpentine may be added.

For Infants and Children.

Injections may be made in the same way as for adults, diminishing the quantity of fluid, and keeping out a portion of the stimulating ingredient, whether salts or senna

CARMINATIVES, or Medicines to Expel Wind.

For Grown Persons.

TEN or fifteen drops of the *Essence of Peppermint*, on a small bit of sugar.

Assafœtida Pills, three at bed-time.

Carminative Injection.—Take of infusion of senna, eight ounces; dissolve in this infusion, assafœtida, a teaspoonful and a half; add peppermint water, one ounce. To be mixed together, and thrown up, pretty warm.

For Infants and Children.

Sweetened Cinnamon or Aniseed Water, one or two teaspoonfuls every half hour

Essence of Peppermint, from one to three drops, every half hour. A tea made of caraway, dill or fennel seeds; or calamus root, chewed and swallowed, as well as hartshorn, 10 drops in water, are all good medicines of this class.

DIURETICS, or Medicines that promote the flow of Urine.

For Grown Persons.

TAKE of cream of tartar, one drachm; borax, half a drachm. Mix. Dissolve in three ounces of tepid water; this quantity to be taken three times a day.

Sweet Spirits of Nitre—A teaspoonful in warm water, four or five times a day.

Acetate of Potass—From twenty grains to a drachm three times a day.

Nitrate of Potass, (saltpetre)—Thirty to sixty grains in a pint of gruel. This quantity to be used as a common drink in the twenty-four hours.

Oil of Juniper—Four drops on white sugar, three times a day.

Squill—One grain in powder, mixed with powdered cinnamon, three times a day; or, two grains of squill, with ten grains of powdered nitre, to be mixed in sugar and water, or molasses, and repeated twice or three times a day.

Diuretic Mixture.—Take of sweet spirits of nitre, one ounce; tincture of squill, two drachms. Mix. Dose, a teaspoonful five or six times a day, given in a teacupful of the following, namely . juniper berries, bruised one ounce; and cream of tartar, half an ounce; infused in a pint of boiling water.

Infusion of Juniper Berries in Cider.—Take of bruised juniper berries, mustard seed, and ginger root, of each half an ounce; grated horse-radish and parsley root, of each one ounce; and infuse them in a quart of hard cider. The dose is a wineglassful, three or four times a day. This infusion has been found useful in cases of general dropsy occurring in patients very much debilitated and unconnected with inflammation, or disorganization of any internal organ. When fever or inflammation is present, it would be improper.

Carbonate of Potash—Half drachm in a pint of water. Dose, two tablespoonfuls every two hours.

Cream Tartar—One ounce in a gallon of water; to be drank during the twenty-four hours.

Oil of Turpentine—One to two teaspoonfuls, three times a day.

For Infants and Children.

Nitrate of Potass, (saltpetre) one drachm; water, eight ounces. Dissolve, and sweeten with refined sugar. Dose, from a teaspoonful to a tablespoonful every three hours, till the water flows freely.

Sweet Spirits of Nitre, one teaspoonful; water, three ounces. To be mixed together, and a little syrup added.

A tablespoonful every two hours.

A drink made by dissolving a drachm of cream of tartar in a quart of boiling water, and sweetening it with sugar, may be used, to increase the urine.

DIAPHORETICS, or Medicines to produce Perspiration or Sweating.

For Grown Persons.

By regulating the doses of the following medicines, and the drink of the patient, as also the quantity of his bed-clothes, we can produce a perspiration more or less copious.

Antimonial Medicines are excellent diaphoretics. A grain of tartar emetic may be dissolved in five ounces of hot water; and a tablespoonful of this solution given every two hours, will generally occasion perspiration. Or James's powder, three or four grains, in honey, jelly, or marmalade, every three hours.

Diaphoretic fever powder.—Take of nitre, powdered, one drachm; tartar emetic, one grain; gum-arabic, half a drachm; for twelve powders; one to be taken every three hours.

A mixture of narcotic and emetic medicines, makes an excellent sudorific or sweating medicine. Such are *Dover's powder,* and the diaphoretic draught, made by adding thirty drops of laudanum to forty drops of antimonial wine, and the same quantity of sweet spirits of nitre, to be taken in an ounce of cinnamon or peppermint water.

These combinations of opium should be avoided when the skin is very hot and dry; but they may be used in rheumatism and other feverish disorders, after the violent excitement is in some degree removed.

Camphor.—Two grains of camphor reduced to powder by the help of a little alcohol or whisky, and half a grain of opium, made into a bolus; to be repeated only once or twice at the interval of four hours.

Diluted Acetate of Ammonia, (Spirit of Mindererus)- from two drachm to half an ounce, in an equal quantity of water, every three hours. (Spirit of mindererus can be made by dissolving a teaspoonful of carbonate of ammonia in a teacupful of vinegar.)

Saline Mixture.—Take of the fresh juice of lemons, one ounce and a half; and of the sub-carbonate of potassa, or chlorate of potash, twenty grains; then add, of white sugar, one or two drachms; of tartar emetic, one grain; and of pure water, an ounce and a half, and the same quantity of cinnamon water. Dose, a tablespoonful every two hours. This is an excellent prescription in most cases of fever.

Diminished doses of *Ipecac* may also be used to promote perspiration.

For Infants and Children.

Antimonial Wine—From four to ten drops, in a teaspoonful of tepid water, every two hours.

Take of tartaric acid, one drachm; carbonate of potass, four scruples. Dissolve each of them separately in an ounce of water, add them together, and, when the effervescence is over, add, syrup, two drachms; cinnamon water, half an ounce; water, four ounces.

Dose, a teaspoonful every two hours.

Diaphoretic Mixture.—Take of sweet spirits of nitre, four drachms; water, two ounces; cinnamon water, two ounces; sugar, four drachms; tartar emetic, one grain. Mix. Dose, a teaspoonful every three or four hours.

A tea made of Virginia snake root, or pleurisy root, or boneset, is good for sweating.

EXPECTORANTS,
To bring Phlegm from the Lungs.

Squill.—The powder of the dried root, one grain night and morning, made into pills with powdered cinnamon and ginger. Or vinegar of squill, a small teaspoonful, with simple syrup, in a litle peppermint water three times a day.

Ipecac.—One grain three times a day, made into lozenges.

Sulphate of Zinc.—One grain, with powdered ginger, twice a day.

The *Steam of Hot Water and Bitter Herbs*, such as hoarehound, sage, &c., inhaled into the lungs.

Expectorant Mixture.—Take of mucilage of gum arabic, four ounces; syrup of squill, four drachms; tarter emetic, two grains; sweet spirits of nitre, three drachms. Mix. A teaspoonful to be taken every three hours.

Brown Mixture.—Take of the powdered extract of liquorice and of powdered gum arabic, of each two drachms; dissolve in four ounces of warm water; then add sweet spirits of nitre, two drachms; tartar emetic, one grain; and laudanum, forty to sixty drops, or paregoric, one teaspoonful. Mix. The dose is a table-spoonful every four hours. This is a good prescription in cases of catarrh, towards the decline of the disease, when a troublesome cough still remains.

Expectorant Pills.—Take of extract of henbane, eight grains; extract o belladonna, one grain; powdered ipecac, four grains; for eight pills Dose, one every three hours.

ABSORBENT MEDICINES,

Or Correctors of Acidity in the Stomach, and of Heartburn.

For Grown Persons.

Carbonate of Potass, or carbonate of soda, from ten to thirty grains, in water.

Prepared Chalk, when the bowels are loose; from twenty grains to two drachms in cinnamon water, or milk.

Calcined Magnesia, when the bowels are costive; to be taken in the same way. The above two articles meeting with an acid in the stomach form a neutral salt: that with chalk is *binding*; with magnesia *laxative*.

Lime Water.—A small wine-glassful, with three table-spoonfuls of milk, three times a day.

For Infants and Children.

Calcined Magnesia, or *Prepared Chalk,* may be given in milk or any liquid, or mixed with the food, in doses of from three to ten grains.

ANODYNES,

Medicines to allay Pain, and procure Sleep.

For Grown Persons.

Opium.—One or two grains. It can be made into a pill without any addition. Opiate pills may be made also by taking equal weights of opium and powdered cinnamon, and forming them into a mass with simple syrup. This mass may be divided, so as to make the pills to contain each one grain of opium.

Paregoric may be given as an anodyne; half to one teaspoonful.

Anodyne Draught.—Take of laudanum, thirty drops; cinnamon water, one ounce. To be sweetened with dissolved jelly or syrup, and taken at once.

Or, sulphate of morphia, two grains; cinnamon water, four ounces sugar, one drachm. Dose, a teaspoonful every three hours.

Anodyne Injection.—To one ounce of olive oil, and three of thin made starch, add thirty, forty, or sixty drops of laudanum, and mix the whole well together.

This injection is particularly useful in cases in which there is great irritation about the lower part of the bowel, bladder or urinary passages, and in dysentery and diarrhœa, after proper evacuations. The relief obtained is sometimes almost instantaneous.

When opium binds the bowels too much, *Henbane* in extract or tincture may be used in larger doses than opium; three grains of the extract, or a drachm of the tincture. A strong tea made of hops, a few spoonfuls every two hours, is also good.

For Infants and Children.

Opiates are so hazardous, that we feel reluctant to sanction the use of any one of them *internally*; but from three to ten drops of paregoric, according to age, given at bed time, may be considered safe.

For *external* use, *Anodyne Balsam,* or the tincture of soap with opium, rubbed on the belly or along the spine, in the quantity of a table-spoonful in many cases allays pain very effectually.

ASTRINGENTS,

Or *Medicines to lessen Discharges of* **Fluids.**

For Grown Persons.

In *Looseness of the Bowels ;* after being sure that they are cleared of all rritating matter, as much as possible.

Astringent Drops.—Take of tincture of rhubarb, two teaspoonfuls; landanum, one teaspoonful. Mix them together. Thirty-six drops to be taken four times a day in a little water.

Chalk Mixture.—Take of prepared chalk, one ounce; refined sugar, half an ounce; mucilage of gum arabic, two ounces. Mix together, and then graduary add, of water. two pints and a half; cinnamon water, two ounces.

Of this, a small cupful may be taken four times a day; and if it be thought necessary to increase its astringent power, ten drops of laudanum, or half a teaspoonful of the tincture of kino, may be added to each dose.

Astringents, in discharges of Blood from the Lungs or Womb.

Infuse a handful of dried *Red Rose Leaves* in a quart of boiling water for half an hour. Strain off the liquor, and add of diluted sulphuric acid, thirty drops; simple syrup, two ounces. A table-spoonful to be taken every two hours, when necessary, during a discharge of blood. Other measures at the same time being employed for the cure of the disease.

Astringents, to be thrown into the Vagina for the cure of Whites.

Thirty grains of *White Vitriol* dissolved in a pint of water.

Or, take of oak bark, two ounces; water, two pints. Boil to one pint; to which, when strained, add one drachm of alum. Inject half a pint up the parts three times a day.

ASTRINGENTS, to check Looseness of Children.

Astringent Mixture.—Take of best Turkey rhubarb, twenty grains; prepared chalk, one drachm; Dover's powder, ten grains; simple cinnamon water, half an ounce; spring water, two ounces and a half. Mix them carefully.

Dose, from one to two teaspoonfuls every six hours. This is found particularly useful in some cases of habitual looseness.

Stronger Astringent Mixture.—Take of tincture of catechu, two drachms; prepared chalk, half an ounce; simple cinnamon water, one ounce; spring water, five ounces. Mix them.

Dose, from two teaspoonfuls to a table-spoonful every three or four hours. This may be rendered still more powerful in checking debilitating looseness, by the addition of a small proportion of laudanum to each dose.

For Grown Persons.

Nut Galls in powder. The dose from ten to twenty grains.

Black-Berry Root. That of the running brier, called *Dew Berry,* is the best. An ounce of it bruised and put in a pint of boiling water. Th dose about half a cupful, three or four times a day.

Alum. Dose five to ten grains : also small doses of *rhubarb.*

White Oak and Chestnut Bark, in substance or strong decoction, are powerful astringents.

Gum Kino and *Catechu* come under this head. The dose of either is from five to ten grains.

Lime Water, particularly when acid exists in the stomach, has a similar effect. The dose is a half teacupful, with an equal quantity of milk.

Common Salt. A table-spoonful has been recommended to stop bleeding of the lungs. The dose may be repeated.

Charcoal in powder, in small doses, has an astringent effect upon the bowels : also when applied to bleeding parts—as the nose, gums, etc.

Cold Water is ranked amongst the most useful of astringents. It is the best application for local bleedings.

TO CHECK VOMITING.

The Effervescing Draughts of Soda and Tartaric Acid.

Toast water taken cold in table-spoonful doses every half hour.

An *Opium* or *Mustard Plaster* to the pit of the stomach.

Lime water, a teaspoonful, with the same quantity of milk every two hours.

TONICS AND BITTERS.

For Strengthening the System.

Peruvian Bark. A teaspoonful three times a day, in milk or port wine.

Sulphate of Quinia, one to two grains before each meal.

The following is a good way of administering the quinine : Take of sulphate of quinia, ten grains ; elixir of vitriol, half a drachm ; white sugar, four spoonfuls ; water, four ounces. Dose, a teaspoonful.

Dogwood Bark and Bark of the Wild Cherry tree, reduced to fine powder, and taken in doses of thirty or forty grains, or drank in strong tea, will be found nearly as good as the Peruvian bark.

Angustura Bark, in doses from five to twenty grains, is by some esteemed equal to the Peruvian.

Columbo Root, in doses from twenty to thirty grains, or taken infused in water or wine—an ounce to the quart—is a powerful tonic, in doses of two tablespoonfuls three times a day.

Gentian Root and *Quassia* are among the strongest bitters. A very common bitter tincture is made of two ounces of gentian, one ounce of orange-peel, and half an ounce of cascarilla bark in a quart of spirit or wine ; and it is a tonic. Dose, one tablespoonful three times a day.

Bitters, in general give tone to the system ; and among those most used are—

Chamomile Flowers, Hops, Virginia Snake Root, Horehound, and *Wormwood*, separately, or in combination made in strong tea, or added to wine or spirit, make agreeable and mild bitters.

Charcoal in powder, in doses of a teaspoonful once in two hours, has been found a valuable tonic.

Nitric Acid. This is a most powerful tonic ; especially in chronic affections of the liver, etc. It is generally given a teaspoonful in the course of a day, diluted in a quart or more of water, and sweetened to render it palatable. To be drank through a quill, to save the teeth.

Iron has long been considered as a tonic, not only when in substance, but when in solution, as in the state of chalybeate waters. The filings of iron were once much used ; but a better form is the *Rust of Iron*, in doses from five to ten grains three times a day.

Green Vitriol. Dose from two to four grains, twice a day.

Tincture of Steel. Dose from fifteen to thirty drops, three times a day.

Chalybeate Waters owe their strengthening qualities to iron, which is dissolved by the agency of fixed air. By putting a few grains of the rust of iron in a bottle, and having it filled with the common *Soda Water* as it is called, as valuable a chalybeate drink may be had as from any of our springs, to which so many resort.

Tonics are not to be given in high fevers, and the following rules should be observed in recovering from sickness:

Let the diet of the patient be accommodated to the state of the system. After recovering from violent disease—it should at first consist of the lightest vegetable matter, as rice, tapioca, arrow-root, and sago, also Irish potatoes. Then eggs, oysters, wild fowl, poultry, and finally beef and mutton, generally selecting the articles most agreeable. The patient should eat *often* and in *small quantities*, and solid food is generally preferable, and should be prepared as plain as possible.

WORM MEDICINES.

EXCEPT Rochelle, Epsom or Glauber Salts, almost any purgative may be used to expel worms.

Take of *calomel*, four grains; powdered *jalap*, four grains; powdered *aloes*, three grains. To be mixed together and given in jelly, honey, or conserve of roses, at bed time, followed next morning by a dose of castor oil.

Five grains of *aloes*, with four of soda, taken at bed time, and followed next morning by a tea-cupful of strong infusion of senna-tea.

A teaspoonful of *common salt*, taken in the morning, when the stomach is empty, will often expel worms.

Two or three cloves of *garlic* may be swallowed in the morning for a length of time, or three grains of *assafœtida* made into a pill.

Injections made of senna-tea, with a teaspoonful of tincture of aloes, may be thrown up the bowels, to destroy the small white worms.

A teaspoonful of *oil of turpentine*, given in syrup, will, in many cases, carry off the tape-worm.

Pink-Root.—The dose is from five to ten grains of the powder, or an ounce of it boiled in a quart of water, of which one or two tablespoonfuls may be given every two or three hours. It is to be followed by a brisk purgative after it has been taken three or four days.

Worm-Seed Oil, extracted from the seed of the Jerusalem oak, in doses of eight or ten drops, taken morning and night for three days, followed by a purgative.

EYE-WATERS.

WHEN there is much inflammation, *decoction or tea of quince seeds*, or *infusion of the pith of sassafras*, applied every three hours. When the inflammation is abated, twelve grains of the *sulphate of zinc*, dissolved in six ounces of rose water, applied every three hours.

Six to ten grains of *acetate of zinc*, in four ounces of rose water, applied every three hours.

Twelve grains of *sugar of lead*, dissolved in six ounces of spring water, with the addition of a tablespoonful of distilled vinegar, is a stimulating eye-water.

The quantity of sugar of lead may be increased if necessary.

GARGLES.

A good domestic gargle for sore throats is made by using *vinegar*, diluted with warm water, and sweetened with honey or sugar.

Infusion of red rose leaves, acidulated with vinegar.

Or a gargle may be made with port wine and a little vinegar, or strong sage tea, with the addition of alum and honey.

Gargles should always be of such a degree of sharpness as to cause a temporary smarting of the throat.

STIMULANTS.

THESE are medicines which excite a general action over the whole system, but of short duration, which is exemplified in the effects of spirituous liquors. A rule respecting them, when they are required, is that they are to be *frequently renewed*, the quantity *gradually increased*, and the kind changed. The most common and the best are our much-abused intoxicating liquors in the various shapes of spirituous liquors, wines, porter, ale, cider, &c. Those which are considered of the strictly *medicinal* kind are:

Spirit or Oil of Turpentine.—The dose is thirty to sixty drops.

Sulphuric Ether.—Dose from one to three teaspoonfuls, mixed in half a tea-cup of water.

Spirit of Lavender.—Dose about a teaspoonful.

Hartshorn. called *Volatile Salts.*—Dose from ten to fifteen grains, made into a pill with syrup. Of the same nature is

Spirits of Hartshorn.—Dose from one to two teaspoonfuls.

Teas of Hops and our garden herbs, as well as of those imported, are also of this class.

Garlic and Onions are stimulants as commonly used.

Opium, and its preparations of Laudanum and Paregoric, in small doses, are equal in stimulating power to spirituous liquors.

Blisters are often used to stimulate the whole system.

Articles to irritate the nose, as volatile alkali, hartshorn, snuff, assafœtida, burnt feathers, and any thing very offensive, are sometimes properly used to rouse a momentary action. Sudden burning, and irritating the skin by whipping, have often done good in rousing the system.

All the *stomachics* and *cordials,* in common use, particularly the various essential oils of peppermint, cinnamon, &c.; all our spices, as pepper, ginger, &c., are often used to rouse the whole system to action, and with very good effects.

ANTI-SPASMODICS

ARE those stimulants which are supposed to remove spasm, or a kind of cramp in parts of the body. Of these the most remarkable are—

Opium and its preparations in large doses, depending on the urgency of the case.

Tincture of Valerian, and Tincture of Hyoscyamus; dose, one half to two teaspoonfuls, three times a day.

Sulphuric Ether—Dose, a small tablespoonful.

Assafœtida—Dose, in substance, eight or ten grains: in tincture, three or four teaspoonfuls.

Musk—From ten to twenty grains.

Essence of Peppermint—A teaspoonful in a glass of cold water

DEMULCENTS

ARE medicines supposed to sheathe or cover parts in a state of irritation; as the mouth, throat, stomach and bowels, in a state of increased sensibility or soreness. Of this class are all the articles which are commonly known to make a mucilage with water, as gum-arabic, and the gums of our orchard trees; teas of elm bark; of the root of the cat tail of our marshes; flax, melon, and quince seed, &c. &c. Oils of the mild kind have a similar effect; especially olive oil.

LOCAL IRRITATING REMEDIES.

NOTWITHSTANDING the term applied to this class of medicines, some of them produce a powerful effect on the whole system; not the least of which may be ranked—

Spanish Flies, or the potatoe fly of this country, universally used for exciting blisters. The fly should be very finely powdered, and mixed with equal quantities of beeswax and tallow, melted together, or with tallow alone; and is to be spread on soft leather or thick linen—or the plaster may be spread with the tallow, and the flies sprinkled on it. In cases where it would be injurious for the flies to adhere to the skin, the plaster may be covered with thin gauze. An ounce of flies in a quart of spirit, forms a good application to irritate the skin.

Mustard Seed, reduced to powder, and mixed up into a paste with vinegar, is also a common mode of irritating the skin.

Nitric-Acid—Two parts with one of water, spread by a feather on the part, speedily destroys the skin, which can be rubbed off in a few minutes, and the raw part kept discharging by irritating ointments.

Burgundy Pitch, spread on leather, and worn on the skin, makes a moderately stimulating plaster; improved by sprinkling on it a little of the dust of the Spanish fly.

Volatile Liniment, made by mixing equal quantities of spirit of hartshorn and olive oil.

Volatile Alkali, or spirits of hartshorn, is frequently used alone, to excite irritation on the surface.

Spirit of Turpentine, Spirit of Camphor, Red Pepper in Spirit, each makes valuable local irritants, and they are often used to relieve rheumatic and other deep-seated pains.

Tartar Emetic, twenty grains in a gill of water, with half a gill of tincture of Spanish flies; and common salt, with or without red pepper, answers a similar purpose.

OINTMENTS.

Those most generally used for common sores are,

Simple Ointment. It is designed merely to sheathe the parts and exclude the air. It is generally made by melting half a pint of olive oil with four ounces of beeswax. But suet alone, or mixed with equal quantities of hog's lard, will answer equally well.

Lead Ointment. This is used for sores of an inflammatory nature. It is made by pounding one drachm of sugar of lead very fine, and intimately rubbing it up with five or six ounces of hog's lard.

Basilicon, or Yellow Resin Ointment. This is used in common sores, requiring a little excitement. It is made by melting one ounce of beeswax and the same quantity of yellow resin, with an ounce and a half of hog's lard.

Red Precipitate Ointment, made by rubbing up one drachm of pow. dered precipitate with one ounce of hog's lard.

Tar Ointment. Valuable for affections of the skin and scald head; made by melting together equal quantities of tar and suet.

CAUSTICS.

These are frequently necessary to destroy the fungous of sores or *proud flesh*, as it is vulgarly termed; and to stimulate them to greater action. That most commonly used is

Burnt Alum. This is common alum deprived of its water, by keeping it on a hot iron until it ceases to boil; it is then powdered and sprinkled on the sores. Powdered rhubarb is a good substitute.

Lunar Caustic. This article, obtained from the druggists, is most used by surgeons. Its application is very simple, the edge of it slightly moistened, the sores are to be gently touched with it.

Nitric Acid. When diluted freely with water, it is very commonly applied as a wash to destroy the worms or maggots of sores in warm weather.

MISCELLANEOUS RECIPES.

Antimonial Wine.—To make this, put twenty grains of powdered tartar emetic in ten fluid ounces of sherry wine, and shake till dissolved : for dose, see tabular list of Medicines.

Rose Water.—Mix five drops of otto of rose with a tablespoonful of powdered white sugar, rubbing well until all the oil has been taken up in the sugar; then dissolve the sugar in one quart of cold water, and strain through muslin.

Spermaceti Cerate.—Melt one ounce of spermaceti, three of white wax, and six ounces of sweet oil, or lard, in a vessel over a slow fire, mixing thoroughly. This makes a good healing application to sores, ulcers, burns, &c.

Resin Cerate, (or *Basilicon Ointment.*)—Resin, five ounces; lard, eight ounces; yellow wax, two ounces. Melt and mix together, and while hot it may be strained through a coarse cloth or sieve. Used the same as the above.

Compound Resin Cerate.—Resin, suet, wax, of each four ounces; turpentine, two ounces; flax-seed oil, two ounces; melt and mix together. This is more stimulating than the two previously given, and is better for old sores and ulcers sometimes, where they have become chronic.

Yeast Poultice, is made by mixing one pound of wheat flour with one pint of yeast, exposing to a gentle heat until it begins to rise; it is then applied.

Simple Cerate, is made by melting four ounces of lard and two ounces of white wax together; it is a good healing application to simple sores, cracked lips, chapped hands, &c.

Charcoal Poultice.—A poultice made from half a pound of oatmeal thickened with water, with the addition of two ounces of finely powdered charcoal, is employed in cases of mortification, to destroy the fetor arising from the dead portions of flesh and offensive discharges.

Leeches.—These are applied to various parts of the body, to draw blood for the cure of disease.

A leech attaches itself to any substance to which it wishes to fix, by an apparatus, constructed on the principle of a leather-sucker, or air-pump, which it has at both ends; the one at the head being like a horse-shoe, with a triangular mouth in the centre, and that at the other end being circular. When they fix on the body, they inflict a small wound of three little flaps, from which they suck blood until they are gorged, or till they are forced to quit their hold; which is best done by sprinkling on them a little salt.

ENEMAS, OR INJECTIONS.—*For flatulent Colic.* Assafœtida, two drachms, and thin gruel, or starch, ten ounces, well mixed together.

Purgative Enema.—Senna leaves, three ounces; glauber salts, one ounce; and boiling water, one pint; when cold, strain.

Common Enema.—Warm water, one pint; sweet oil and molasses, of each one ounce; common salt, one drachm.

Anodyne Enema.—To relieve pain in the lower part of the bowels, constant straining, as at stool, or profuse diarrhœa. Thin starch, half a pint; olive oil, one ounce; opium, from a half to two or three grains for an adult; smaller doses, in proportion, for a child.

Warming Plaster.—This plaster forms an excellent local irritant, in cases in which the action is wished to be kept up for a long time, without exciting a blister. It is composed of Burgundy pitch and Spanish-fly cerate, seven parts of the first and one of the latter, melted together, and then spread on leather.

Warner's Cordial.—This is an excellent purgative in persons troubled with a weak stomach, flatulence, or tendency to cramps of the bowels. It is composed of an ounce of rhubarb, two drachms senna, a drachm of coriander, and the same quantity of fennel-seed bruised; red saunders, two drachms; saffron and liquorice, of each, half a drachm; stoned raisins, half a pound, and diluted alcohol or whisky, three pints; to be steeped together for two weeks, then strained through paper: dose, half an ounce or an ounce.

Turner's Cerate.—This cerate is a very useful dressing to produce the healing of simple ulcers, excoriations, slight burns, blisters, &c. It is made by melting together half a pound of yellow wax and two pounds of lard, and stirring into the mixture, while fluid, half a pound of prepared carbonate of zinc on an empty stomach.

Kentish Ointment.—This ointment, made by mixing together two ounces of basilicon ointment, and two drachms of turpentine, has long been celebrated as a dressing for burns and scalds. Care should be taken to prevent its contact with the sound skin surrounding the burn or scald; for though a soothing application to the latter, in the surrounding parts it will be very apt to produce severe inflammation.

Gregory's Powder.—This is a useful laxative powder, composed of equal parts of calcined magnesia, powdered rhubarb, and ginger; it was a favorite prescription of the late Dr. Gregory, of Edinburgh. He considered it possessing various good properties, the magnesia correcting acidity, the rhubarb acting as a tonic and laxative, and the ginger being a good aromatic for the stomach, and preventing griping of the bowels. The dose of this compound powder is one or two drachms, or a heaped teaspoonful; and it may be taken in water, in gruel, milk, or any vehicle, that may be most convenient. It generally operates easily and effectually, and may be taken at any time of the day. The dose may be repeated after an interval of four or six hours, if the first dose does not produce its proper effect. Sometimes it may be advisable, especially for children, to omit the ginger, and to give simply a mixture of rhubarb and magnesia, in the dose of a small teaspoonful.

Effervescing Draught.—The effervescing draught is made by dissolving a drachm, or a drachm and a half of carbonate of soda, of potash, or of ammonia, in an ounce of water, and mixing with this an ounce of lemon-juice, with a little water and sugar; or if lemon-juice cannot be procured, dissolving a drachm of crystallized citric or tartaric acid in an ounce of water, and adding this to the alkaline solution. The two solutions when they meet, occasion, by their mutual action, an effervescence, in consequence of the escape of the carbonic acid; and should be swallowed while this action is going on. The medicinal virtues of the effervescing draught are to check vomiting, and to determine the blood to the skin; hence it is very useful in a variety of diseases, especially feverish and dyspeptic complaints. The materials for making the effervescing draught are kept in the shops under the name of soda powders, and directions are given for their use. They are thought to give relief in the symptoms of indigestion which follow over-indulgence in eating or drinking.

Coxe's Hive Syrup.—The hive syrup is made by boiling squill and seneca roots, of each four ounces, in four pints of water, until the whole is reduced one-half; the product is then to be strained, and two pints added of clarified honey, when it is to be boiled down to three pints, and forty-eight grains of tartar emetic, dissolved in a little water, added to the residue. The dose is from ten drops to a teaspoonful, according to the age of the patient. It is principally employed as an emetic and expectorant, in cases of croup, or in the lung and throat affections of children.

Neutral Mixture.—This mixture is one of the most agreeable, mild diaphoretic sweats we possess, in case of fever. It is made from recent lime juice, or lemon juice, one ounce and a half, saturated with subcarbonate of potash, with the addition of a drachm or two of white sugar, and three ounces of pure water, or mint-water. The dose is a table-spoonful every two or three hours. Its powers are decidedly augmented by the addition of half a grain of tartar emetic to the mixture; or when this is not thought advisable, by adding a drachm or two of sweet spirits of nitre.

Lenitive Electuary.—This is a very gentle and agreeable laxative in cases of simple costiveness. It is made by rubbing together in a mortar an ounce of senna leaves, and half an ounce of coriander seeds; then sifting ten ounces of the powder through a sieve. The remainder, with the addition of three drachms of liquorice root, and two ounces of figs, is to be boiled in half a pint of water, until the whole is reduced to one-half. The liquor being pressed out and strained, is to be evaporated to one gill, and to this is to be added four ounces of sugar, and a syrup made in the usual manner by boiling; one ounce of the pulp of prunes, the same quantity of tamarinds, and of senna, being well mixed together in a mortar, are to be added to the syrup, and the whole well combined with the sifted powder. The dose is a portion of the size of a nutmeg, or a table-spoonful.

Aloetic Pills; aloes and castile soap equal parts, made into five grain pills; for costiveness without any peculiarity of symptoms: two pills for a dose at bed-time.

Lady Webster's, or *Lady Crespigny's Pills,* are made of equal parts of rhubarb, aloes, and gum mastich. This last ingredient is not of much virtue in itself, but makes the solution of the others in the bowels gradual and equal. The dose of these pills, which have not received any particular name, is two or three, and the time for taking them is immediately before dinner; they then mix with the food, prevent flatulency, and are usually found to operate next morning after breakfast.

Disinfectant.—One-half pound of copperas dissolved in a bucket of water, poured down the sink three or four times, will completely destroy the offensive odor. As a disinfecting agent to scatter around premises affected with any unpleasant odor, nothing is better than a mixture of four parts of charcoal and copperas, by weight. All sorts of glass vessels and other utensils may be effectually cured from offensive smells by rinsing them with charcoal powder, after the grosser impurities have been scoured off with sand and soap.

Blue Vitriol. As much of it as any given quantity of water can dissolve, is frequently applied to old sores.

Although many remedies have been mentioned, do not understand me as recommending them for *general use.* Annexed is a list of such as we think ought to be in every family, at least in every neighborhood, especially in country places, also for steamers, ships, etc.

Your great dependence for the cure of disease should be on the *most simple means;* you should give a decided preference to local, instead of general remedies, in most cases.

A LIST OF MEDICINES FOR A MEDICINE CHEST,

For Ships, Families, etc.

	Ounces.		Ounces.
Powdered Jalap	½	Gum Ammoniac	4
Powdered Rhubarb	1	Gum Assafœtida	4
Magnesia, calcined	1	Balsom of Peru	2
Tartar Emetic	½	Buchu Leaves	2
Powdered Ipecac	½	Cardamon Seed	2
Powdered Aloes	1	Prepared Chalk	4
Laudanum	2	Powdered Cinnamon Bark	4
Paregoric	2	Essence of "	2
Blistering Plaster	2	Extract of Colocynth	1
Camphor	4	Syrup of Squills	4
Columbo Root	4	" Ipecac	4
Sugar Lead	1	" Rhubarb	4
White Vitriol	1	Dover's Powders	2
Blue Vitriol	1	Gum Guaiacum	1
Powdered Nut Galls	1	Carbonate of Ammonia, or Harts-	
Spirit Nitre	4	horn	2
Ether Sulphuric	4	Tincture of Iodine	2
Senna and Manna, each	8	Gum Kino	1
Flowers of Sulphur	8	Compound Spirits of Lavender	4
Chamomile Flowers	8	Blue Mass, for " Blue Pills "	2
Powdered Peruvian Bark	8	Powdered Gum Myrrh	2
Epsom Salts 2 pounds, and Cream		Gum Opium, powdered, 2 drachms.	
of Tartar ½ pound.		Quassia	4
Oil of Turpentine, and Spirit of		Rochelle Salts	8
Hartshorn, each one bottle.		Glauber Salts	4
Castor Oil, and Olive Oil, each		Oil of Worm Seed	½
one bottle.		Quinine	1
Essence of Peppermint	2	Tincture of Hyoscyamus, or Hen-	
Virginia Snake Root, ½ pound.		bane	4
Also a little Alum, Nitre, Borax		Tincture of Valerian	4
and Basilicon Ointment.		Wine of Colchicum	4
Tincture of Aloes	8	Court Plaster, and Adhesive Plas-	
Antimonial Wine	2	ter.	
		Uva Ursi Leaves	4

LIST OF MEDICINES—THEIR DOSES AND USES.

The following comprises a list of medicines adapted to a work of this kind. The dose should be regulated according to the age as follows: 14 to 16 years of age, give two-thirds the dose ordered for a grown person; 7 to 10 years, one-half the dose; 4 to 6 years, one-third; 3 to 4 years, one-fourth; 1 year old, one-eighth the dose for a grown person:

Medicines.	DOSES.		How Taken.	Uses.
	Grown Persons.	Children from 2 to 5 years.		
Æther	30 drops to 1 drachm.	8 to 10 drops.	Flaxseed tea.	Asthma, Cramp, and Flatulence.
Aloes	10 to 20 grains.	2 to 6 grains.	In pills or powder.	Obstinate Costiveness.
Aloes, tincture	3 to 6 drachms.	1 to 2 drachms.	In cinnamon water.	Costiveness and Worms.
Alum, powdered	3 to 10 grains.		In pills, 3 times a day.	Flooding, Chronic Dysentery.
Ammoniac Gum	10 to 15 grains.		Made into pills, twice a day.	Chronic Cough, Asthma, &c.
Antimonial Wine	2 to 4 drachms.	1 to 2 drachms.	Mix in water.	Emetic.
Antimonial Wine, as an alterative	12 to 20 drops.	4 to 8 drops.	In water, twice a day or every hour or two	To produce moisture on the skin.
Assafœtida, tinct. of	30 to 60 drops.	8 to 12 drops.	In water, every 3 or 4 hours.	Nervousness.
Assafœtida Pill	10 to 15 grains.		In Pills, twice a day.	In Nervous Diseases.
Balsam Copaiva	20 to 60 drops.	6 to 8 drops.	In sugar, syrup, honey.	Whites, Gravel, Cough, &c.
Balsam of Peru	5 to 10 drops.		In sugar or honey.	Flatulence, Asthma, Coughs.
Bark, Peruvian, pow'r.	20 to 60 grains.	6 to 10 grains.	In a little mint-water, 4 or 6 times a day.	Ague, Indigestion, and general Weakness.
Bark, Peruvian, decoc.	3 to 4 tablespoonfuls.	1 to 2 tablespoonfuls.	3 or 4 times a day.	Relaxation and Debility.
Bark, Peruvian, tinct.	2 to 4 drachms.	40 to 60 drops.	3 or 4 times a day.	Ague. Fever. &c.
Buchu Leaves, Tincture of	2 to 3 teaspoonfuls.		3 times a day, in decoction of marsh mallow root or flax-seed tea	Irritation in Bladder.

Calomel..	10 to 20 grains.	3 to 5 grains.	In powder, mix with molasses.	Derangements of the Liver, to be followed by Oil or Salts.
Camphor	2 to 4 grains.	1 to 2 grains,	In a pill, twice a day.	Whooping Cough, Nervousness.
Cardamon Seeds, tinct.	1 to 3 drachms.		In water, 3 times a day.	Indigestion, Flatulence, Cramp.
Castor Oil......	From ½ to 3 ounces.	1 teaspoonful to a tablespoonful.	Cold coffee or mint-water.	Colic, Costiveness, Worms, &c.
Cascarilla Bark, tinct.	1 to 3 drachms.	20 to 30 drops.	In water, 3 times a day.	Indigestion and Debility.
Chalk, prepared......	10 to 15 grains.	4 to 6 grains.	In mint-water, 4 times a day.	Looseness or Sourness of the Stomach.
Camomile Flowers, powdered...	10 to 20 grains.	6 to 10 grains.	In mint-water 2 or 3 times a day.	Dyspepsia, Worms, Weakness of Stomach.
Cinnamon, powdered.	5 to 10 grains	2 to 4 grains.	In water, 3 times a day.	Indigestion, Flatulence, &c.
Cinnamon, essence of..	3 to 10 drops.	1 drop.	In water.	Flatulence, Colicky Pains, &c.
Cinnamon, tincture of.	3 to 4 drachms.	20 to 30 drops.	In water.	Diarrhœa.
Columbo Root, pow'd.	10 to 20 grains.	3 to 5 grains.	In mint-water, times a day.	Dyspepsia, Chron. Diarrhœa, &c.
Columbo Root, tinct..	1 to 3 drachms,	10 to 20 drops.	In mint-water, times a day.	Dyspepsia, Debility.
Colocynth, extract of..	10 to 15 grains.		Pills.	Costiveness.
Cream of Tartar......	1 to 4 drachms.	20 to 30 grains.	Honey, every morning, or in warm water sweetened.	Inflammations, Eruptions of the Skin, and Cooling Purge.
Cubebs, ground......	15 to 25 grains.		3 times a day, in water.	Whites, Diseases of Kidney, &c.
Cubebs, tincture of...	2 to 3 teaspoonfuls.		3 times a day, in water.	Diseases of Kidney, &c.
Dover's Powder......	10 to 20 grains.	3 to 6 grains.	In water.	Rheumatism, recent colds, and to produce perspiration, sleep.
Elixir of Vitriol......	10 to 12 drops.	3 to 5 drops.	In a glass of water.	Flatulence, Weakness of Stomach, Night Sweats, Loss of Appetite, &c.
Epsom Salts.	4 to 8 drachms.	1 drachm.	In water, or in mint water.	Costiveness.
Gum Guaiacum......	5 to 15 grains.		In pills, twice a day.	Chronic Rheumatism and Gout.

Medicines.	DOSES. Grown Persons.	Children from 2 to 5 years.	How Taken.	Uses.
Hartshorn, spirit of...	5 to 10 drops.	2 to 3 drops.	Honey and water, 3 times a day.	Hysterics, Convulsions, Heartburn.
Hoffmann's Anodyne Liquor..........	30 to 40 drops.	6 to 10 drops.	Water, twice a day.	Nervous Fever, Asthma, and Neuralgia.
Iodine, tincture of....	10 to 30 drops.		Flax-seed tea or lemonade, 3 times a day.	Scrofula and Wen.
Ipecac, powder........	20 to 30 grains.	5 to 10 grains.	In water.	When Ipecac is given in small doses, it produces sweating—valuable in fever, inflammat'n.
Ipecac, wine of........	4 to 8 drachms.	2 to 3 drachms.	In water, warm.	Costiveness.
Jalap, powder.........	20 to 30 grains.	4 to 6 grains.	In mint-water.	Costiveness.
Kino, Gum, tincture of.	2 to 3 drachms.	15 to 20 drops.	In mint-water, 3 or 4 times a day.	Diarrhœa.
Lavender, spirits of...	30 to 80 drops.	10 to 20 drops.	In water.	Lowness of Spirits, Nervousness.
Lobelia, tincture of...	40 drops to 1 teaspoonful.		In water, 2 or 3 times a day.	Asthma, Bronchitis.
Madder, powder.......	10 to 60 grains.		Mint-water, 2 or 3 times a day.	Chlorosis or Green Sickness.
Magnesia, calcined....	20 to 40 grains.	5 to 10 grains.	In common or in mint-water.	Heartburn and Sourness of the Stomach.
Manna...............	3 to 6 drachms.	1 to 2 drachms.	Mint-water or tea.	Costiveness; a good medicine for children.
Mercurial, or Blue Pill.	6 to 12 grains.	2 to 4 grains.	In pills, twice a day.	Diseases of the Liver.
Muriatic Acid. Diluted	10 to 30 drops.	3 to 6 drops.	In a glass of water, twice a day.	Scrofula, Eruptions of the Skin.
Myrrh, powdered	5 to 10 grains.	2 to 4 grains.	Mint-water, twice daily	Green Sickness and Weakness.
Myrrh, tincture of....	1 to 2 drachms.	10 to 15 drops.	In water, 2 or 3 times a day.	" "
Nitre, sweet spirit of..	20 to 60 drops.	6 to 12 drops.	In water, every 2, 3 or 4 hours.	Difficulty of Urinating, Fevers.

	Dose.	Dose.	How given.	Uses.
Nitric Acid. Diluted.	12 to 30 drops.	4 to 6 drops.	In a glass of water, 2 or 3 times a day.	Diseases of kidneys and Scrofula
Opium......	1 to 2 grains.		A pill.	Costiveness, Acute Pain, Diarrhœa, &c.
Opium, tincture of, or Laudanum.	10 to 35 drops.	3 to 5 drops.	In water.	To produce sleep, when in pain.
Paregoric, elixir......	1 to 2 drachms.	15 to 20 drops.	In water.	Cough, Asthma, and Cramp.
Peppermint, essence of.	3 to 15 drops.	1 to 2 drops.	In water.	Colicky Pains and Flatulence.
Poppies, syrup of.....	2 to 4 drachms.	1 teaspoonful.	In water.	Spasms, Acute Pain, and Cough.
Quassia, tincture of...	30 to 60 drops.	10 to 12 drops.	In ginger tea, 2 or 3 times a day.	Dyspepsia and Flatulency.
Rhubarb Powder.....	20 to 30 grains.	5 to 8 grains.	In mint-water.	For Costiveness.
Rhubarb Tincture.....	4 to 8 drachms.	1 to 2 drachms.	In mint-water.	Costiveness.
Rochelle Salt.......	6 to 12 drachms.	2 to 4 drachms.	In water.	Costiveness.
Roses, infus, or tea of.	1 to 2 ounces.	3 to 4 drachms.	2 or 3 times a day.	Dyspepsia, Flooding, &c.
Roses, conserve of.	1 to 2 drachms.	½ drachm.	2 or 3 times a day.	Cough.
Salts, Glauber's.....	6 to 12 drachms.	1 to 2 drachms.	In mint-water.	Costiveness, &c.
Salts, Epsom, purified.	6 to 8 drachms.	1 to 2 drachms.	In water.	Costiveness, &c.
Sarsaparilla Powder...	20 to 60 grains.	5 to 10 grains.	In mint-water, times a day.	Scrofula, and all impurities of the blood.
Sarsaparilla, comp'd decoction of......	3 to 4 ounces.	1 to 2 ounces.	3 or 4 times a day.	Scrofula, and all impurities of the blood.
Scammony, powder...	10 to 20 grains.	2 to 4 grains.	In mint-water, 3 or 4 times a day.	Obstinate Costiveness.
Senna, infus, or tea of.	2 to 3 ounces.	2 to 3 drachms.	On an empty stomach.	Costiveness and Worms.
Senna, tincture of.....	6 to 12 drachms.	2 to 3 drachms.	In water.	Costiveness.
Squill, powder.......	1 to 2 grains.	¼ to 1 grain.	In a pill, twice a day.	Dropsy, Asthma, Chro. Cough.
Squill, tincture of.....	15 to 30 drops.	6 to 10 drops.	Mint-water, twice a day	Dropsy, Asthma, Chron. Cough.
Sulphate of Quinine .	2 to 8 grains.	½ to 1 grain.	In a pill, or in water, 3 or 4 times a day.	Ague, and general debility.
Sulphur, flowers of.....	1 to 2 drachms.	10 to 20 grains.	In honey, once a day.	Diseases of Skin, Piles, Worms.
Tartar Emetic.......	1 to 6 grains.		In warm water.	Emetic and to produce sweating.

Medicines.	Doses. Grown Persons.	Doses. Children from 2 to 5 years.	How Taken	Uses.
Tincture of Cantharides....	10 to 30 drops.	6 to 8 drops.	Barley water, or in gruel, twice a day.	Whooping Cough, &c.
Tincture of Gentian, compound, Stoughton's Bitters....	1 to 2 drachms.	12 to 30 drops.	In water, 3 times a day.	Dyspepsia and Flatulence.
Tincture of Gum Guaiacum....	30 to 60 drops.	5 to 6 drops.	In honey, twice a day.	Rheumatism and Gout.
Tincture Henbane (Hyoscyamus)...	20 to 60 drops.	3 to 5 drops.	In mint-water.	Spasms, Acute Pains, Nervousness.
Tincture of Iron, muriated....	10 to 30 drops.	3 to 6 drops.	In a glass of water, twice a day.	Indigestion, Rickets, Debility.
Tincture of Jalap....	2 to 4 drachms.	1 drachm.	In mint-water.	Obstinate Costiveness.
Tincture Opium, Laudanum....	10 to 30 drops.	3 to 5 drops.	In water.	Spasms, Acute Pains, Diarrhœa, &c.
Tolu, tincture of....	30 to 60 drops.		Water, sugar, or honey.	Chronic Cough.
Turpentine, Spirits of.	15 to 20 drops.	10 to 40 drops.	In honey, twice a day.	Gravel, Rheumatism, Worms.
Valerian, tincture....	1 to 3 drachms.		In water, 3 times a day.	Nervous Headache, &c.
Wine, Antimonial....	4 to 8 drachms.	1 to 2 drachms.	In water.	Emetic, &c.
Wine of Colchicum Seeds....	5 to 50 drops.		In mint-water, twice a day.	Rheumatism and Gout.
Wine of Ipecac....	4 to 8 drachms.	2 to 3 drachms.	In water.	Emetic.
Wine of Rhubarb....	6 to 12 drachms.	1 to 2 drachms.	In mint-water.	Costiveness and Dyspepsia.

Mint Water is made by dropping a few drops of Essence of Peppermint into water, and is designed to destroy the unpleasant taste of the medicine. Cinnamon Water is made in the same way by dropping a few drops of Essence of Cinnamon into water, and can be used the same as the Mint Water by those who prefer it, as convenient and pleasant to take medicine in. A *drachm* is about one teaspoonful; a *grain* is the weight of a grain of black pepper or allspice

THE MOVEMENT CURE.

This new curative treatment, so called, is nothing more than *extending into minutiæ* the old principle of the "*dumb-bells,*" for giving greater strength to the muscles of the arms, and thereby taking off local irritation existing in the lungs, etc.

While we can not admit the wonderful cures said to be performed by this treatment, still we have seen cases greatly benefited by it, and under favorable circumstances is worthy a fair trial, as it can be done in the patient's own room, either alone, or in some cases having a friend to assist.

The following is about the plainest description that can be given for popular use.

ACTIVE MOVEMENTS.—The purpose of an active movement, is to convey to, and concentrate upon a selected point, the nutrition and energies of the system. Such a movement may accomplish a two-fold purpose, that of supplying a part, and of relieving another part more or less distant.

The mode of effecting this purpose is as follows:—The person to receive the application, is placed in an easy unconstrained position, sitting, lying, half lying, kneeling, or any convenient position that will suitably adjust all parts of the body to the purpose. The body is fixed either by the hands of an assistant, or by means of apparatus, so as to prevent as much as possible any motion of all parts of the body, except the *acting* part. The patient is in some cases directed to *move* the free part in a particular direction, the effort to do so is *resisted by the operator*, with a force proportionate to the exertion made—very nicely graduated to the particular condition of the part, and of the system at large. The resistance is *not* uniform, but varies according to the varying action of muscles, as perceived by the operator. In other cases the operator acts while the patient resists. The action is the same, but in one case the patient's acting muscles are shortened; in the other, lengthened. The operation is a sort of *wrestle*, in which a very limited portion of the organism is engaged. The motion must be much slower than the natural movement of the part engaged, which fact strongly fixes the attention and concentrates the will. The act is repeated two or three times with all the care and precision the operator can command, being cautious not to *induce fatigue*. A perfect rest in the lying position succeeds, of some ten or fifteen minutes. The changes of matter induced by the movement continue for that length of time, producing an afflux of power and nutritive material to the part, provided the patient remains quiet. If, however, other actions be engaged in, it detracts from and diminishes the effect of the movement. If movements succeed each other rapidly, very much of the peculiar effects are lost and the operation becomes to a certain extent gymnastic.

The effect of a movement if properly applied and received, is to transmit the available force of the system, together with the conditions for its production to the *acting part :* this part receives what the whole system by the process is made to contribute. Thus a lax, weak, bloodless region is reinforced with fresh supplies contributed by the whole system. Every portion of the body is, in turn, and at proper intervals, subjected to similar operations. Such as the fingers, arms, legs, feet, etc.

The great point is to commence these manipulations *gradually,* and end in the same way ; and always follow the partial movements by thorough friction with the hand or a rough towel, and afterwards wrap up the part in a flannel cloth ; the under garments should always be flannel or woolen, also.

POISONS AND THEIR ANTIDOTES.

The effects produced by many poisonous substances, take place with such promptness, that but little time is allowed for the exhibition of remedies, and the patient is often destroyed before the physician arrives ; whereas, had the proper treatment been *immediately* instituted, the fatal result might have been prevented.

We shall in this place confine ourselves to a brief account of the leading effects produced by the introduction into the stomach of the various classes of poisons, the antidotes proposed for the principal articles of these classes, and the general medical treatment demanded in cases of poisoning.

Poisons may be divided into the *corrosive* or *acrid* (which destroy the parts to which applied), the *narcotic* (or stupifying), and those acting both as corrosives and narcotics.

The symptoms resulting from the first class, or acrid poisons, in addition to the particular taste of the article itself, are heat, irritation, or an extraordinary and sudden sensation of dryness, constriction (tightness) and roughness at the root of the tongue, and in the gullet; these are succeeded by violent efforts to vomit, and sharp pains in the stomach and bowels ; there is also great thirst, copious discharges by vomiting and stool, attended with much straining, and followed by hiccup ; a sense of constriction across the diaphragm (or chest), and difficulty of breathing; pain is generally felt about the kidneys, followed by strangury (difficulty in urinating), convulsions at length come on, or cramps of the hands, trembling of the limbs, extinction of the voice, repeated fainting, cold sweats, and usually a hard and irregular pulse.

The narcotic poisons produce the following effects : stupor, numbness, a great inclination to sleep, coldness and stiffness of the extremities, a cold sweat of an offensive or greasy nature, swelling of the neck and face, protrusion of the eyes, with a haggard cast of countenance, thickening of the tongue, frequent vertigo (giddiness), impaired or depraved vision, delirium, general debility, palpitation of the heart, the pulse at first full and strong, afterwards becomes unequal and intermittent ; there is also paralysis of the lower extremities, retraction of the lips, general swelling of the body, and swelling of the veins. At the conclusion of the disease slight convulsions and pain are sometimes present.

The effects of the narcotico-acrid poisons, such as belladonna, aconite, etc., are distinguished by a combination of several of the symptoms of both the foregoing classes. There is generally agitation, pain, acute cries, sometimes stupor, and convulsive motions of the muscles of the face, jaws and extremities, and occasionally extreme stiffness of the limbs, and contraction of the muscles of the chest ; the eyes are red and starting from their sockets, the pupils frequently dilated ; there is often great insensibility to external impressions ; the mouth is full of foam ; the tongue and gums livid, or purple, with nausea, vomiting, frequent stools, etc. Often these symptoms attack in paroxysms, and between them the patient is left comparatively easy for a few moments.

It might appear easy, from an attention to the symptoms we have recited, to distinguish the nature of the poisonous article under the effects of which the patient is laboring ; but nevertheless, nothing is generally more difficult. Substances very different in their nature, *produce similar effects ;* as, for example, cantharides (Spanish flies), certain acrid vegetable substances and the caustic minerals. The difficulty is increased by the circumstance, that articles of ordinary food, perfectly innoxious in themselves, so far as regards any poisonous property, in certain conditions of the stomach, and in certain constitutions, when eaten, sometimes cause the most alarming symptoms. Roasted cheese, fish, crabs, lobsters, clams, mushroons, or even apples and cherries, have been known to produce the most alarming symptoms and cause a suspicion of poison having been taken. A variety, also, is frequently observed in the symptoms caused by the same

poison in different persons. Many circumstances may con
duce to this, such as

The mode in which the article has been taken. When
swallowed in a liquid form, the effects of a poison are generally more prompt and marked than when it is taken in a
solid state.

If the article be taken on a full or empty stomach, its effects
will vary ; being much more rapid and certain in the latter
case than in the first.

The circumstance of vomiting occurring immediately, or
not until after a considerable time, will produce a difference
in the effects of the poison. In the former case the article
may be rejected from the stomach before it has had time to
produce any injurious effects. Thus large doses of arsenic
have been taken intentionally as a poison, but in consequence
of copious vomiting instantly following, the lives of the individuals have been preserved.

To distinguish cases of poisoning from accidental affections
of the stomach, produced by other causes, demands a judicious
and cautious examination of every circumstance relative to
the character and disposition of the patient ; the possibility of
his having taken poison ; the article of which he had last eaten
or drank ; the vessel in which it had been contained ; the
patient's own confession, if able to speak, &c. The diseases
and symptoms most likely to be mistaken for the effects of
poisons, are probably those produced by idiosyncrasy (pecul -
arity of constitution) and indigestion, and cases of sudden an l
unexpected illness. But the most striking cases of resemblance
to the effects of poisons, probably occur in those who, after
being long accustomed to a particular species of food, for the
first time use another kind. The treatment, in cases of
poisoning, varies according to the individual articles taken.
As a general rule, in those cases in which the corrosive an l
acrid poisons have been swallowed, the indications of cure are,

First, to endeavor to *discharge* the poison as quickly as
possible from the stomach.

Then endeavor to destroy its poisonous properties, by the
administration of antidotes. And afterwards to prevent or
subdue inflammation.

The first indication is to be effected by the administration
of an active emetic ; or, if vomiting has already occurred, in
general by the copious administration of diluent drinks ; or we
may attempt to remove the article from the stomach, by an
appropriate pump, if it can be had.

Arsenic.—This is an article very frequently made use of to destroy life; it is, also, often taken in mistake for other articles, nearly resembling it in their external appearance, either of food or medicine. Arsenic may be taken in such quantity as merely to produce disorder of the stomach and system, without necessarily destroying life; or it may be taken in such quantities as to produce death at a later period than twenty-four hours; or, lastly, the quantity may be such as to induce death within twenty-four hours. When taken in the slightest portion, the symptoms produced by arsenic, are uneasiness at the stomach, with a sense of heat. When the dose is somewhat greater, but not so great as to produce death, violent vomiting is commonly the first symptom; although, in some instances, it is preceded by a sense of heat in the tongue and throat; in other cases, these sensations are not felt during the whole course of the disease. When the vomiting is immediate, and the poison has been taken on a *full stomach*, the patient seems to owe his escape to the poison being discharged before it has had time to act.

The next symptom claiming attention is purging, sometimes of blood; but purging is less frequent in the slight cases, than in those where the degree of poisoning is greater. In the region of the stomach and bowels, pain is frequently felt: it is often, however, rather an *unsupportable uneasiness and oppression* than pain, properly speaking. A sense of coldness, especially of the extremities, with cold sweats, seems nearly always to be present, with general paleness of the face and surface; and in some cases languor, faintness and a tendency to sleep. In this degree of poison, convulsions are not frequently observed; and thirst and fever are seldom present.

In the second degree of poisoning from arsenic, when the patient lives a day or two, the first sensations are heat and thirst, vomiting, or inexpressible distress. The first is less frequent than the two others; purging is not often present; convulsions generally take place, however. In the third degree of poisoning, when death takes place within a few hours, the symptoms succeed each other rapidly, or occur at the same time; fainting and general debility almost invariably precede the vomiting, which occurs in most cases, as well as purging and griping, and death seems generally to proceed from exhaustion and rapid sinking of the vital powers.

The indications in the treatment of poisoning from arsenic are:—To remove the poison. To protect the stomach and bowels from its effects; and if the patient survive sufficiently long, to diminish inflammation.

The removal of the poison is to be attempted by emetics of

sulphate of zinc, **or** if vomiting be present, by the **aid of** diluent drinks, or a vegetable emetic (ipecac being the best.) Tartar emetic *should never be administered.* But when vomiting does not quickly ensue from these means, the urgency of the case demands a resort to more direct remedies. The stomach may be washed out by means of taking large drinks of sweetened water, and then causing vomiting by tickling the throat with a feather, &c.; in this manner, a quantity of liquid is to be used, **so** as to dilute or **suspend** the poison, and by means of a stomach pump, when to be had, the whole may be withdrawn. By this procedure, we may, in many cases, **succeed** in saving the life of the patient.

The second indication may be effected **by** means **of** milk, lime-water, **soap-suds**, and drinks sweetened with sugar or honey. Fatty **or** oily substances are of doubtful utility.

In a case reported by Joseph Hume, life **was** saved by administering freely, after vomiting had ceased, retching and pain, however, remaining, the carbonate of magnesia twenty grains, with twenty drops of laudanum, in water; one or two cases subsequently reported are in favor of this practice.

The third indication is to subdue inflammation by the same remedies as in ordinary inflammation of the stomach.

For arsenic we unfortunately possess no antidote, strictly speaking. A **preparation of** iron has been vaunted, but it is of doubtful **efficacy**; if **either** this **or** the stomach pump is used, it will **be safest in medical** hands. White arsenic is not the only preparation of the metal **by** which poisoning occurs; the coloring substances known by the name of King's yellow and Scheele's green are **both** compounds of arsenic, and being frequently and **culpably** used in confectionery, have proved fatal. Similar **symptoms** occur and similar treatment is **to be** followed as **after** poisoning by white **arsenic.** Whether in poisoning by arsenic, or by any other **agent, the** vomited matters should always be carefully preserved in a vessel by themselves, **for** medical inspection; and if there is any suspicion of foul play, some responsible person should place them under lock and key. The chemist can detect the smallest amount of arsenic, even after years have elapsed.

Corrosive Sublimate. — Besides the ordinary symptoms caused by corrosive poisons, the present article produces **a** peculiar sense **of** stricture and burning heat in the throat and gullet, increased when attempts are made to swallow; there **is** also dysentery, bloody vomiting, and sometimes diminished or **even** suppressed secretion of urine. The treatment of poisoning **from** this article **is** to administer **an** emetic, or if vomiting **is** present, as large a quantity of the whites of eggs, well mixed with water, as the stomach can contain. By the experiments of Orfila, it is proved that albumen decomposes corrosive sublimate, forming a triple compound, consisting of albumen, muriatic acid, and calomel. Dr. Taddei, of Italy, has recommended wheat flour, starch, or gluten, mixed with water, as an antidote to corrosive sublimate; hence, when the whites of eggs are not at hand, either of the latter should be employed as directed above.

The plentiful use of mucilaginous drinks is also very proper as an accessary remedy. The *antidotes* to corrosive sublimate are, therefore, albumen (white of eggs) and vegetable gluten (flour, starch, &c.)

Tartar Emetic.—This substance, in large doses, is undoubtedly a poison. It is by no means, however, so destructive as either of the foregoing. The remedies are, if vomiting be present, to wash the article from the stomach by large draughts of tepid drinks; if vomiting be not present, to excite it by tickling the throat, and by the administration of large quantities of warm water, sweetened. If, notwithstanding these means, vomiting be not induced, we are to resort to antidotes. These are decoctions or infusions (teas) of any astringent vegetable substances. The following may be employed: a tea of Peruvian bark; strong green or black tea; a decoction of galls, of oak bark, or of any of the other astringent roots or barks. The above articles are named in the order of their efficacy. From the experiments of Berthollet, the Peruvian bark would appear most certainly efficacious, and when it can be procured, should invariably be preferred. When the vomiting is excessive, opium may be administered.

The Salts of Copper.—These, in certain doses, are all poisonous. Verdigris, or the impure carbonate, is the one most commonly employed. The symptoms are the same as in the case of other corrosive poisons. We are to endeavor, when it has been taken, to get it out of the stomach by the same means as have already been mentioned. Sugar was once considered as the antidote for this poison. Subsequent experiments, however, have lessened the estimation in which it was at first held, and have pointed out *albumen* as the article most to be depended upon; hence, the whites of eggs mixed with water are to be administered at first; their operation being aided by the use of large quantities of sugar and water. Should inflammatory symptoms remain after the presumed evacuation of the poison, apply mustard or leeches to the stomach. For the removal of the spasmodic affections that are apt to remain, laudanum or opium will be required.

Sulphate of Zinc.—When taken in an over-dose, vomiting should be excited by copious draughts of warm water, e mollient drinks of slippery elm, flax-seed tea, &c. *Milk* is the proper antidote. Inflammation is to be prevented by the ordinary means, and irritation allayed by opium.

Muriate of Tin.—The treatment is the same as in the former article : milk is also its proper antidote.

Nitrate of Silver.—When accidentally taken in an over-dose, a solution of *common salt* in water is to be administered ; at the same time, the patient should take plentifully of emollient and mucilaginous drinks.

Nitrate of Bismuth.—The same general treatment as in the case of other corrosive poisons, with milk and mucilaginous drinks plentifully administered.

The Salts of Lead, when taken in large quantities, produce poisonous effects, and when gradually introduced into the system, they produce a peculiar species of colic, which has been already treated of. When taken in an over-dose, the proper treatment is to endeavor as speedily as possible to empty the stomach by the ordinary means. The sulphate of soda, (Glauber's salts,) or of magnesia, (Epsom salts,) is the most effectual antidote for lead ; it should be given in strong solution ; at the same time, mucilaginous drinks and purgatives are to be administered.

Sulphuric Acid.—Taken in an undiluted state, or in large quantities, it produces all the symptoms attendant upon violent inflammation of the throat, gullet and stomach, or when concentrated, it may destroy at once the lining membrane of those parts.

Large quantities of calcined magnesia in milk, syrup, molasses, or as little water as possible, must be instantly administered ; or, if not at hand, soap and water, chalk and water, or lime-water. The caustic must be neutralized, or the patient is inevitably lost. The subsequent treatment will depend upon the degree of inflammation present. Demulcent drinks barley-water, gum-water, whey, milk diet and injections oi thin gruel, will always be proper.

Nitric Acid.—When taken in excess, the treatment is to give freely carbonate of soda, or magnesia, or calcined magnesia, or lime-water, and the other general treatment recommended above.

The Alkalies, such as potash, ammonia, caustic soda, sale

ratus, &c. For these, when taken in excess, vinegar and lemon-juice are the most valuable remedies; they are to be aided by the plentiful use of mucilaginous drinks and emollient injections. The remaining treatment will depend upon the degree of inflammation.

Barytes.—All the salts of this earth, except the sulphate, are poisonous in certain doses. When taken, vomiting is to be excited, and the plentiful use of a solution of sulphate of soda, (Glauber salts,) or magnesia, (Epsom salts,) commenced with early. These decompose the poison and produce the insoluble sulphate, which of course is inert.

Nitrate of Potash, (salt petre,) when taken in excess, is a poison producing inflammation of the stomach, &c. Treatment, vomiting, mucilaginous drinks, and mustard to the stomach, according to circumstances.

Muriate of Ammonia.—The treatment is the same as that directed in the last case.

Acrid Vegetable Poisons.—The treatment for poisoning from these is, to dislodge the article from the stomach as speedily as possible, by vomiting, and then to administer large quantities of mucilaginous drinks, emollient injections, &c. To overcome violent irritation and spasm of the stomach and bowels, give frequent doses of opium and landanum.

Narcotic Poisons—Opium.—When opium or any of its preparations are taken in a large quantity, so as to act as a poison, the following symptoms are usually perceived within a short period: insensibility and incapacity of exercising muscular motion; breathing scarcely perceptible, and a small and feeble pulse, which usually becomes full and slow. As the effects of the poison increase, the state of stupor becomes more complete; swallowing is suspended; the breathing is occasionally laborious; the pupils are insensible to the application of light; the countenance is livid or pale and death-like, and the muscles of the limbs and trunk are in a state of relaxation; vomiting sometimes supervenes; death is often preceded by convulsions. In cases of recovery, a weakness will sometimes be left in the lower extremities, nearly approaching to paralysis, and the bladder will be unable to retain its contents.

The following are the directions for treating a case of poisoning from opium:

Induce vomiting, if possible, with sulphate of zinc, sulphate of copper, (blue vitriol,) or tartar emetic. In endeavoring to induce vomiting, great quantities of watery fluids will be improper, as they dissolve the opium and promote its absorption. The vomiting should, therefore, be accomplished without the administration of any more liquid than is neces sary to dissolve the emetic.

The operation of the emetic may be accelerated by tickling the throat with the finger, a feather, &c., but as one of the effects of this poison upon the stomach is to render the latter *insensible* to the impression of emetics, much time should not be lost in vainly waiting until they shall operate, when by the aid of the gum elastic tube and syringe, the contents of the stomach may be pumped out and fluids afterwards injected, so as entirely to wash out every portion of the poison. The patient should not be allowed to *remain quiet* in one position, but should be *moved about* between two assistants; stinging with nettles, or even the application of a cowskin has been proposed, and put in practice, under these circumstances, with good effect.

Mustard plasters on the extremities should never be neglected. The effusion of cold water is also a remedy of considerable efficacy in rousing the system from the state of stupor in which it is thrown by the effects of narcotic poisons, particularly the one under consideration ; large pitchers or buckets of water should be splashed from a height over the head and shoulders of the patient, or over his whole body, and persevered in until the patient indicates a return to a state of animation.

Now administer alternately, water acidulated with any vegetable acid, and a strong warm infusion of coffee. The experiments of Orfila have shown that the exhibition of vegetable acids *previously to the evacuation of the opium, is highly improper*, as they accelerate and aggravate the action of the poison ; *after*, however, the latter has been entirely discharged from the stomach, water acidulated with vinegar, lemon-juice, or other vegetable acid, tends to diminish and correct its effects upon the system, to which, also, the infusion or decoction of coffee is admirably adapted.

In about ten or twelve hours, administer an injection, and let the arms and legs of the patient be well rubbed with the flesh-brush, soft coarse flannel, or some stimulating application. Dr. Beck states that he has known the most happy results at this particular juncture, and during the latter stage, from repeated injections of a strong watery solution of assafœtida. So long as any of the opium is suspected to be retained in the bowels, purgative injections should be continued.

The above treatment for poisoning from opium or its pre

parations, and acrid vegetable poisons, is adapted in al essential particulars, for poisoning from the various kinds of poisonous plants, herbs, &c.

Iodine and Iodide of Potash.—Give freely of starch or wheaten flour and water, combined with ipecac and warm water.

Creosote, is coagulated and rendered comparatively harmless, by white of eggs, or if not to be had, starch or flour and water, given freely, with ipecac and warm water, sweetened.

Green Vitriol, or Copperas.—Give freely of carbonate of soda and water, followed by ipecac and sweetened water, and then give cold coffee, to strengthen the system and quiet the nervous pain.

Poisonous Fish.—Give an emetic of ipecac, and warm sweetened water, or tickle the throat to produce vomiting. Then give a purge of castor oil or Rochelle salts or Epsom salts, and if necessary, injections of soapsuds up the bowels. Afterwards give a drink of vinegar, sugar and water, frequently. A dose of laudanum may also be needed to produce sleep.

ERYSIPELAS.

This disease is also called "*St. Anthony's Fire,*" or "*The Rose,*" and is an inflammatory affection of the skin alone, or of the skin and cellular tissue or soft substance beneath the skin. Like other inflammations, it varies in degree and extent, in different cases. When it affects merely the external surface of the skin, in which case the latter is red, not sensibly swollen, soft and without fluctuation, the disease is termed erythema. The cases to which the term erysipelas is more generally applied, are marked by the same symptoms as above, but of a more intense grade; there is greater redness, considerable swelling, a peculiar burning pain, and an effusion takes place beneath the skin, raising the latter in the form of blisters of various sizes; very generally there is effusion also in the subcutaneous cellular tissue. The most aggravated form of the disease is termed phlegmonous erysipelas; in this both the skin and cellular membranes are inflamed, and extensive collections of matter and sloughing of the cellular structure are quickly produced.

Erysipelas usually affects the face and limbs; less frequently, especially in adults, the surface of the chest and abdomen. In a few instances, the disease has been known to pervade the entire surface of the body.

The disease is confined to no particular sex or constitution. It is more common, however, in infants and young children, as

well as in the aged, than in persons about the middle period
of life. It likewise more frequently attacks females than
males; and persons of a sanguine and irritable temperament,
and of luxurious and intemperate lives, are more liable to its
attacks than any others.

Erysipelas is confined to very small spaces sometimes, in
others it extends over the whole head and face, or occupies
an entire limb. It not unfrequently commences at a point,
and gradually extends in every direction, until it involves a
very large portion of the skin. In other instances, in the part
first attacked, the inflammation runs through its various
stages and disappears, while it subsequently extends over a
new surface to pursue the same course. In this manner, it
may travel gradually from the head to the feet. In other
cases, again, the erysipelas may suddenly disappear from the
part primarily affected, to reappear in another and remote
part of the body. It occasionally happens, that when the disease very suddenly disappears from the skin, some internal
organ, and particularly the brain, in cases of erysipelas of the
face, suddenly present all the symptoms of inflammation.

The part affected with erysipelas in its simplest form, presents the ordinary symptoms of inflammation, namely, swelling, heat, and redness. The swelling, however, is softer, more
irregular and diffused than in common inflammation; the
heat is more intense, and the accompanying pain is a *burning*
or *smarting*, similar to that from the application of mustard
or scalding water, in place of being throbbing as in the
latter; the redness is *brighter* and more *intense* and *disappears
upon pressure*, but returns the moment the pressure is removed. When erysipelas attacks a limb, in general the whole
circumference of the latter becomes enlarged, and the skin
presents a kind of smooth shining appearance, and a somewhat doughy feel, as though a fluid was effused beneath it,
which is, in fact, the case in most instances, when the disease
is of any violence or extent. When the face is the seat of
erysipelas, the features become deformed; the mouth is often
drawn towards one side; the nose is enormously enlarged, and
the eye-balls, becoming swollen, close up the eyes entirely.

In a few days, the period differing in different cases, vesi-
cations, (blisters,) varying in size, arise upon different parts of
the inflamed surface, especially towards its centre. They are
of an irregular form, and filled with a fluid, at first clear and
watery, becoming subsequently straw-colored. The skin after
a time, gives way, allowing the fluid of the blisters to escape,
which generally drying upon the skin, covers it with this

scales. About the eighth or ninth day of the disease, but occasionally much later, the redness of the affected surfaces changes to a brownish or yellow hue; the vesicles entirely subside, and the skin dries and scales off.

In general, the disease is preceded and accompanied with fever, varying in its character according to the constitution, age, and general state of health of the patient. In the young, the robust, and those of full habits, we have a very decided attack of fever, and often of considerable intensity. When the face or scalp is the seat of erysipelas, there are often pain and oppression of the head, inclination to sleep, or delirium. The tongue becomes dry and brown; the pulse rapid and feeble, with great loss of muscular strength. In other cases, the heart and nervous system are less affected, but we have pain in the stomach, foul tongue, a bad taste in the mouth, nausea and costiveness of the bowels.

This disease is produced by the same causes as other inflammations. It may result from cold and various irritants applied to the skin, or it may be produced sympathetically from irritations seated in the stomach and bowels. Phlegmonous erysipelas very generally results from wounds, bruises, extensive ulcerations, or from the influence of cold.

The treatment of this disease, when it is possible, should be entrusted to a good physician, but in many cases this is not the case. The following plan of domestic treatment is recommended: Always begin by giving an emetic, (vomit,) of ipecac; then clear out the bowels by the following mixture: mix ten grains of calomel, ten of Dover's powders, ten of rhubarb; divide into three parts, give one part every two hours; to be followed by a seidlitz powder, if the bowels are not freely opened in ten hours. If this mixture is not to be had conveniently, a dose or two of castor oil or Rochelle salts, or other good purgative will answer, giving at the same time nauseating doses of ipecac, (about half to one grain of the powder every two hours.) In weak or delicate persons these nauseating doses can be dispensed with, after one vomit in the beginning.

The local treatment is also of much importance; the best application to the inflamed part is caustic, nitrate of silver. Erysipelas tending to spread, may be stopped by surrounding the affected part *entirely* with a cauterized ring. The parts to be touched must, in the first place, be shaved, if covered with hair, and the skin must always be thoroughly cleansed from its natural oily secretion, by washing with soap and water. It must then be moistened all round, and the stick of caustic drawn slowly and gently over it, so as to make a line of demarcation at least a quarter of an inch broad; but this line must be entire throughout—deficiency in one spot may permit the inflammation to extend by the outlet. It is not asserted that in every case this caustic line will inevitably stop the disease, but it will do so in the majority, if care be taken that it is efficiently done, and that it includes, without doubt, every portion of the affected skin. When the solution of caustic is to be used to spell the disease, it should be used of the strength of forty grains to two drachms of water. The inflamed surface must be gently cleansed by soap and warm water, and the solution applied all over it by means of a camel-hair brush or a feather. The practice now recommended is *perfectly safe*, is very efficacious, both as a preventive against the extension, and as a cure of this formidable disease, and might be quite justifiably employed by an intelligent person in the absence of medical assistance.

There is a very common and often a good and comfortable local remedy in mild cases of erysipelas; hot fomentations, either of simple water or a decoction of poppy-heads or hops, applied *continuously for many hours*, by means of flannel, give much relief in some cases; or a lotion composed of twenty grains of sugar of lead, a drachm of laudanum, and sixteen ounces of water, may be used slightly warm, and applied by means of linen cloths, with much advantage. A poultice of charcoal, slippery elm, and hops, frequently renewed, is also a good remedy in mild cases.

Erysipelas is the most formidable enemy which can gain a footing in a surgical hospital; and for the same reason, when the disease occurs in private houses, caution should be observed that persons suffering from wounds do not come into close contact with the affected; and indeed in any case, the same precaution should be adopted in erysipelas as in contagious diseases generally. This is doubly requisite in a house in which a confinement is expected, or has recently taken place, for there is an undoubted close connection between erysipelas and fatal childbed inflammation, and under such circumstances too much care can not be exercised.

Telling Diseases by the Tongue.—Much could be said about this "wondrous little member," but in the present case we are only taking it into consideration in a medical point of view, in what is to be regarded both with reference to its own disorders, and to the indications it affords of disorder in other parts of the system.

The condition and appearance of the tongue are indications almost always consulted by a physician in investigating a case of disease, and most valuable guides they are at times, when experience, observation, etc., have given the power of reading them aright. When the appearances of the tongue, however, are admitted as evidence, consideration must always be given to the *natural* state of the organ in the individual, for some never have a clean tongue, while in others it scarcely becomes furred, even when considerable disorder is going on in the system. In chronic disorders, especially of the digestive region, the most valuable indications are often afforded by the tongue, immediately after the night's sleep, *before food has been taken.* Persons who sleep with their mouths open, generally have a dry tongue in consequence, but in most persons in health, the mouth should be pleasantly moist on awakening in the morning; if it is the reverse—if the tongue is dry, or clammy, or vivid, and covered with fur, there is usually disorder of the digestive organs, permanent or temporary, from some indiscretion in food, and especially in the use of stimulants. In feverish conditions of the system generally, the tongue is liable to become dry. The appearance of the fur on the tongue varies greatly: it may be thick and dirty-white, as it is in stomach and febrile disorders, and especially in sore-throat; it may be a thin, creamy-looking white, as in inflammatory disease within the abdomen; or it may be yellow, as in biliary disorder. It may be patchy, as in scarlet fever; or, the centre and sides of the tongue being preternaturally red, as in some forms of intestinal (bowel) irritation, may contrast with the white fur in other parts. Further, the tongue may be morbidly clean and red also in intestinal irritation, and in haemorrhage; in the former case, perhaps feeling sore, as if scalded. Again, partaking of the general debilitated condition of the system, the tongue may be pale, when it is also usually broad and fat, indicating general want of tone and power in the muscular fibres. The motions of the tongue, moreover, when it is protruded, give a clue to the state of the nervous system especially; thus in paralysis it is drawn to one side; in delirium tremens and nervous affections, it is tremulous; in the low stages of fever, it cannot be protruded

at all. Persons, therefore, who hear physicians ask a patien
to put out the tongue, should bear in mind that it means some
thing more than "putting on professional airs."

Symptoms of Diseases.—As it is of the very greatest
importance to ascertain at the beginning the nature of every
disease, we give here a few of the leading symptoms of dis
eases, as they ordinarily make their appearance. In observ
ing and forming deductions from symptoms, the first questions
ought to be—do they indicate an acute attack? have they
supervened suddenly? and if so, to what can the attack be
traced? Has there been exposure to cold and wet, or to
checked perspiration?—those fruitful sources of inflammatory
and rheumatic affections. Has there been exposure to conta
gion in any form, or to malaria of any kind; or is there any
prevailing epidemic? Can any violence, at no very distant
date, account for the attack? Careful consideration of the
" history " of the affection will often throw much light upon
its nature. Again, if the usual symptoms of fever indicate
inflammatory affection, it is to be considered whether pain or
uneasiness in any part, or disordered function of any organ,
indicate that the disease has localized itself. If inflammatory
symptoms are absent, the spasmodic character of pain, or the
nervous character of the disorder, become questions for con
sideration. Should the symptoms of ailment be chronic, the
same consideration of the history and of the hereditary ten
dencies ought to be entered into, and attention particularly
directed to the fact of there having been progressive loss of
flesh, habitual complaint of cold, unusual lassitude, alteration
in the complexion, difference in sleeping, etc.

By systematizing inquiries and observations, a much clearer
idea will be gained of the state of an individual who is an
object of care and solicitude, than by making them at ran
dom. Thus beginning at the Head, attention should be direct
ed to any unusual sensations complained of by the person, or
any unusual manifestations apparent to others. These are:
pain, giddiness, affection of the senses, confusion of thought,
or impairment of mental power; flushings, twitchings, draw
ing of the features to one side; disturbed sleep; moaning
grating of the teeth; sleeplessness, or too great sleepiness
Passing downward to the Organs of Respiration (breathing)
alterations in the character of the voice; in the respiration
as to the frequency or otherwise; in the power of lying in
any or every posture, are all matters for observation; also
any habitual cough, and its character When the Digestive

Organs are disordered, the period of their chief disorder, as connected with taking food, is an important symptom: whether the uneasiness comes on quickly after a meal, or not for some hours; whether it is worse after long fasting, or the reverse; whether there is habitual vomiting, etc. With respect to the Bowels, the nature of the motions or stools is to be inquired into, and especially the fact of thorough daily relief. In inquiry into the state of the urinary organs, the amount of the secretion, its nature as to color, or its tendency to deposit sediments immediately after being passed, or when it becomes cool, are principal objects. If the calls are too frequent, it is to be noticed whether this depends on increased quantity or on diminution, which causes irritation from greater concentration. In this way, by carefully and *systematically* considering a case, even an unprofessional person may acquire very considerable knowledge of its leading features, sufficient probably to enable him to refer to those articles in this work from which he will derive proper information; in many cases, sufficient to open the eyes to a condition of health that calls for the prompt submission to proper medical advice. When this is determined on, the observation of symptoms, either in his own case or that of another, such as a child, will enable any individual to furnish a physician, even at a first interview, with such a history as will afford him much assistance in forming his opinion. That makes a work of this kind of great value to every family, in fact to every individual.

Salivation.—It is to be hoped that the day for " dosing with mercury " for the many ills that flesh is heir to, has past. The first symptoms of the constitution being affected by mercury, or of approaching salivation, is a sense of fulness and tenderness of the gums; the teeth feel as it were elongated, and the person cannot bite any firm substance, such as crust, as well as usual; coincident with these symptoms, the breath acquires a peculiar fetor (bad odor), which, once smelled, cannot be forgotten, and the gums, if examined, are seen to be slightly swollen, and of rather a purple hue. Sometimes the face swells to an enormous size, and presents hideous appearance. There are some constitutions so susceptible of the action of the medicine, that the smallest dose cannot be taken without its producing free, or even violent salivation. Unfortunately, but little can be done to cut short, or even alleviate greatly, a course of mercurial salivation; cold, of course, is to be avoided; the alum wash for the mouth, or tincture of myrrh, or camphorated spirit in water used to rinse the mouth, afford some relief. A lotion made with two teaspoonfulls of ether, or spirits nitre, to eight ounces of water, is also serviceable, and diminishes the fetor ; a so-lution of common salt, in the same proportions, will have the same effect. A few leeches may be applied under the jaw. Seidlitz powders, or Epsom salts, largely diluted, may also be given with advantage, if the patient can swallow them, and is not in a very reduced state. A tablespoonful of a mixture of equal proportions of flowers of sulphur and cream of tartar, given once a day, on a fasting stomach, is often beneficial. Also a wash of white-oak bark, (tea,) cold, used freely every two hours, is beneficial. Cold green tea, also, and strong coffee, cold, are good remedies.

Influenza.—This disease consists of a peculiar feverish attack, accompanied with catarrhal affection of the air-tubes of the lungs, and great prostration of strength. It is not un-common to call various forms of cold and catarrh, influenza ; but the true influenza is a very distinct disease, and sel-dom occurs but as an epidemic, attacking large numbers at once. The symptoms of influenza are those of general fever ; coming on suddenly, there is shivering, loss of appetite, per-haps vomiting, heat, and thirst, with cough, headache, and generally great depression and languor. The feverish symp-toms may last from one day to ten, but their general duration is from three to five, or even seven days, the cough usually remaining a variable time after the acute symptoms are gone, according to exposure and circumstances, such as a predispo-sition to cough, etc.

To the strong and healthy, influenza is but a trifling dis-ease. It certainly prostrates even them for a few days, and leaves them weak ; but it is in almost all cases perfectly de-void of danger—*with ordinary care*—and requires little or no medicine. A few days in bed, according to the severity of the case, with low diet, a gentle purgative and diluent, drinks such as flaxseed tea, lemonade, with gingers, tea, etc., and the feet in hot water, being all that is required. If the catarrhal symptoms are severe, treatment similar to what is recommended for catarrh or cold may be had re-course to.

To the weakly and the aged, influenza is, on the other hand, a comparatively fatal disease ; and, from the almost univer-sal nature of its attack, carries off more, perhaps, of these classes than many more apparently severe and more dreaded disorders. The attack of influenza, in the description of per sons above mentioned, should be the signal for medical at tendance. Weakening medicines, especially, must not be resorted to ; confinement to bed, and the use of warm teas, will be required ; broth, strong or weak, must be allowed, ac cording to circumstances ; if the strength is deficient, wine may be requisite. and stimulant expectorant medicines, espe cially in the aged, if the expectoration is abundant, viscid, and difficult to be got up. In such cases, the following will be found useful :—Take of carbonate of ammonia, thirty to forty grains ; tincture of squill, one drachm ; wine of ipecac, forty drops ; water or camphor julep, sufficient to make an eight ounce mixture, of which two tablespoonfuls, or one-eighth, may be given every few hours. If the cough is very irritating and troublesome, two drachms of paregoric may be added to the above, but the opium rather tends to check the free expectoration. which is so desirable. Demulcent drinks, such as barley-water, etc., should not be neglected, and a mus tard plaster or blister to the chest will do good. In severe forms of the disease, with difficult breathing, if the strength is much reduced and the appetite bad, two doses of decoction of Peruvian bark may be given during the day.

Persons who greatly suffer from delicate chests, should be ware of allowing the effects of influenza to hang about them ; as the debility and cough are very apt, if predisposition exists, to lay the foundation of consumption. The strong and healthy may trust to the domestic management of influenza ; but the weak and aged ought to have proper medical advice, if it is within reach. Many persons of frail constitution, who might have lived for years with proper care. have fallen vic tims to effects of influenza, colds, etc.

Change of Climate, or Acclimation.—It is a fact worthy of remark, that the air and its temperature are largely concerned in the process of acclimation ; the former is so much more rarified in hot than in cold climates, that in the vital process of respiration, a comparatively much smaller quantity is habitually consumed ; less oxygen is taken in, and the process of oxidation or combustion, which is continually going on within the body, is slower ; we reasonably conclude that by this process of combustion, the animal heat, in part at least, is maintained ; but of course, in a hot climate, a less active condition is sufficient to keep up the average tempera ture. The process of oxidation or combustion effected on the one hand by the oxygen inspired, is supported on the other by some of the elements—carbon and hydrogen—of the food. It is evident, therefore, that if an individual who has become resident in a hot climate, makes a practice of consuming as much nutriment as he used to do, without injury to health, in a cold one, he must take more than is requisite; consequently the blood becomes overcharged with a quantity of noxious matter, which the rarified air and inactive habits of warm countries do not tend to remove; and if the course be continued, an attack of illness, probably of a biliary nature, is the consequence. Even in temperate climates, the difference between the consumption of oxygen, in winter and in summer, is considerable. How great must be the difference to those who permanently settle in tropical heats! certainly sufficient to require much alteration in habits of living. The abundant animal diet, the fats and alcoholic drinks of the colder climes, all of which contain carbon and hydrogen in abundance, and assist materially in sustaining temperature, must give place to the fruits, vegetables, etc., of warmer regions; *vice versa*, on going from a warm or temperate country to a colder—as the experience of all arctic travelers testifies—a larger proportion of animal diet, and that of a more fat or oily character, is requisite to maintain health and strength, and those only who are capable of consuming and digesting this full allowance, are fit for encountering the cold of the north. From what has been said, it is evident how important due regulation of the food is to safe and speedy acclimation; it is the main element, and the most under man's control. Modern science and discovery will render him much assistance, but study of the natural products of the soil and of native habits is essential.

The great increase of the functions of the skin which takes place on removal to a warm climate, requires attention. It renders the constitution more susceptible to the influences of a damp or chill air, such as frequently occurs in evening. The best preservative is *woollen clothing* of some kind, be it ever so thin, worn next the skin. Persons who, from a warm climate, of which they are either natives, or to which they have become accustomed, come to reside in a variable or cold country, are peculiarly liable to affections of the chest or lungs, and not unfrequently become the subjects of consumption, for the want of a little timely medical advice, which a work like this is intended to give. Such persons should by all means go more warmly clad than those who have been raised or acclimated to a northern country—putting on an extra flannel under-shirt by all means.

Chilblain.—This disease, though not very frequent, is quite troublesome; it is an inflammatory affection of the skin, more particularly of the fingers or toes, caused by alternations of cold and heat, and is characterized rather by irritating and troublesome itching than by pain. Persons of fine skin, scrofulous constitution, or languid circulation, are most liable to suffer from chilblains, and old people and children more than those of middle life. The sudden exposure of the skin when very cold to a high state of temperature, is generally and justly considered to be an exciting cause of the affection; but one quite as frequent is keeping the surface in a state of artificial warmth, by the use of sleeping-socks and hot applications in bed, or of fur-lined shoes and foot-warmers in the day time. All these applications keep the skin in a continual state of unnatural perspiration, weaken its tone, and so render it more susceptible of the effects of cold when exposed to it. To prevent chilblains, in the predisposed, the feet ought to be regularly bathed with cold, or (in the case of the aged) tepid water, or salt water, every morning, and afterward well rubbed with a rough towel, exercise being employed to preserve the warmth of the extremities rather than artificial heat. When chilblains have formed, and the skin is unbroken, stimulating applications are requisite; many different ones are used—spirit, such as brandy, camphorated spirit, paregoric, or turpentine, will any of them be of service, applied by means of a piece of linen, or gently rubbed on. When the skin of a chilblain breaks, an ulcer is the consequence, which discharges a thin slimy fluid, and is often difficult to heal. In this case, the inflammation should be subdued, in the first place by means of a poultice, and afterward an ointment used, made with ten grains of sugar of lead, to the ounce of lard. Of course all friction or pressure from boots or shoes must be guarded against. Frequently anointing the parts with sweet oil or lard may be beneficial.

Nervousness, or Nervous Diseases.—The term " Nervousness " is a kind of undefined expression, after a manner of speaking; and yet, call it what we may, the disease is very prevalent. Females are much more liable to nervous disorder than males, independent of hysterical affection, which constitutes one of the most marked phases of the malady, and many of the remarks on which apply to the present subject.

In nervous disorders, there is usually great susceptibility to external influences, and at the same time mental emotions,

whether of joy or grief, fancied or real, exert much influence
over the body and its functions. The heart palpitates, the
hand trembles, the face flushes under the most trivial excitement. Much of this is undoubtedly due to constitutional
timidity; but it is also notably increased in debilitated states
of the constitution, and those who have never been what is
called "nervous," are apt to become so in some particular
condition of impaired health. The affection is, indeed, very
nearly akin to hypocondriasis (lowness of spirits); it is essentially a disorder of *weakness*, and is relieved by whatever increases, temporarily or permanently, the power of the nervous system. The temporary relief to nervous sensations
which is afforded by alcoholic stimuli, is very apt to lead
those who suffer from them to put too much trust in, and to
resort too habitually to the use of those palliatives—a practice which must be followed by pernicious consequences;
sometimes, too, opiates are habitually made use of, and are no
less injurious.

Undoubtedly, when properly employed, alcoholic stimuli,
and even opium, are valuable in the treatment of nervous disease, but they must never be substituted for more permanent
means of invigoration, particularly *regular and sufficient exercise in the open air*, on foot or horseback, good nourishing
diet, with a sufficient amount of animal food, and attention
to the bowels and the state of the skin. The producing cause,
whether excessive mental exertion, sedentary employment,
late hours, or excess of any kind, must of course be modified
as much as possible. The shower bath is often recommended,
and often useful in these affections, but some persons cannot
bear the shock; when this is the case, the cold or tepid bath
down the back does much good, particularly if there be any
tenderness of the spine on pressure, a fact which should always be investigated in those who suffer much from nervous
disease; it very commonly exists and is overlooked. When
the tenderness is at all marked, it will require special treatment by counter irritation, by the use of liniments, friction,
etc. In addition to the regulation of the bowels, by proper
purgatives, or by injections, quinine (one grain three times a
day), and the preparations of iron, are the most generally
useful remedies; tincture of valerian may be used as a palliative during an aggravated attack, but should be sparingly
resorted to; or tincture of valerian and hyoscyamus used together in equal parts (one teaspoonful three times a day) may
be tried.

Fullness of Blood, or Plethora.—It is not unusual to see what are called robust, strong, full-blooded persons, that look to be the picture of health, and yet they are anything but well; being always exposed to danger from the too great fullness of the blood vessels, and richness of blood. Individuals of the sanguine temperament, while leading a life of mental activity and anxiety, have greater powers of activity than most others, but they, in many instances, border upon plethora, and if they become so placed that their former activity is either uncalled for or interfered with, provided there is not much mental anxiety, they quickly become plethoric; the vessels are overloaded with rich blood, and instead of the former power of exertion, oppressive languor and inactivity succeed; in fact, the whole of the functions, and the nervous system especially, are weighed down and clogged—there is mental sluggishness, heavy sleep, and inaptitude for exertion.   This last symptom is too often mistaken for weakness; the person laboring under the mistake resorts to additional food and stimulants—it need scarcely be added, only to increase the evil.   An individual in this condition, it may be said, is ripe for inflammation ; if cold be taken, it is very likely to light up inflammatory action somewhere, and once lighted up, the action is very liable to be of the severest kind.

Should febrile disease of any kind—as for instance, small-pox, or erysipelas, or rheumatic fever, be excited in the constitution, the symptoms run high, and the case is very likely to become one of danger.   For similar reasons, accidents are not well borne : at least their after effects are often such as to put life in danger.

Persons who are in a plethoric condition, not unfrequently get relieved by some natural effort; piles show themselves and bleed, or the nose bleeds, or spontaneous diarrhœa comes on, and instead of the individual being weakened, he feels stronger than before.   The evil results of mistaking a state of oppression from plethora, of *false* debility for one of weakness, must be evident to all.   Even the pulse is liable to deceive, and in these states of oppression to seem low and weak, but it is essentially different from the pulse of debility.   The latter, if the finger is pressed even slightly upon it, is extinguished at once; but the pulse of oppression seems rather to resist the pressure, to become stronger, and to beat up against the finger, rather than to give way.   As might be expected, plethoric individuals are often the subjects of apoplexy.   A state of plethora must always be one, if not of danger, at least of hazard, and ought to be guarded against.

If a person suffering from plethora is threatened with an immediate attack, such as apoplexy, the condition cannot be too soon or too actively removed. Bleeding in some way, free purging, and low diet, are the immediate remedies; but in the absence of any threatened attack, it is not advisable to invoke the aid of these powerful agents; the system should be reduced *gradually* and *steadily* by the formation of, and perseverance in, modes of living suited to counteract the tendency. When a man suffering from the effect of plethora gets rid of his unpleasant symptoms by a "coup," such as the loss of a basinful of blood, by a few calomel pills and black draughts, he is probably highly pleased to be so easily rid of his enemy, and by means which involve no self-restraint or giving up of indulgences; so, trusting to the repetition of the same remedial measures, he puts no check upon himself, and when the plethora again reaches a certain height, he again bleeds and purges, and this goes on until he is overtaken some day with an apoplectic attack, or until he becomes the subject of organic disease.

Persons who have a tendency to plethora, must have exercise—they must use up their blood and muscle in active motion; but in doing this, especially at first, they must beware of *over-doing it.* It will not do for a plethoric man to commence a new system of *living for health,* with violent exertion—otherwise he might precipitate the very evil he dreads. Plethora, to be reduced, must be so *steadily,* but *gradually;* active exercise, increased as the ability to take it increases, must be balanced with food proportioned to the amount taken, and animal food in very moderate proportion used. Early hours, and curtailment of the time devoted to sleep, is desirable. In most cases, tepid bathing is preferable to either hot or cold, and, either by it or by sponging, the skin must be kept active. The bowels require especial attention, and are better rather lax than otherwise; any slight tendency to plethoric oppression being counteracted by acting upon them by proper purgatives, such as Epsom salts, or by seidlitz powders. If the kidneys are inactive, spirits nitre, twenty to forty drops three times a day, in a wineglass of water, or a teaspoonful of cream tartar in same quantity of water, about three or four times a day, will be beneficial.

Old Age.—"The youngest of us will be old some day, if we live long enough," sounds a little on the Hibernian order, but the idea intended to be conveyed is not a bad one, and we should not forget the aged and the infirm. With old age increases the liability to such hereditary diseases as gout, gravel, rheumatism, apoplexy and paralysis, and in women especially, to cancer. Now the effects of excesses and dissipation in early life, which may have been unfelt during the vigor of manhood, too often add to the natural infirmities. Whatever may have been the previous modes of living, it is always a dangerous experiment to make material or sudden change in them; after age has begun to tell upon the constitution, it should not be done, but for important reasons, and under medical advice. The weakened digestion of advanced life should be considered in the food, which, while it is nutritious, ought at the same time to be lightly cooked, and everything like hardening avoided. Where the teeth are deficient, meat should be well divided, either by mincing before cooking, or by the knife after. The meals should be light, not at too long intervals. If the dinner be *early*, as it ought to be for the aged, who are not obliged to hurry off to business, supper, though a light one, should always be taken. The skin of old people is often most shamefully and disgustingly neglected, and no point in their management is more closely connected with their comfort and health; it should frequently be sponged with tepid water, and well rubbed afterward with a rough towel, to promote reaction. It ought at the same time to be carefully protected by woollen clothing: old people are most injuriously susceptible of the changes of external temperature, particularly cold. Exercise by the old should be continued as long as they are able to take it, but never extended to fatigue. Sleeplessness, so frequently and so loudly complained of by aged people, is, in some respects, natural; as life advances, nature would seem to require less of the soft restorer. It is not well to endeavor to overcome it by narcotic medicines. If possible, the time of sleep should, by habit, be kept to the early hours of the night; and, in summer especially, the tedium of the early morning may be relieved by reading, knitting, sewing, or some other light employment. In advanced life, the urinary organs require the greatest care; the call to relieve them should never on any account be delayed; on the slightest symptoms of derangement, proper medical advice ought to be taken at once; it may prevent evils which too often render the latter year miserable. It is most important for old people to give themselves time to *empty* the *bladder thoroughly;* they do this with more difficulty than the young. The medicines prescribed for the aged should be, whenever it is possible, of a warm character, to counteract the tendency to flatulent distension; large doses of mercurials, neutral salts, and strong purgatives, are all to be avoided. Pills, especially if at all hard, are apt to pass through the bowels *unchanged*. When a mild purgative is required by an old person, none is more suitable than a moderate dose of infusion of senna, to which a little ginger. or a teaspoonful of Peruvian bark, or of gen tian root, is added. Six to eight drachms of the compound decoction of aloes answers well, if there is no great tendency to piles. When the bowels are habitually constipated, injection of a pint to a pint and a half of warm soap-water, must be given occasionally as required; this counteracts the great tendency to accumulation in the bowels. The doses of medi cine ought always to be *diminished* after the period of incipient old age, as old persons do not bear the effects of medicines so well.

Flatulence.—This is commonly the result of indigestion, but it is often also the effect of nervous disorder. In the former case, it is probably chiefly due to the extrication of gas from the badly-digested food-mass in a state of partial fermentation. In the latter, it is only possible to account for the enormous quantities of "wind" which are discharged by its formation (secretion) in the bowels. Toward the termination of fever and other acute diseases, flatulent distention of the bowels, or "tympanitis," as it is called, is always an unfavorable symptom.

Persons who suffer from flatulence require sedulously to avoid most kinds of vegetable food and fruits. Individual experience, however, is the best guide on this head. When a severe attack of flatulence comes on, carminatives, such as chewing a piece of calamus or ginger root and swallowing the juice, a few drops of peppermint in water, or eating some peppermint drops, are generally resorted to, and often prove useful. A drink of water as warm as it can be taken, is often a very good and simple remedy. But in many cases, particularly in nervous individuals, with pale tongues, the mineral acids will often be of more service—either twenty or thirty drops of dilute nitric acid in a wineglassful of infusion of orange peel, or some other warm bitter; or, better still, aromatic sulphuric acid, in ten drop doses, in a wineglassful of water. As preventatives, keeping the bowels well open, and the use of the flesh-brush over the stomach and bowels twice a day, and plenty of out-door exercise, are the best. Persons subject to this complaint should be careful to eat *slowly* and *chew their food well.*

Noises in the Ears.—These are often compared to the sounds of "bells ringing," "simmering of water in a kettle," "singing of insects," "roaring of the sea," &c. They are often extremly troublesome, and may arise from many and different causes. Mere temporary derangement of the digestive organs will in some persons produce them. They are often indicative of determination of blood to the head, and when accompanied by symptoms of this tendency, ought not to be neglected. Partial obstruction of the Eustachian tube by cold, or accumulation of wax in the external ear-passage, are apt to occasion these noises, and they are accompanied with some degree of deafness. Of course the remedy must vary with the cause. If the digestive organs are deranged, they must be regulated; if cold be the cause, the symptoms may be left to pass away with the temporary ailment. In some cases of chronic or continued noise in the ears, regularly bathing the head with cold water every morning will sometimes remove it, when other remedies fail.

Weakness of the Bladder in Old Persons.—An inability to perfectly retain the urine, and a weakness of the bladder, is a frequent disorder of aged persons. It often commences with and is accompanied by imperfect emptying of the organs, either through carelessness or weakness. Sponging the lower parts of the abdomen, &c., &c., with vinegar and water, or salt water, may be of service. Some physicians recommend the use of the tincture of the ergot of rye in these cases, a teaspoonful three times a day, but as a general rule they should be placed under regular medical attendance. The same may be said of that very troublesome complaint of old age, catarrh of the bladder, in which large quantities of thick mucus are discharged. A tea made of *Uva Ursi*, and *Buchu* leaves (in equal parts), taken in the quantity of a small wineglassful three times a day, with the addition of twenty drops of spirits of nitre, is beneficial in diseases of this kind.

Bile.—This is secreted by the liver, is of a brownish-yellow color, and has a very bitter taste. Its composition is complex, and it undoubtedly fills more than one important function in the body. Bile is separated by the liver from dark blood, which, passing through that gland, on its way to the heart, from the abdominal organs, is thus purified of noxious matters, containing a large amount of carbon, before re-entering the general circulation. The separated bile is discharged into the duodenum (top of the bowels), and mixing with the digested food, appears to assist in fitting certain of the constituents for absorbing into, and assimilation or transformation in the body. A large proportion of the constituents of bile are along with the food re-absorbed into the system, and are probably intended and adapted to support the processes of respiratory combustion in the lungs. It is chiefly the coloring matter of the bile which is discharged from the bowels in health, which gives color to the discharges from the bowels.

Biliousness. or Biliary Derangement.—When persons feel unwell and do not know exactly what is the matter, it generally turns out they have made up their minds that they are "bilious"! It is true that biliary disorders are very common. In fact biliary derangement is so frequent an ailment in civilized life—its history is so intimately connected with the general principles of health, and the prevention, or at least alleviation, of the disorder is so much under individual control, that it has special claims upon our attention.

It has been shown, under the subject of Bile, that in ordinary health there must be a certain balance maintained between the secretion and ultimate destination of the bile, the assimilation of food, and the functions of respiration; that in the excreted bile the blood is freed from certain principles containing a large amount of carbon, which could not be retained in it without injury to health. That further, the bile having been separated from the blood by the liver, and thrown out into the general tract of the bowels, performs an important part in the function of assimilation (digestion); and that, lastly, a considerable proportion of the bile—without the coloring matter—is *reabsorbed* into the system, with the nutriment, in such a state as to fit it, or rather its carbon, for union with the oxygen which enters by the lungs, so that while heat is generated, the carbon, by taking the form of carbonic acid, is fitted for excretion by the lungs or skin. Upon these facts hinge the causes of one at least of the most prevalent biliary disorders, that which depends upon the introduction into the system of a proportion of carbon aliment too great to be removed by the oxygen obtainable through the lungs, and which has its ordinary termination in the attacks which are termed "bilious attacks," "sick headaches," "bowel complaints," "cholera morbus," &c., according to the manner in which the patient is affected.

The second form of biliary disorder depends upon torpidity or inactivity of the liver itself. The third form is the reverse of the first; the gland itself may be sufficiently active, but the blood does not afford sufficient material to work upon, and bile is deficient. This is most frequent in children.

The first form of biliary disorder, that dependent upon the accumulation of carbon, or of the elements of bile in the blood, must evidently be owing to one of the following causes, or a combination of them; either too much food, espe￭ially of a highly carbonized character, such as fats, oils, sugar, &c., is habitually consumed, or the habits are too physically inactive to keep the functions of respiration, animal heat, and circulation, in healthy action, or the external atmosphere is so temporarily or permanently rarefied by heat that the individual cannot obtain the full supply of oxygen in respiration; lastly, the excretory function of the skin (sweating, &c.), may be impeded. Now, though it is unquestionable that some persons have a much greater ten￭dency to biliary disorder than others, it is also unquestiona￭ble that all have it in their power in a great degree, if not entirely, to control or obviate that tendency, by attention to, and practical application of, the above principles. In those who suffer habitually from sick headaches—which depend generally upon the presence of bile in the stomach—and from other forms of biliary disorder, there is generally traceable great error in diet. Fats, as found in ham and bacon gene￭rally, melted butter, pastry, meat, malt liquors or wine, and other highly carbonized articles of diet, are taken too freely, or at least are too regularly indulged in, while at the same time very little active exercise is taken; the blood becomes overloaded with carbon; languor, sleepiness, headaches, giddi￭ness, loss of appetite, furred tongue, depression of spirits are the consequences, and continue until at last the system is relieved, wholly or partially, by an excessive secretion of vitiated bile, which passes off either by vomiting or purging. That deficient exercise has much to do with such a state of system, is evident from the much greater prevalence of such attacks among females, who take little exercise, than among men; and indeed they would be still more prevalent among the former, were it not for the monthly sickness. Habitual neglect of the skin, also, by impeding the excretion of car￭bonic acid from its extensive surface, undoubtedly assists the evil.

From what has now been said it is evident how much the avoidance of biliary disorder is under individual control the question is in reality not one of medicine, but of *diet* and **regimen**; medicine certainly may be required, but not by any

means to the extent it is often used. Those who are habitu
ally liable to biliary disorder ought most strictly to regulate
the diet; fats of all kinds (except, in some cases, bacon),
must be avoided; butter either entirely avoided, or used in
very small proportion, and never when *melted;* animal food
may be taken in moderation, but should never be consumed
at night; much sugar, strong tea or coffee, malt liquor and
wines, are all bad. In addition to plain meat, bread, well
boiled vegetables, farinaceous preparations, such as rice,
potatoes, &c., and fruits, ripe or cooked, are the best articles
of diet. Exercise regularly in the open air *must* be taken,
and the skin kept clear and in an active state, by frequent
bathing, friction, &c. If the bowels are confined, a pint of
warm water, used as an injection, will be most suitable. It
is much better not to trust to medicine. When, from any
cause, the languor, sleepiness, furred tongue, &c., give notice
of an impending billous attack, eat sparingly of the mildest
liet and take a purgative of some kind. Having thus cleared
the system, it is better to trust to diet and regimen than to a
repetition of the dose as a corrective of indulgence. There is
reason to believe that an emetic of ipecac or warm water
taken on the first approach of the disease, would break it up
suddenly in most cases.

Sciatica.—This is a kind of *Neuralgia,* or *Nervous
Rheumatism,* which affects the Sciotic nerve, the great nerve
of the lower extremities. This nerve, the largest in the body,
passes down the back of the thigh to the ham, a little above
which it divides into two main branches. The nerve sometimes becomes the seat of severe neuralgic pain, felt down its
entire course, or perhaps in the hip only, or sometimes in the
foot and ankle; the pain comes on in paroxysms, and is generally increased by exercise; in some cases pressure upon the
course of the nerve causes pain.

Sciatica is often attended with so much suffering, that it
affects the general health to a considerable degree; moreover,
it is frequently most difficult to get rid of. Leeches and
cupping, in the first instance, down the course of the nerve,
especially in phletoric, robust subjects, followed by blisters,
are useful; or heat and moisture may be used with advantage in the form of a bran poultice, followed twice or three
times a day by the application of a liniment. Kerosene oil
frequently applied is an excellent application, to be followed
by laudanum and turpentine, well rubbed in, and then
covered with flannel.

The bowels being cleared by a purgative, if there is no ten dency to fever, drachm doses of carbonate of iron, given three times in the twenty-four hours, often cures quickly; or turpentine, in doses of fifteen drops, given in milk, three times a day, may be tried; or quinine, in two-grain doses, every eight hours. There is considerable uncertainty in the effect of remedies in Sciatica, even in skilful hands In all cases of Sciatica, perfect rest of the limb is essential. Sulphur baths are also recommended.

Gonorrhœa, As this book has been written for all classes, a short statement giving the symptoms, causes, treatment, &c., of the various forms of sexual, or (as they are usually called,) venereal diseases, is necessary. In the first place, it should be understood, that Gonorrhœa, or Clap, and the Venereal disease, are two entirely distinct diseases. A person, for instance, having Gonorrhœa cannot give another person the Venereal disease, nor can one having the Venereal disease give another the Gonorrhœa, or Clap. Neither can one disease change into the other; but both diseases may exist in the same person at the same time, and both may be communicated at the same time, or only one, as may happen. The important point to consider is that they are two distinct diseases. This is very important, because Gonorrhœa is simply a local disease, while the Venereal disease, or Pox, is a constitutional disease, which poisons the blood. This is an infectious disease; it is seated in the urinary passage, from the orifice of which there issues a discharge of matter, attended, more or less, with pain and heat in making water. There is no outward sore or ulcer in Gonorrhœa; neither is there any within the passage. If it were possible to take an inside view of the urinary passage, nothing would be seen there but redness and a little swelling, and the discharge, whether much or little, would be seen issuing from the red and swollen state of the membrane which lines it, just as matter is sometimes seen issuing from an inflamed eye. Venereal disease, on the contrary, is not seated in the urinary passage, but externally, on some part or other of the genital organs, and consists in one or more little ulcers or sores, which do not heal like others, but continue unhealed, and, at length, assume a peculiar character. Gonorrhœa consists in *a discharge from the urinary passage*, attended, more or less, with pain and heat in making water, without ulceration or sores; whereas Venereal disease consists in external ulceration or sores, without discharge from the urinary passage, and without pain or heat in making water. No one, it is presumed, after this can possibly mistake these two diseases.

A person having been exposed to this infectious malady (Gonorrhœa), at some period within seven days, if he should have taken the disorder (which we will suppose to be the first attack), will feel a little pain and heat in making water. This leads him to make an examination, when he will find the orifice of the urinary passage looks a little redder than usual, its lips rather pouting or swollen, and on squeezing it, an uncommon dampness or thin mucous will be seen. In the course of a few hours, or by the next day, the pain and heat in making water will have increased, the discharge also increased and of a purulent or white creamy appearance. For some days these symptoms go on increasing in severity, the matter becoming of a yellowish color. Sometimes the pain and heat in making water will be exceedingly great, amounting to a sensation of absolute scalding; at this time the discharge generally turns to a greenish color, and instead of being about as thick as cream, is thin and watery. Under these circumstances, there is also a painful sensation along the whole length of the urinary passage. Gonorrhœa always begins with pain and inflammation, more or less, which after a time abates very much or entirely subsides. This is called the inflammatory stage of the disease.

The degree of pain and inflammation is not the same in all cases; the oftener a person has had the disease, as a general rule, the less will be the inflammation, so that persons who have had the disease several times may feel scarcely any pain, the only notice of it being the soiling of the clothes with matter. After a time this inflammation subsides, whether remedies have been used or not.

In the commencement, the seat of gonorrhœa is just within the orifice of the urinary passage; but the disease soon progresses, until it reaches the neck of the bladder. How long a time it may occupy in its passage through the urinary canal, is uncertain, and differs in different individuals, and under different circumstances of constitution, mode of life, etc. The painful inflammatory symptoms, however, seem principally to attend the invasion of the disease, and most of all, its effect upon the first portion of the passage; when these are subsided, we may consider it has advanced to its ordinary distance.

After a time, or if there has been much pain, after this has subsided, another stage of the disease commences. This stage consists simply of an infectious discharge issuing from the orifice, with little or no pain. This discharge is more or less in quantity in different persons, and is much increased by whatever may excite, or by irregularities of any kind. This, specially when the discharge is slight, and quite unattended with pain, is what is properly called " Gleet," and when once suffered to commence, there is no telling how long it may continue, or how it is to terminate ; but let it ever be remembered that, as long as this discharge continues, the disease may be communicated by infection. This, then, is the *Second* or *Chronic Stage* of Gonorrhœa. Sometimes there is great swelling of the foreskin and prepuce, which is not free from danger, as mortification sometimes takes place, on account of a kind of ligature being formed of the skin around the parts, called Phymosis when the skin cannot be brought back, and Pariphymosis when it cannot be brought forward. Swelling of the glands in the groin, which often takes place, is the result of inflammation.

In every part of the body, there are certain little vessels, called absorbents; and these, in their course, run through absorbent glands, of which there are several in each groin. Now it happens that inflammation in the urinary passage is communicated through the medium of these vessels to the glands in the groin, which become painful and enlarged in consequence. It is, however, worthy of remark, that in gonorrhœa, glandular swellings seldom go on to form an actual bubo (which is an abscess in the groin), as they are apt to do in Venereal Disease, so that they are comparatively trifling ; after a little while, they generally subside and disappear. Swelling of the testicle sometimes takes place, and generally the pain is great, and the necessity of rest and lying in bed is urgent. Much, however, may be done to prevent it, if early attended to, and therefore the symptoms which precede and lead to it should be noticed carefully.

The first thing which precedes a swelling of the testicle, is a sense of pain running down the whole length of the urinary passage, accompanied with a feeling as if a drop or two of urine were remaining in the passage after making water. To this there succeeds a dull pain in the groin, going on to affec the testicle, which presently feels heavy, and very tender to the touch ; or, the pain and tenderness of the testicle may set in at once, without any of these premonitory symptoms. No time should be lost. It should be attacked instantly, in order, if possible, to arrest the inflammation before fully developed.

Inflammation of the Bladder is sometimes caused by gonorrhœa. It may be slight, causing merely the inconvenience
of too frequent urinating, with difficulty of restraining it,
or, the bladder may empty itself every five or ten minutes,
passing only an ounce or two of highly-colored urine, with
great pain, and followed by blood, the desire to urinate
remaining unrelieved. It is then attended with fever. This
affection oftener appears in a mild or sub-acute form, but may
set in with great severity, and is then a most painful and harassing addition to the gonorrhœa.

The question is sometimes asked, "Does gonorrhœa ever
wear itself out?" It may possibly do so; but it is running
a most terrible risk, for it may, in the meantime, produce
gleet, and stricture of the urethra.

In regard to the treatment of gonorrhœa, I would by *all
means* recommend those who have been so unfortunate as to
contract this disease, to apply at once to the best physician
they can find; but where that can not be done, the following
course of treatment is recommended:—As soon as it is known
by the symptoms that the disease has been contracted, the
patient should live on low diet, eating no *meat* for several
days; a dose of Epsom salts, or Rochelle salts should be
taken every second morning, on an empty stomach. The
parts should be frequently and thoroughly bathed in cold
water several times a day; and one heaping teaspoonful of
powdered cubebs should be taken in a glass of cold water
half an hour before each meal; or a tablespoonful one hour
before meals, should be taken of the following mixture:
Balsam copaiba, spirits nitre, spirits of turpentine, of each
three ounces; essence of peppermint, one tablespoonful. The
powdered cubebs should always be tried *first*, when they can
be had; if not better, after a few days, use the mixture given
above. It is well to change the medicines every few days,
taking first the cubebs, then the mixture, and being careful
by all means to keep the parts scrupulously clean, by bathing
and the free use of soap and water. In nine cases out of ten,
the above course of treatment will be found successful.
Swelled testicle is to be treated by *rest, absolute rest*, free
purging with salts once a day, and the application of poultices, cooling washes, lotions, and if need be, leeches; the
parts should be supported by a suspensory bandage until
cured.

The symptoms of gonorrhœa, in females, are a burning or

smarting pain of the privates, which is rendered much worse by urinating; there is also a kind of numbness or heaviness of the adjacent parts, down the thighs, across the lower part of the abdomen. pain in the back, etc. These symptoms pass off in a few days, and there succeeds a discharge of matter similar to that witnessed in males, and which may gradually run into Chronic Lencorrhœa, or Whites.

Often the confiding and faithful wife gets this disease from a heartless husband. If this disease in females is neglected, it is very apt to run into what is called the " Whites," which is very often hard to cure under such circumstances. As females may get this disease from contact in water-closets, or using a towel that had been used by a diseased person, they should not let the disease run on, but have it cured at once.

There are other diseases which resemble gonorrhœa. The natural secretion of the parts accumulating beneath the foreskin (in males), often irritates and excoriates the inner surface of the foreskin, giving rise to uneasiness and swelling, accompanied by an offensive discharge. By reading the symptoms of gonorrhœa, the difference is easily told—the discharge coming from the urinary passage, instead of the foreskin.

Free and frequent bathing in cold water, or cold green tea, and thoroughly cleansing the parts with soap and water, and keeping them clean, is the best treatment. Symptoms resembling those observed in gonorrhœa, may be communicated by females having the " Whites," or female weakness. This has often been the cause of great unhappiness; the husband believing his wife to have given him a loathsome disease—that she has proved untrue in her marriage vows—and a separation sometimes being the result, although the wife may have been perfectly innocent. A little medical knowledge on the subject would have solved the "mystery," and saved all this unhappiness. A woman with Whites is unfit for anything of this kind while in that condition, and should be thoroughly cured as soon as possible.

The disease in the male resulting from Whites in the female, is to be treated the same as gonorrhœa, as it is a local inflammation of about the same nature.

A discharge occurs in female children sometimes, which often creates serious apprehensions in the minds of parents; it is generally the result of pin worms in the privates of the child, but sometimes is produced by inflamed gums. Of course an examination into the case will show its cause, and

naturally suggest the proper remedy, by removing the
cause.

Bubo, or swelling in the groin, is an enlargement of the
gland in that region, and is characterized by swelling, redness, and pain. As soon as it makes its appearance, apply to
a physician, if possible; when that can not be done, apply
tincture of iodine twice a day, until the swelling subsides
The treatment is the same, whether the bubo has been pro
duced by gonorrhœa or syphilis, so far as can be done by any
but the physician. If suppuration (coming to a head) takes
place, it may require poultices until it breaks and runs, or .a
opened.

Syphilis This, as we have before stated, is
an entirely distinct disease from Gonorrhœa. It commences
with minute erosive ulcers on some parts of the genital
organs. These are termed *chancres*. As the chancres gradually spread and deepen, they secrete a virus, which, being
absorbed, may occasion a constitutional taint or cachexy,
called secondary or constitutional syphilis. The first stage of
a chancre is called the vesicular. A small red point is first
noticed, which soon becomes elevated into a pimple; this is
gradually changed into a vesicle or sac, containing a turbid
humor covered by a scab, which is eventually detached,
becoming an open ulcer. The time elapsing from the first
appearance of the morbid point to the complete ulceration,
may be a few hours, or several days.

The second stage is that of *ulceration*. The sores are
generally of a roundish form, involving the whole skin and
the adjacent parts with an indurated (or hardened) base and
edges. The size of the ulcerated surface varies from that
of a pin's point to a dollar. Sometimes the centre and borders of the chancre are elevated so as to form a fungous
(proud flesh) projection, constituting what is called the
raised chancre. Chancres in close proximity often become
united into a single ulcer. This stage of ulceration may last
for several weeks or months. In some cases, it has continued
for years.

The third stage is that of *cicatrization* (or healing.) When
a chancre is about to heal, it passes into the condition of a
wound. The difference between an ulcer and a wound consists in the difference of the membrane secreting the humor
which covers the two solutions of continuity. When this
cicatrizing membrane appears, the edges of the ulcer which
are detached sink, and approach the base, to which they unite

The inequalities of the base, the indentations of the borders, disappear, and are replaced by fleshy granulations (or healing points), resembling those in suppurating (running) wounds. As the healing of the ulcer progresses, its circular form becomes changed and angular at several points of its circumference, because its edges are drawn inward, and in an unequal manner, by the membrane. The cicatrix (or scar) which succeeds to chancre, like that of every solution of continuity with loss of substance, is shrivelled and depressed; sometimes, instead of being depressed, it is, on the contrary, elevated like a honeycomb. On the mucous membrane these marks at length completely disappear.

Once the healthy process commences, its progress is rapid, for it is now no longer that of a chancre, but a wound. Cicatrization is not complete while there remains a single point of the grayish base. This point may extend, become developed, and the chancre itself be renewed; or, to speak more correctly, become enlarged, and invade the cicatrized portion, and the adjacent integuments, which were not before attacked. But when the cicatrization has once covered the whole surface, a return is no longer possible; before another chancre can appear, there must be a new inoculation—another attack of the disease.

The chancre, as described in the preceding article, is called *regular* When the sores extend irregularly, and progress rapidly over the surface, in the form of a malignant erysipelas, they have been termed *phagedenic*. The term *chancre* signifies a gnawing, corroding affection; but the disease is denominated *phagedenic* only when it is unusually erosive, irregular, and spreading.

Another form is called *gangrenous.* It implies a destruction or death of small portions of tissue beyond the boundaries of the ulceration. The phagedenic and gangrenous varieties do not depend on peculiar modification or virulence of the original affection, but on the morbid conditions and habits of the patients themselves, or upon the unfavorable circumstances in which they are placed.

Another variety has been denominated *diptheritic* or *pultaceous.* It is generally noticed in old and feeble persons. The ulcer is deeply colored, disposed to bleed, and the surrounding tissues are thickened and indurated. The ulcer is coated with a false membrane, and bleeds if this is removed. This membrane or coating is, however, immediately reproduced. In some cases, the ravages of the disorganizing process are terrible. The genital organs are rapidly destroyed.

and the adjacent parts often present a disgusting mass of cor
ruption. The indurated chancre is defined as a small, round,
callous ulcer.

In the treatment of this disease, as we mentioned when
treating of gonorrhœa, also, we would by *all means* advise
those who contract this terrible malady to seek the very best
medical advice at once. It is too dangerous a disease ever to
be treated by any but a skillful physician. However, as this
book may, and doubtless will be, the fireside companion of
many who may be so situated that they cannot have the
advice of a physician in person, we propose giving the best
advice possible, to be used under such circumstances.

The first thing to be done is to heal up the sores or chancres
by the application either of blue stone (sulphate of copper),
or nitrate of silver. A stick or piece of blue stone inserted
in a goose quill, or any similar contrivance, should be moistened with water, and applied to the sores every day until
cured. A simple piece of dry lint being applied all the time
to absorb the discharge, and which should be frequently
changed in the twenty-four hours.

Or, if you decide to use nitrate of silver (lunar caustic),
apply the solid stick, moistened in water, once every two
days. It should be applied thoroughly to every part of the
chancrous surface which exhibits the least trace of the grayish, or yellowish, or dusky-white matter. A thin film or pel-
licle instantly forms over the surface, which is cast off in a
day or two, when the caustic should be re-applied, if there is
the least appearance of the erosive matter aforesaid. After
a few applications—provided the proper hygienic measures
and regimenal directions are attended to—the ulcerated surface will present a smooth, florid, healing appearance, when
the caustic may be dispensed with. Should the ulcer, however, at any future time, take on the virulent character, the
caustic treatment is to be resumed. It sometimes happens
that a large chancrous surface will present the appearance of
healing granulations at some points, and the erosive action at
others. Here it is necessary to cauterize only the virulent
points. The sore should be thoroughly cleansed with tepid
water before the application of the caustic, and afterwards
covered with a little dry lint. The lint should be kept in constant contact with the secreting surface, so as to absorb the
matter; and very frequently changed, especially when the
ulcer is large, and the secretion of matter copious. Cerates,
plasters, astringent lotions, medicated washes, etc., are useless
or injurious. The essential conditions on which we are to

predicate the safe treatment and prompt cure of all forms and degrees of chancrous diseases, are : disorganization of the ulcerous surface, by means of the caustics referred to ; local cleanliness, and bodily purification, by means of plenty of soap and water.

Besides this local treatment, medicines must be used for the constitutional disease. Begin by giving three grains of blue pill once or twice a day, until the *gums begin to get sore.* Then leave off the use of the blue pill until all the soreness of the gums is gone ; then, say after about ten days, use the blue pill as before, until the gums begin to get sore the second time, when it should be left off altogether. In the meantime, a teaspoonful or two each of cream of tartar and flowers of sulphur should be given in a glass of water, once every one or two days. After this course of treatment, as a general thing, the patient will be cured. *Be careful* that none of the poisonous matter from sores of this kind gets on to a towel, or any place where the peculiar contagion may be propagated, as it is highly contagious, and may get into the eyes and destroy sight, in the ears, nose, etc.

After taking the medical treatment just recommended, the system may be weak ; at any rate, use half a pint of sarsaparilla tea (cold), three times a day, on an empty stomach. In this peculiarly loathsome disease, persons may desire further directions and advice, for it often assumes some aggravated form, which requires a physician's attendance. It should never be forgotten that in gonorrhœa and syphilis every possible pains should be taken to prevent others from getting these diseases, as they are contagious in the extreme. Any of the discharge getting into the eyes, ears, mouth, nose, fundament, etc., or into any part where the skin is off, or a sore, will produce the most frightful result in many cases; therefore the greatest cleanliness is necessary. Especially be careful of the disease being propagated by others using a towel that has been used by any one having these diseases. Great care is necessary on this account in going into privies and water-closets.

Gleet—Is almost always the consequence of badly treated gonorrhœa ; it is simply a continuance of the discharge, though it is thinner and in much smaller quantities. It is principally dangerous on account of its often producing stricture of the urinal passage, and should be attended to at once. Internal remedies, as a general thing, will not cure it. Injections of a mixture made by putting a teaspoonful of sulphate of zinc into

Stricture of the Urethra or Urinal Passage.

half pint, or one pint of cold water; or same quantity of
sugar of lead and water, used from three to five times a day,
will be best. It may be advisable to change the injection
every fourth day; using, as may be needed or convenient, a
tea made of blackberry root, white oak bark, or alum water,
gradually discontinuing the injections as soon as the discharge
ceases. The bowels should be kept freely open, and diet not
too stimulating; all kinds of spirituous liquors being avoided
in this as in gonorrhœa, and all diseases of this kind.

Stricture of the Urethra or Urinal Passage—
Consists in chronic inflammation, usually following gonorr-
.œa and gleet, causing a narrowing of the passage, and thereby
interfering with the voiding of the urine. The disease is
caused also by masturbation, and anything that produces a
chronic irritation in the passage. The diseased condition exists
during the early or first stage of stricture, only as a soft,
swollen or puffy state of the delicate skin lining the passage
In the next or second stage of stricture, an important change
has taken place. That portion of the passage which was
before in a merely tumefied or swollen condition, has now
acquired a certain firmness; it resembles a band encircling
the passage, narrowing it at this point, and actually reducing
ts capacity. It now offers a decided impediment to the urine,
and if a moderate sized instrument (Bougie) is introduced, an
evident resistance is met; though with a gentle pressure it
yields, and the instrument pretty easily passes on.
In a still more advanced or third stage, the dilatable condition has disappeared; the stricture has become firm, or callous, as it is commonly called; the contraction is unyielding,
and an instrument meets an abrupt positive resistance, the
calibre of the passage being frequently so diminished as
hardly to allow the urine to pass at all, or even an instrument
of the smallest possible size.
There is no natural cure for Stricture. When it once begins,
it is sure to go on increasing, either rapidly or slowly. Fortunately, however, we have the means of curing it by art.
Among the very early symptoms of a Stricture, there is one
in which great confidence may be placed. It relates to the
manner in which the last few drops of urine pass, which is by
dribbling away. This is a symptom of great value in determining any doubtful case of Stricture. Another early symptom of stricture is a scattering of the stream of urine in
making water, or splitting into two or three small streams. A
certain hesitation in commencing to urinate, although the

stream flows fully and easily enough when once started, may be added. It is very common to have this hesitating or waiting longer than natural, then a full stream, then again the dribbling away afterwards.

The remedy for Stricture is Bougies; they are to be introduced up the urinary passage, commencing with a small one, and using a larger one from time to time, until the Stricture is dilated, and the passage becomes of its natural size again. This treatment should never be attempted by any but a medical man, as it is of too delicate a nature to be undertaken by any other.

Albuminaria, or Bright's Disease of the Kidneys.—This disease is an affection of the Kidneys, and was first described by Dr. Bright, of England. Its most distinguishing symptom is the presence of the serum (or watery portion) of the blood in the urine, so that when the latter fluid is heated to near boiling, the albumen becomes coagulated, like the white of an egg, causing merely a cloudiness if in small proportion, but sometimes existing in such quantity as to form a nearly solid mass. This condition of the urine is always to be looked upon seriously. It sometimes comes on slowly, more particularly in those addicted to the excessive use of ardent spirits; or it may be the immediate consequence of severe cold and repressed perspiration; it is not an unfrequent sequel to scarlet fever. The *sudden* development of this condition of urine is accompanied with feverish symptoms and dropsical swelling of the face, with stiffness of the eyelids, swelling of the extremities, and if it proceed far, of the trunk of the body also. It ought at once to be submitted to the treatment of a physician. In the absence of this assistance, should sudden swelling, as above described, come on, and with it symptoms of general fever, a portion of the urine may be heated in a metal spoon to boiling; if it becomes thick or cloudy, and if it is not cleared by the addition of a few drops of vinegar, it may safely be concluded that the kidneys are suffering. Blood, according to the strength of the patient, may be taken from the loins by cupping, the patient confined to bed, and a bath of the temperature of 90° taken for half an hour, once in twenty four hours. A diaphoretic (or sweating) mixture, is to b given, and the bowels purged with cream tartar and jalap, or some other good purgative. The diet must be kept low as long as fever continues. The case ought not to be trusted to domestic treatment further than is unavoidable; for even m the most skillful hands, it is beyond doubt a formidable disease.

Poison-Vine Eruption.—This Vine, a species of Sumach, and one or two other plants, cause by contact in some persons an inflamed eruption, or small blisters, which is in some cases very painful. The hands and face are its most common localities, but it may appear on any other part of the person. Its duration, in severe cases, is likely to be from one to two weeks. In some cases it is very mild, and does not last over a day or two, and only in a mild form. Much depends on the state of the constitution, temperament, etc.

The treatment of this complaint is simple. Apply a mixture of one teaspoonful of sugar of lead, dissolved in half a pint of cold water, every two hours, with a camel's hair brush, or soft linen mop. Fluid extract of Virginia snakeroot is said to be almost a specific, applied over the affected parts freely and frequently. When nothing better can be had, apply sweet oil or common lard.

Masturbation.—In all ages of the world, of which we have any account, there has prevailed in the youth of both sexes a most destructive habit, by reason of which many lives are lost every year, besides in other instances the laying the foundation for disease and a debilitated condition, which the best efforts of a life-time do not counteract. This habit is known by the name of secret habits of youth, or masturbation, and is one of the most prevalent and most universal of any that afflicts the human race. One reason why the vice or habit is so very destructive to youth, is from the fact that it is continued in secret from day to day, and from year to year, without a knowledge of its real consequences. There are very few parents who ever think to warn their children of the dangers of this habit, either by speaking to them, or putting into their hands a treatise on the subject, and even do all in their power to keep their children from getting hold of any books on the subject. This kind of false modesty has been the cause of many a blooming youth filling a premature grave; and we hope to see the time when parents, teachers and guardians will treat those entrusted to their care with more confidence, and the exercise of better judgment.

This habit, began in youth, is often continued many years. It is not *always* the result of initiation, or contracted by contact with those who indulge in the habit, but this is the cause in most cases, and one bad boy will ruin twenty good ones. There are those who think their children can only contract bad habits by contact, and among the rest the habit of masturbation, and think if they can keep their families from vicious company they are safe. This is generally true, but there are exceptions to the rule; for no matter how strict a parent may be, the very innocence of a child makes him a ready tool for the designing, when they chance to fall into their company. This habit is oftener learned at school than elsewhere.

The following are some of the consequences of this destructive habit: Consumption is often induced by this habit, by debilitating the system, and causing tubercles to be developed in the lungs. Loss of memory is among the most common effects produced by it. Insanity is sometimes produced by this habit; the unusual and unnatural excitement produced by the very frequent repetition of the act, sooner or later reaches the brain, the great nervous centre, unless abstained from; and this weakness continues in a ratio with the extent to which the practice has been carried, until it no longer controls the body, and there is lowness of spirits, a disposition to commit suicide, restlessness, discontented mind, and an exceedingly unhappy irritability of temper, causing the patient to make himself, and everybody around him, unhappy. An uneasy, aching pain, heaviness, and weakness across the back and loins, is a frequent symptom, especially in the morning. Palpitation of the heart, shortness of breath, and nervousness, also result from masturbation. A nervous, aching pain in the head, bones and muscles, resembling rheumatism, is a frequent consequence. By weakening the general powers of the system, such persons have not the ordinary powers of resisting diseases, and for this reason when they expose themselves to an atmosphere tainted with ordinary epidemic or miasmatic poisons, they are more liable to the disease. Stricture, or narrowing of the urinal passage, is produced by this habit.

In females, the following diseases are produced. Whites, falling out of the hairs and eyebrows, bad breath, loss of the natural voice, barrenness, falling of the womb, epileptic fits, ulceration of the neck of the womb, hysteries, a desire for seclusion from society, etc. The *most frequent disease* produced by masturbation in males, is called spermatorrhœa, or involuntary emission during sleep. These at first are rare, happening once in the course of two or three weeks, and takes place during a dream of a lascivious character. They soon become more frequent, and if accompanied by a dream, they do not wake the person, and finally they happen almost or quite without any sensation.

The treatment in this disease will be of no avail, unless the vice which produces it is abandoned entirely. It is a hard matter for a patient to do this, but it *must* be done, or he cannot be cured. To accomplish this, he must avoid all conversation or reading calculated to excite the imagination on such subjects, and go into the society of cultivated and virtuous females ; cultivate a taste for music ; and keep himself busy always at some agreeable employment. Tonic medicines may be necessary, and good digestible and nourishing food should be used. Keep the bowels regular, avoid sleeping on the back, and bathe the parts in cold water frequently. I would advise parents not to let their children grow up in ignorance on this subject ; tell them of its dangers, and advise them to shun the " first false step." There is a great deal too much mock-modesty on this subject on the part of parents.

Hay-Asthma, or Summer Bronchitis—Hay-Fever.

—This disease is so called on account of its occurring during hay-time, or summer, and is thought to be caused by the odor of new-mown hay ; but it may be caused by other strong odors. It does not differ very much from the ordinary asthma, except perhaps there is not so much difficulty of breathing, and the attacks last longer in the hay-asthma ; **the lining membrane of the nose is also much more inflamed and the throat irritated in the latter disease.**

The best thing to do is to remain within doors and keep quiet for a few days ; take a few doses of Rochelle salts or rhubarb, also a teaspoonful of paregoric at bed-time for two or three nights, and live on light diet. A dose or two of quinine (one grain) may be beneficial, night and morning.

Hysteric Fits, or Hysteria

—Is a peculiar affection, and is the source very often of much unhappiness. A fit of hysterics may assume different forms, but, generally, the female becomes *apparently*, of a sudden, partially insensible, t may be, falls down, but more generally has sufficient warning to seat herself on a chair. The eyes are closed, the lids tremulous, the limbs are stretched out, and spasmodically and suddenly contracted, at intervals ; or there is violent struggling, the chest heaves, the heart and vessels of the neck beat violently, and the face is more or less flushed

Frequently the patient puts the hand to the throat and neck, as if to dispel some uneasiness, and not uncommonly gives utterance to incoherent or disconnected sentences, generally in a peevish or distressed tone of voice. In most cases the power of supporting the body when seated, remains, unless it is worked off the chair in the struggles. At length the attack, having lasted for a longer or shorter period, from a few minutes to some hours, terminates, probably with a fit of sobbing and crying; the patient recovers consciousnes, but is left exhausted and fatigued with the efforts and struggles, and, perhaps, falls into disturbed or heavy-snoring sleep. When the fit has terminated, or even during its progress, if continued, the kidneys act very freely, and large quantities of urine, almost resembling pure water, are voided.

Such are the leading features of a "fit" of hysteria, but they may be greatly varied. The struggles, especially, being so violent as to require the assistance of two or three strong men to restrain a comparatively feeble female, and to prevent her injuring herself, and sometimes, though not commonly, those around her.

Such are the outward manifestations of a fit of hysteria; but before it comes on, many patients complain of a sense of general oppression or uneasiness, with coldness or numbness of the limbs. Just previous to the accession, the characteristic hysteric "globus," or ball in the throat, is probably felt. It seems as if a ball commenced rolling upward in the bowels, generally from the lower left side, and as if it kept gradually ascending toward the throat, which it seems entirely to fill up, causing those sensations which induce hysteric patients so often to carry the hand to, and pull at the forepart of the neck or throat. During the continuance of a fit of hysteria, little either need or should be done, beyond preventing the patient hurting herself during the struggling. Cold water dashed upon the face may be useful, or it may be poured in a stream upon the head for a few minutes at a time; a mustard plaster on the lower part of the neck may be applied. If there is much flatulence, a half teaspoonful of essence of peppermint in a wine glass of water will give it relief. It must be remembered, that in most cases, the patient is sensible of what is going on around, and may, in the excited state of the nervous system, be painfully alive to any unguarded or unfavorable opinions uttered by those in attendance. For this reason, it is not to be recommended that, as sometimes is done, severe and violent remedies should be proposed within hearing of the patient, with the view of frightening her out of the fit. Such a course has had the opposite effect, causing an aggravation of the symptoms. This is a different thing from threatening severe remedies while the patients are comparatively well. Such a plan of treatment, it is well known, has often succeeded in putting a stop to the spread of hysteria (by imitation) through schools, or such like collections of young females.

The exciting causes of hysteria are, remotely, whatever tends to exalt the influence of the nervous system. Among the moderately-fed and hard-working population in the country, hysteria is comparatively rare, but it is not unfrequent in servants who remove from the poor living of their own homes to the stimulating diet of a rich man's house. Most generally, hysteria, although in some degree the result of constitutional tendency, is connected with debility, and irregularity of the usual conditions of female health—all these being aggravated by emotions of the mind, particularly those which are connected with the affections; these, too, when in direct excitement, as well as inordinate physical exertion, which produces exhaustion of the nervous system, must be ranked as amid the most general directly originating causes of the hysterical fit itself. But little can be done for this disease during the "fit;" about the best thing to be done is to give an emetic or vomit of Ipecac, or if nothing better, drinking every few minutes of warm water, until vomiting occurs. Also injections of warm water or soap suds up the bowels, to produce an evacuation, is advisable.

The attack is to be prevented by keeping the bowels open, avoiding overloading the stomach, and plenty of exercise—doing a good day's washing occasionally or scrubbing the floors being of more use, a great deal, than medicine.

Clergyman's Sore Throat.—So called from the fact that it is oftener witnessed in ministers of the Gospel than others. It is a peculiar affection of the throat and organs of voice, to which public speakers are liable. According to Mr. Macready, actors who have to assume feigned tones, are more liable to it from that cause. The seat of the disease is the mucous follicles (little lumps) scattered over the membranes of the throat, larynx (vocal box) etc., being extended to the latter from the former. The commencement of the disease is insidious; it begins with an uneasy sensation, as if there was something in the throat which required to be hawked up or swallowed down; at the same time the mucous secretion is viscid (tough). As the larynx becomes affected the voice is changed, becomes hoarse, unequal in tone, or quite extinguished; there may be slight pain about the parts, but not much cough in the earlier stage of the disease. All the symptoms become aggravated by cold, by vicissitudes of temperature, or by exertion of the voice in reading, speaking, &c. The above sources of aggravation are of course to be guarded against, and the general health attended to; but the cure of the disease, which consists chiefly in the repeated application of a strong solution of lunar caustic to the parts affected, must only be entrusted to a physician. Dr. Green, of New York, has the credit of first accurately describing the disease, and of prescribing the treatment above mentioned, which is very successful. But when you do not have the opportunity of having that done, I would state that Medicated Inhalation, as recommended under the head of "Catarrh in the Head," is very beneficial in this disease, assisted by gargles of tar-water, strong green tea, cold water, &c., and the external application of kerosine oil to the throat till it produces pimples on the skin, to be applied and re-applied if necessary. The stooping position required by ministers who *read* their sermons instead of *preaching* them, is one great *cause* of this difficulty. The best plan is to do like Christ, the Saviour, did while on earth going about doing good, "preach from the *heart*," instead of *talking* from *paper*—it will be better for both soul and body, for preacher and for people.

Frost-bitten Parts.—Usually it is the ears, toes, nose and fingers. In such cases, as in warming the body to avoid aching, the restoration to heat must be very gradual; a limb has been frozen perfectly stiff, and by being rubbed in snow, afterwards in cold water, and very slowly warmed, its life has been preserved. The sudden application of heat never fails to occasion inflammation, and mortification quickly follows. When the heat has been *gradually* restored, and action and sensation perceived, the part should be rubbed with spirits (gin, whisky, or Bay rum), the patient be put to bed and kept comfortable, perspiration excited by warm drinks, and perfect rest enjoined until the effects subside. Bathing the feet in cold water night and morning, and afterward rubbing with a coarse towel or flesh-brush, will harden the feet against the effects of cold.

Common Cold.—"What is best to do for a common cold?" is a question often asked, and as it is one of the most prevalent ailments to which frail man is subject, we propose answering the question in a common-sense way. The inhabitants of every climate are liable to take cold at different seasons of the year, particularly when the changes of the atmosphere are sudden. The persons most subject to this disease are those of a delicate and irritable constitution, and whose employments expose them to sudden transitions from heat to cold.

A cold is attended with a weight or uneasiness in the head, fullness and oppression at the chest, a sense of distension an stopping up of the nose, followed by a secretion of thin mucous from the nose, watering of the eyes, soreness of the throat, cough, with expectoration of mucous, cold shiverings, succeeded by flushes of heat, and pain in different parts of the body—very frequently the chest.

This disease is not generally attended with danger when appearing in a mild form, and early precautions are used. If the symptoms should be highly inflammatory, and the constitution of the patient delicate and irritable, the most vigilant attention is demanded in order to arrest the progress of the disease; otherwise the most serious consequences may ensue. It is by such neglect that two-thirds of the cases of consumption and other pulmonary affections in this country, arise.

When a person finds himself much indisposed from exposure to cold, he should *at once confine himself to the house;* use a spare, mild diet; drink barley water with lemon juice in it, or any other warm diluent, or mucilaginous drinks, and particularly avoid eating or drinking anything stimulating. Previous to going to bed, put the feet in warm water for a few minutes, then wipe dry and rub them well with a flesh-brush or rough towel, put on the stockings again, and get into bed at once (between blankets being the best in cold weather); then from half to one teaspoonful of paregoric, and about an equal quantity of syrup of ipecac should be taken in half a cup full of warm tea of any kind. The following morning take a dose of Rochelle salts or Seidlitz powder, or some other good purgative. The mucilaginous drinks should be taken through the day. Repeat the paregoric two or three nights if necessary. Any good liniment or a mustard plaster applied to the chest (front and back) and also the neck will be of service. A violent fit of tickling cough may be relieved at once by taking a lard pill or gargling with a spoonful of melted lard, or sweet oil. The course of treatment above advised will cure the great majority of common colds, when strictly followed. To avoid taking *fresh cold,* gargle the throat with cold water frequently during the day, especially before going into the cold air; bathing the neck in cold water is also a good remedy.

Continued Common Cough.—This is generally the effect of a cold neglected, or improperly treated. Sometimes t arises from indigestion, or irritable matter in the stomach and bowels.

A common cough ought always to be viewed as a serious disease, and should never fail to excite the fear and anxiety of the patient and friends. By early and vigilant attention to this disease, thousands of lives are saved which otherwise are destroyed by pulmonary consumption. As a general thing it will be found more advantageous to employ *external* than *internal* remedies. The bowels should be first attended to by giving two or three good purges, a day or two apart; then ∩ub the throat and chest *thoroughly* two or three times a day with some stimulating liniment; the author has found the common kerosine oil about the best. If it proves too irritating, omit it for a day or two or more, and rub the parts with sweet oil or melted lard. Red flannel should be worn next the skin if it can be tolerated; if too irritating it can be lined with silk. In addition to this treatment I would advise Medicated Inhalation, by all means (see Catarrh in the Head). The diet should be nourishing, but be careful to avoid everything that disagrees with the stomach. Gargling the throat with green tea (cold) or cold water frequently through the day and night is often beneficial.

Propagation of Diseases by Infection.—" Is it Catching?" is a question very often asked by those whose kindly natures have prompted them to perform a Christian duty by " visiting the sick." To all such, and all others, the following remarks on the subject will be found important and interesting. Contagious diseases may be communicated only by actual contact of individuals, as in the case of itch, &c.; by inoculation as in the case of cow pox; or in addition to both or either of these modes of transmission, through the atmosphere by infection, as in the case of small pox, &c.

This power of propagation through the atmosphere, does not, independent of epidemic and other influences, extend far from the patient. Certain circumstances influence the extent of contagious diffusion. Of these, the most distinctly ascertained are atmospheric impurities; for it is ever observed, and we believe it may be predicated of every disease possessing the property of remote contagion, that its contagious matter is pre

pagated to greater distances in a dirty, crowded, and ill-ventilated apartment, than in one of which the air is pure. The
same principle applies to articles of dress and furniture ; those
which are contaminated by animal secretions and effluvia
being much more readily impregnated with contagious matter than those which are clean. Peculiar atmospheric conditions, certainly, also favor the propagation of disease by
contagion ; sometimes these conditions are inappreciable, at
others they are evidently connected with a superabundance
of warmth and moisture, and also, we have good reason to
conclude, with certain states of electrical disturbances. The
discovery of the new agent, or modification of the known
existing agent oxygen—named ozone—may probably shed
some new light upon the subject of contagion. Actual contact,
however, or even immediate vicinage, to a person laboring
under a contagious disease, is not requisite for its propagation
to others. This may be effected by means of substances to
which the contagious matter clings. These substances, which
go by the name of *fomites*, are more generally clothing and
stuff furniture which have been about or near the bodies of
those laboring under the disorder. These fomites are apt to
be impregnated with the poison in a very concentrated condition, and are capable, not only of retaining it for a long
period, but of transporting it from place to place. A sofa on
which a patient laboring under scarlet fever had lain has been
known to propagate the disease six months afterward ; and
clothes which have been about the sick are constantly ascertained to have been the medium of conveying fever, &c., to
distant localities. Wool and cotton seem particularly apt to
attract and retain contagious emanations ; but, indeed, all
loose textures have the property ; while on the other hand,
polished and hard surfaces and substances are much less likely
to act as fomites, if they do so at all. Everything of unnecessary drapery or clothing should be removed from the
chambers of those sick of contagious maladies, or indeed of
any malady ; for a sick chamber must always, in a lesser or
greater degree, have an atmosphere containing unhealthy
emanations, which it is expedient, both for the good of the patient and of others, should find no unnecessary attractions or
lodgments. Further, it is advisable to have the furniture as
much as possible of hard and polished substances, and the
dresses of those in attendance upon the sick, especially if ha
bitually so, might with advantage be made with a glazed
surface. Those substances which have necessarily become
the fomites of contagious matter ought to be scrupulously

freed from it by complete and lengthened exposure to the open air, by washing, or by exposure to the fumes of chlorine in a close apartment ; or by all three, the chlorine fumigation being first resorted to. Those persons under whose management a case of contagious disease has occurred, ought, as a Christian duty, to make sure that every article of stuff, furniture, clothing, &c., has been fully and carefully purified before others, either in the way of social intercourse or in occupation, particularly the washerwoman, come in contact with them. The following systematic course of action should be pursued when the generation of contagious matter has ceased in an apartment, either by the death or recovery of the patient, premising, of course, that throughout the illness measures have been (or ought to have been) resorted to to preserve purity. During the day, *the door being shut*, the windows *should be open* to their full extent, and the infected articles freely exposed to the air ; during the night, the windows and door being *closed*, chlorine should be well diffused through the apartment. This having been repeated, if possible, for two days and nights, all textile fabrics and the like should be removed ; those that are capable of being washed put into cold water, and the others placed in the open air.

All articles of furniture left in the room, also the floor and oil-painted wood-work, should be well scoured. If the chamber be a whitewashed or colored one, it should be " re-done ;" if papered, it is only a safe precaution to re-paper it. The bed requires the greatest amount of care ; if of wool, it is better destroyed altogether ; if of hair or feathers, these should be exposed to the heat of rebaking, that is, at least to a temperature of 210° Fahr. ; and the ticking either thoroughly fumigated and washed, or entirely renewed. These directions may appear minute and troublesome, but they are far from being too much so when put in comparison with the fearful scourge of a contagious disease which has established itself in a household or community, and which perhaps might have been checked at the outset by the adoption of prompt and vigorous measures. The poor and the ignorant cannot or will not adopt, in most instances, effective precautions ; it remains for the rich, for the well-informed, to point out their necessity, and lend a helping hand to their fulfillment, not only as an act of Christian charity, but as a means of safety for themselves. It is not a necessary character of contagious disease that it has itself sprung from contagion ; some of the most virulent and spreading fevers, such as those of the ship, or of the old jails, had no such commencement, but had their origin in the decomposing emanations from the bodies of numbers of individuals confined in unventilated and insufficient spaces. In addition to the disinfectants already mentioned—air, water, and chlorine—many others are and have been used, such as the vapor of vinegar, of pitch, of tobacco, or camphor, or burnt sugar; large fires also used to be a favorite method; but none of these last mentioned are to be relied upon solely. The vapor of muriatic acid and the absorbing properties of newly-slaked lime, may be resorted to, in the absence of chlorine, with advantage. In many instances, particularly in the case of clothes and other textures that will not wash, heat might be used more extensively than it is at present as a disinfectant. A little trouble or labor in time may save much suffering and many lives.

Convalescence, or Getting Well.—This is often a time of peculiar trial to the attendants upon the sick, and *what* to do, and *how* to do it, are matters of great importance. The commencement of convalescence, or the point at which the characteristic symptoms of disease cease, is sometimes distinctly marked, more especially after acute disorders; frequently, however, the tendency toward health, particularly after chronic disease, is much more insensibly established. In the latter case, too, the progress of the convalescence is slower than it is in the former. Its rapidity or protraction is much influenced by age, and the nature and treatment of the previous malady. Children convalesce rapidly, old people the reverse. In no case, perhaps, is convalescence more tardy and unsatisfactory than after illness, in which much loss of blood, or of its constituents, has taken place.

When convalescence from acute disease commences, the previously quick pulse falls to the natural standard, the tongue begins to clear, the skin becomes cool, sleep is refreshing, the mind acquires a more healthy and hopeful tone, and the person *looks* better. There is nothing which more assures a physician of the condition of his patient than the look, the expression of the countenance, to which the first glance, as he enters the room of sickness, is almost instinctively directed. The look of convalescence is tranquil and placid, not the heightened color and bright eye of hectic, which so often deceive the inexperienced with delusive hopes. When the brain has been much affected, however, the condition of the mind, and consequently the countenance, assumes its natural look more slowly.

The management of convalescence is extremely important

Errors in this respect frequently expose the already weakened patient to attacks of other disorders, or induce relapses to the diseased actions which had just been cast off. The convalescence after some particular diseases is more liable to such accidents than it is in others. That after fever is peculiarly so; and after scarlet fever, the tendency to cold and its consequences, dropsical swelling, and affection of the kidneys, is so very common, and so frequently fatal, that the greatest possible care is requisite. During convalescence from acute disease, and especially of an eruptive character, many of the disorders characteristic of the scrofulous constitution show themselves: the eyes become the seat of chronic inflammation, purulent discharge from the ears occurs, and chronic eruptions show themselves upon the skin, of the head especially. These disorders, now, perhaps, for the first time apparent, are apt to continue even after convalescence, properly so called, is over. Relapse in convalescence often occurs from too soon employing actively the previously affected organ; the liability to this mishap must be evident to the common sense of every one. In the case of the eye, it is evident to the senses, after inflammation of that organ, its undue exercise, or even its exposure to full daylight, will often be followed by a return of the disease. Such is the case elsewhere; and whether it be the eye, or the brain, or the stomach which has been affected, return to the ordinary exertions of health must be made with the *greatest caution.*

The clothing of a convalescent patient requires particular attention; there is much susceptibility to cold and to atmospheric vicissitudes. General exercise is to be resumed cautiously, and should never be carried to the extent of *fatigue.* Diet, however, is the great source both of error and mischief—the greatest difficulty which the physician has to contend with; that is, in getting it properly attended to, and his orders properly carried out, particularly among the poor. While a disease is in progress and alarm is felt, directions are tolerably well, or strictly, obeyed; but no sooner does the patient begin to get better, than irregularities commence. The popular idea seems to be that convalescence must advance in proportion to the amount, and often to the stimulant qualities, of the food given; *and many a hopeful case sinks back into fatal relapse from the willful and injudicious kindness of friends.* The point is one which requires to be strongly enforced, that in *diet,* as in everything else, convalescence must be *gradual,* and that nothing is more dangerous, more likely to induce relapse, than the injudicious use of solid animal food or of stimulants. Milk, and the various farinaceous preparations with which it is usually combined, such as arrow-root, sago, rice, bread, &c., is perhaps the most generally useful article of diet in convalescence; next come the broths made from fowl, mutton, veal, or beef, alone, or mixed with crackers or bread; next in succession, are eggs lightly boiled; and lastly, solid meats, of which tender mutton is probably the best, are to be permitted. Ripe fruits in their season, if not contra indicated by the nature of the previous disease, and if they do not occasion flatulence or diarrhœa, are both grateful and serviceable. The patient must be careful not to overload the stomach on any account. In whatever form nourishment is given to the convalescent, it should be in *small quantity at a time*, but as frequently repeated as the natural appetite requires. The atmospheric purity of the chambers occupied by persons recovering from sickness requires great attention, and the temperature ought to be kept as nearly as possible about 58° Fahr. Lastly when convalescence has reached a certain point, there is no remedy which so surely promotes perfect recovery and confirms health as change of air; but care must be taken that in so doing there must not be injudicious exposure during unsuitable weather.

Falling out of the Hair, or Baldness.—Those who lose this great natural ornament before the "head is whitened over with the frost of many winters," feel the loss very keenly, as well they may, for the hair is not only *ornamental* but *useful*. Falling of the hair occurs from weakness, either of the body generally, or of the hair-bulbs, or "follicles," themselves. Various local stimulant applications are used in such cases, of which Balsam of Peru—a drachm stirred well into an ounce of simple cerate when melted—is said to be a good application.

Baldness, or loss or deficiency of the hair or parts usually covered by it, is sometimes seen in infants. It frequently occurs in adults of the male sex, even in the prime of life, and almost universally, in a greater or less degree in old age. The direct occasion of baldness is defect in the hair follicles from which the hair is developed; and this defect may arise from diseases affecting the skin itself, from acute general disease, as fever; or chronic constitutional disease, such as consumption; it may also arise from constitutional peculiarity, or the diminished circulation of blood, such as occurs in advanced life. Some families appear to be peculiarly liable to become the subjects of baldness, even in early life; those who perspire much about the head are often bald. Generally, however, whatever occasions a diminished supply of blood to the scalp or skin, gives the hair a tendency to shed, and the treatment must be directed to stimulating the skin as much as possible by proper and timely applications. After acute disease, if the hair falls off, shaving the part two or three times in succession will probably strengthen the growth. In other cases, much covering upon the head, which causes perspiration, and consequently weakens the skin, must be avoided: and the head should be well washed with cold water every morning, and afterward rubbed and brushed to promote reaction. Various applications are recommended to prevent or cure baldness; they are all stimulant. Those of which cantharides, or Spanish blistering flies, form an ingredient, are generally most serviceable. A drachm of the tincture of cantharides, rubbed up with an ounce of lard, will form a sufficiently stimulating ointment. The infusion of the leaves of the *Asarum Europæum*, a plant which occurs wild in the woods in England, is a very efficacious stimulant to the hair follicles; the infusion may be used as a lotion to the scalp. Falling off of the hair, which is occasioned by eruptive disease, or which is accompanied with inflammation of the skin, of course requires a different and more soothing treatment.

In the baldness of early life, the hair drops off without the previous change of color which occurs in age; in the latter case, of course, no treatment is either likely to be resorted to or to be of service. Still everybody must *try*, so let them try.

INDEX.

www.ingramcontent.com/pod-product-compliance
Lightning Source LLC
Chambersburg PA
CBHW020902210326
41598CB00018B/1755